Institute of Mathematical

LECTURE NOTES–MONOGRAPH SERIES

# Current Issues in Statistical Inference: Essays in Honor of D. Basu

Malay Ghosh and Pramod K. Pathak, Editors

Volume 17

Institute of Mathematical Statistics

*Lecture Notes–Monograph Series*



The production of the *IMS Lecture Notes–Monograph Series* is managed by the IMS Business Office: Jessica Utts, IMS Treasurer, and Jose L. Gonzalez, IMS Business Manager.

Library of Congress Catalog Card Number: 92-53163

International Standard Book Number 0-940600-24-2



Printed in the United States of America

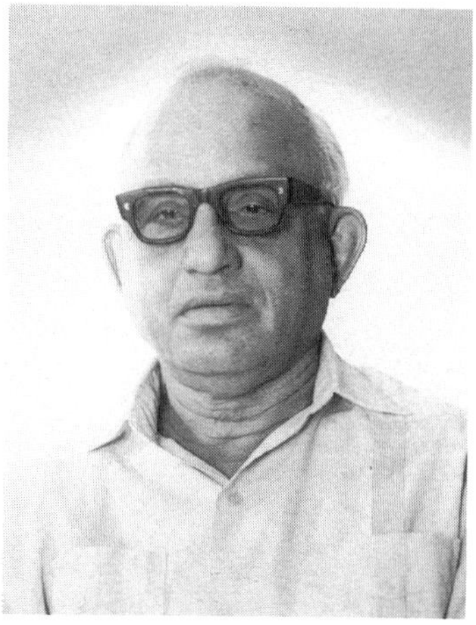

D. Basu

## PREFACE

Dr. D. Basu's pioneering contributions to statistics started at the beginning of the fifties. For about four decades, Dr. Basu, in many of his fundamental writings, has examined critically the foundations of statistical inference, concepts such as information, likelihood, invariance, ancillarity, randomization, fiducial probabilities, logical foundations of survey sampling, and many related concepts. His research has led to some path-breaking results such as independence of ancillary and boundedly complete sufficient statistics, characterization of sufficiency in finite population sampling, the design independence of Bayesian inference procedures in sample surveys, to name a few. His research has influenced several generations of statisticians, and will continue to do so for years to come. Most of Dr. Basu's critical essays are now collected in a Springer volume entitled *Statistical Information and Likelihood*, thanks to the efforts of Professor J.K. Ghosh.

Professor Basu was born on July 5, 1924, in Dacca, now in Bangladesh. He received a Master's Degree in Mathematics from Dacca University around 1945, and taught there briefly from 1947 to 1948. He moved to Calcutta in 1948, where he worked as an actuary with an insurance company for some time. In 1950, he joined the Indian Statistical Institute as a research scholar under Professor C.R. Rao. In 1953, he submitted his Ph.D. thesis to the Calcutta University and went to Berkeley as a Fulbright scholar. His associations with Neyman at Berkeley and with Fisher at the Indian Statistical Institute in 1955 gave him a deep insight into both the Neyman-Pearson theory as well as the Fisherian theory of ancillarity and conditionality. He knew and understood these paradigms better than most of his contemporaries. His critical examination of both the Neyman-Pearsonian and the Fisherian modes of inference eventually forced him to a Bayesian point of view, via the likelihood route. The final conversion to Bayesianism came in January, 1968, when Basu was invited to speak at a Bayesian Session in the Statistics Section of the Indian Science Congress. He confesses that, while preparing for these lectures, he became convinced that Bayesian inference did indeed provide one with a logical resolution of the underlying inconsistencies of both the Neyman-Pearson and the Fisherian theories. Since then, Dr. Basu became an ardent Bayesian and, in many of his foundation papers, pointed out the deficiencies of both the Neyman-Pearsonian and the Fisherian methods.

Professor Basu was on the Faculty of the Indian Statistical Institute for many years. His passion for travel has taken him to universities all over the world as a visitor, e.g. UNC at Chapel Hill, University of Chicago, University of New Mexico, University of Sheffield, University of Adelaide, Iowa State University, to name a few. He was a Professor of Statistics at Florida State

University from 1976 until his retirement in 1986. Throughout his professional career, he has maintained strong ties with the Indian Statistical Institute. Now in his retirement, when he is not abroad, he loves to return to the ISI to look around the classrooms, the flower-beds, and the rose gardens which he so painstakingly helped created during his association with the Institute.

In his fruitful research career spanning nearly four decades, Dr. Basu's emphasis has always been on the foundations and the underlying concepts rather than on the technicalities. In keeping with his philosophy, essays in this festschrift volume, dedicated to Dr. Basu on the occasion of his 65th birthday, place the major emphasis on the foundational issues of statistical inference. Most of the papers in this volume are review articles written by his friends and colleagues in those areas of statistics that have interested Dr. Basu most during his active research career. This monograph differs from other festschrift volumes in yet another respect. It is written in a narrative style which has typified so much of Dr. Basu's own writings in statistics. We believe that this is a fitting tribute to a scientist whose simplicity of exposition has earned him a special place in the evolution of contemporary statistics.

We take this opportunity to thank all the authors of this volume who spent so much time writing and rewriting their articles. We would also like to thank the referees (names arranged alphabetically): J. Berger, A. Bose, G. Casella, R. Christensen, L. Kuo, D. Lane, G. Meeden, R.V. Ramamoorthi, B.K. Sinha, J. Srivastava, and W.J. Zimmer for their selfless service. Special thanks are due to Professor Robert J. Serfling, Editor of the IMS Lecture Notes Monograph Series for agreeing to publish this collection of essays. The project would never have been completed without his active encouragement at different stages of its preparation. We also thank Jose L. Gonzalez, the IMS Business Manager for his valuable advice at the final stages of the preparation of this volume.

Finally, we wish to thank Ms. Cindy Zimmerman for her patient and careful typing of all the manuscripts in a unified format.

Malay Ghosh
University of Florida,
Gainesville

Pramod K. Pathak
University of New Mexico,
Albuquerque

TABLE OF CONTENTS

# CONDITIONAL INFERENCE FROM CONFIDENCE SETS

George Casella, Cornell University

**Abstract**

Ideas of inference using conditional confidence have grown out of many different schools of statistical thought. The development of these ideas is traced, starting with some original ideas of Fisher. The influence of other researchers, such as Basu and Buehler, is also discussed. The development is traced to the present, through the work of Pierce and Robinson, to current work in conditional inference.

## Introduction

The development of conditional inference, in particular that based on confidence sets, has followed many paths. There are now several inferential methods that use this name. For example, the likelihood based methods of Hinkley (1980), or Cox and Reid (1987), are conditional inference methods. The attempt of Kiefer (1977), to merge conditional ideas with frequentist theory is also conditional inference.

The one common factor in the different conditional inferences is the requirement of reasonable (coherent) post-data inference. That is, inferential statements made after the data have been seen should have some logical consistency. Another approach to conditional inference, one that gained structure through the work of Buehler (1959) and Robinson (1979a,b), provides an objective framework for assessing post-data validity. It is this version of conditional inference, based on confidence sets, on which we will concentrate.

The different versions of conditional inference have a common origin in ideas of Fisher. These ideas of Fisher are somewhat intuitive, and leave some gaps in development (but not to Fisher!). The origins in Fisher were later refined by Basu, who relied on ideas of Bayesian inference to close any gaps. This is where our review begins.

## 1. The seeds of conditional inference

Many influential ideas in statistics can be attributed to Sir Ronald Fisher. One of the most elusive, perhaps, is that of conditional inference. In Fisher (1959, page 78) we find the ideas of a *reference set*:

---

This paper was written in honor of Professor D. Basu on the occasion of his $65^{th}$ birthday. Research supported by National Science Foundation Grant No. DMS089-0039.

> In attempting to identify a test of significance ··· with a test for acceptance, one of the deepest dissimilarities lies in the population, or reference set, available for making statements of probability.

Interpreting Fisher, we find that he is concerned with the range of the inferences, that is, with the set in the population to which the inference should apply. In this sense, he is concerned with conditional inference, inference conditional on some subset of the sample space. The exact nature of his concern is not, at first, clear. It does emerge in some later statements, again from Fisher (1959, page 81). In talking of inference from Student's t distribution, he says

> The reference set for which this probability statement holds is that of the values of $\mu$, $\overline{x}$ and $s$ corresponding to the same sample ··· there is no possibility of *recognizing* any subset of cases ··· for which any different value of the probability should hold. (my italics)

In this statement we see one of the keystones of conditional inference. There should not be a subset of the sample space (a *recognizable* subset) on which the inference from a procedure can be substantially altered. If such subsets exist, then inference from the procedure is suspect.

If such a recognizable subset existed, then Fisher would no doubt find it, however, there does not seem to be any general methodology used. Although ideas of estimating and eliminating nuisance parameters are used, and also ideas of ancillarity are used, no general scheme is defined.

One famous example is Fisher's criticism of Welch's solution to the Behrens-Fisher problem. If $\overline{x}_i$, $s_i^2$, $i = 1,2$, are the sample mean and variance from samples of size n from independent normal populations with unknown parameters $\mu_i$ and $\sigma_i^2$, Fisher (1956) derived the following fact. Under the hypothesis $H_0$: $\mu_1 = \mu_2$, for any value $t$,

$$P\left(\left|\frac{\sqrt{n}(\overline{X}_1 - \overline{X}_2)}{\sqrt{S_1^2 + S_2^2}}\right| > t \mid S_1^2 = S_2^2\right) = P(|\, T_{2(n-1)}| > \tau t)\ , \tag{1}$$

where $T_{2(n-1)}$ has Student's $t$ distribution with $2(n-1)$ degrees of freedom, and $\tau$ is an unknown parameter satisfying $0 \le \tau \le 1$. Thus, conditional on $S_1^2 = S_2^2$, the random variable $\sqrt{n}|\overline{X}_1 - \overline{X}_2|/\sqrt{S_1^2 + S_2^2}$ is stochastically greater than $|T_{2(n-1)}|$. Fisher used this fact to show that Welch's solution suffered from the property that the probability of rejecting a true $H_0$, given that $S_1^2 = S_2^2$, was bounded below by the nominal level. Thus, on the recognizable subset $\{(s_1^2, s_2^2)\colon s_1^2 = s_2^2\}$, Welch's solution has an actual error rate greater than the nominal level.

This conditional behavior would be even more disturbing if the set $\{(s_1^2, s_2^2)\colon s_1^2 = s_2^2\}$ is taken as a reference set, i.e., a set on which the conditional inference should be applied. Fisher's argument for conditioning on this set, or more generally on the ratio $s_1^2/s_2^2$, is elusive. The fact that Fisher considers this a reasonable reference set appears again in Fisher (1959), where he discusses his solution to the Behrens-Fisher problem.

The fact remains, however, that the mechanism of choice of a reference set is elusive. Although concepts of ancillarity and elimination of nuisance parameters are considered, a general mechanism for choosing a conditional reference set is not known.

## 2. Basu's refinement

In doing conditional, or post-data, inference the evidential meaning of the inference becomes increasingly important. Fisher's idea of a reference set has some meaning, i.e., it defines a part of the sample space on which inference is to be restricted. On the other hand, the connotation of a recognizable set does not carry this distinction.

A recognizable set is only a set that is in the sample space, and may give no meaningful inference base. Poor conditional (post-data) performance of a procedure on a recognizable set is taken as criticism, but if this recognizable set is not a meaningful reference set, then the criticism may be vacuous.

Fisher had the intuition to choose recognizable subsets that were also meaningful reference sets. Thus, when he leveled criticism (or praise) of the conditional performance of a procedure using a particular recognizable set, this set was also a meaningful reference set. One of the major clues left to us by Fisher, on how to chose these reference sets, is that they should use ancillary information.

Alas, many of us are not possessed with Fisher's intuition in choosing reference sets. When Basu started to think about this, he realized that basing conditioning sets on ancillary information was not, in itself, a reasonable technique in general. In Basu (1964, page 17, Statistical Information), he says

> The ancillary argument of Fisher cannot be extended ⋯. We end this discourse with an example where ⋯ the ancillary argument leads us to a rather curious and totally unacceptable 'reference set'.

Basu then gives an example to illustrate his point. The point that we should be concerned with is that the choice of the reference set is not automatic. Of course, Basu does not give us a recipe for choosing a reference set, but rather argues that the only reasonable procedures are free of conditional defects.

## 3. Conditional and unconditional inference

Inference made conditional on the data must, necessarily, connect a statement about the unknown parameters to the data actually observed. This fact separates conditional confidence inference from unconditional, or pre-data, confidence inference. This latter inference, that of the frequentist (Neyman-Pearson) school, need not apply in any way, to the data at hand. A frequentist inference merely states how the procedure will perform in repeated trials, even if such a statement is ludicrous in the face of the observed data.

This dichotomy, between conditional and unconditional inference, most often results in a statistician choosing one stand and rejecting the other. Fisher rejected unconditional inference in favor of conditional. Basu, although starting in the Neyman-Pearson camp, ultimately rejected unconditional inference in favor of Bayesian conditional inference. Indeed, perhaps Basu stated his belief most elegantly in Basu (1981, page 173, Statistical Information)

> With $E_x$ as the (Neyman-Pearson) confidence set corresponding to the observed sample $x$, can any *evidential* meaning be attached to the assertion $\theta \in E_x$? Suppose on the basis of sample $X$ one can construct a 95% confidence interval estimator for the parameter $\theta$, then does it mean that (the random variable) $X$ has *information* on $\theta$ in some sense?

Of course, Basu gave examples of 95% Neyman-Pearson confidence intervals with no information at all about $\theta$. For example, if $\theta \in [0,1]$, and $X \sim U(0,1)$ ($X$ is $\theta$-free), then for any fixed set $B \subset (0,1)$, the set

$$E_x = \begin{cases} B & \text{if } 0 < X \leq .05 \\ (0,1) & \text{if } 0.5 < X < .95 \\ B^{c} & \text{if } .95 \leq X < 1 \end{cases}$$

is a 95% unconditional confidence set for $\theta$. But, of course, we cannot attach any evidential meaning to the statement "$\theta \in E_x$." (We note, in passing, that the conditional behavior of this set is wretched. For example, $P(\theta \in E_x | 0 < X \leq .05) = P(\theta \in B)$ and $P(\theta \in E_x | .95 \leq X < 1) = P(\theta \in B^c)$. One of these two probabilities must be smaller than .95. Further, $P(\theta \in E_x | .05 < X < .95) = 1$, showing that the post-data inference can be moved all over.)

As we trace the development of conditional inference, we will see that Basu's teachings are there. Many papers take the approach of verifying good conditional properties by verifying Bayesianity. However, this might be a case where some good can come out of greed. Why should we be satisfied with only good post-data behavior or good pre-data behavior? Why can't we try for both? The answer is that we can not only try for both, we can sometimes attain it. The procedures that do can be acclaimed by both camps – conditional and unconditional.

**Formalizing Conditional Inference**

The work of Buehler (1959) was a landmark attempt in examining post-data validity of Neyman-Pearson procedures. Buehler's work is pioneering for two reasons. One, he examined post-data behavior of frequency based rules (not necessarily Bayes rules) and two, he developed criteria for carrying out this evaluation in an objective manner. Buehler's work was based on other seminal ideas of Tukey (1958) and Stein (1961), and was ultimately generalized and formalized by Robinson (1979a,b). We briefly describe Robinson's set-up.

The random variable $X$ has density $f(x \mid \theta)$ and, based on observing $X = x$, a confidence procedure $< C(x),\gamma(x) >$ is constructed. A confidence procedure consists of a set $C(x)$ and a probability assertion $\gamma(x)$. The validity of $\gamma(x)$ as a confidence assertion is measured by the ability of $< (C(x),\gamma(x) >$ to maintain its confidence even when evaluated conditionally. To be specific, we consider $\gamma(x)$ to be an evaluation of the coverage properties of $C(x)$ in the sense that

$$E_\theta \gamma(X) \approx P_\theta(\theta \in C(X)) . \qquad (2)$$

Suppose now that a recognizable subset, $\mathcal{A}$, of the sample space, and an $\epsilon > 0$ exists such that

$$E_\theta(\gamma(X) \mid X \in \mathcal{A}) - P_\theta(\theta \in C(X) \mid X \in \mathcal{A}) \geq \epsilon . \quad \forall \theta \qquad (3)$$

Then, we have qualitatively changed the confidence behavior. On the set $\mathcal{A}$, our conditional assertion is suspect: The asserted probability, $\gamma(x)$, is, on the average, uniformly greater than the actual conditional coverage.

In Robinson's terminology, (3) is a special case of a relevant betting function, defined as follows:

**Definition 1:** A function $k(x)$, $-1 \leq k(x) \leq 1$ is *relevant* for $< C(x),\gamma(x) >$ if

$$E_\theta \left\{ \Big( I(\theta \in C(X) - \gamma(X) \Big) k(X) \right\} \geq \epsilon E_\theta |k(X)| \qquad (4)$$

for all $\theta$ and some $\epsilon > 0$. If $\epsilon = 0$, $k(x)$ is *semirelevant.*

For statistical purposes, the most interesting forms of functions $k(x)$ are indicator functions. Such functions reduce (4) to forms like (3), and allow interpretations in terms of conditional coverage probabilities. If $k(x) < 0$ is relevant, it is called *negatively biased.* If $k(x) = -I(X \in \mathcal{A})$ then (4) would reduce to (3). *Positively-biased* sets can similarly be defined. In the previously mentioned criticism by Fisher of Welch's solution to the Behrens-Fisher problem, Fisher identified a *negatively-biased relevant* subset.

Buehler and Fedderson (1963) identify, in a special case, a positively-biased relevant subset for the one-sample $t$ interval (they also attribute a similar result to Stein, 1961). Later, Brown (1967) generalized this result to any one-sample $t$ interval. For a random sample $X_1,\cdots,X_n$ from $n(\mu,\sigma^2)$, Brown identified constants $K$ and $\epsilon$ so that

$$P\left(\mu \in \bar{X} \pm tS \,\big|\, |\bar{X}|\,/\,S < K\right) \geq 1-\alpha+\epsilon \quad \forall\mu,\sigma^2 , \tag{5}$$

where $t$ is the cutoff yielding a nominal $1-\alpha$ interval. This can be interpreted as saying that the conditional coverage of the $t$ interval, after accepting $H_0{:}\mu = 0$, is uniformly greater than the nominal level.

Identification of semirelevant subsets is less interesting than identification of relevant subsets, as most procedures with a frequentist guarantee will allow them. For example, from (5) we can deduce

$$P\left(\mu \in \bar{X} \pm tS \,\big|\, |\bar{X}|\,/\,S > K\right) \leq 1-\alpha \quad \forall\mu,\sigma^2 , \tag{6}$$

identifying a negatively-biased semirelevant set for the $t$ interval. However, Robinson (1976) showed that the $t$ interval allows no negatively biased relevant sets. This led him to conclude that elimination of negatively-biased semirelevant sets was too strong a conditional criterion, but elimination of negatively-biased relevant sets was about right. (The elimination of positively biased sets is of lesser concern, as this corresponds to being conservative. However, note there are situations when this direction of error can be important.)

An interesting set of papers are those by Olshen (1973), and Scheffé (1977) with a rejoinder by Olshen (1977). In the 1973 paper, Olshen established a result like (6) for the Scheffé multiple comparisons procedure. Specifically, Olshen showed that the conditional coverage of the Scheffé procedure, given that the ANOVA F test rejects $H_0$, is less than or equal to the nominal level. Thus, Olshen generalized Brown (1967) in one direction, identifying a negatively biased semirelevant set for the Scheffé intervals. Scheffé took exception to this criticism, and answered Olshen in the 1977 article.

The connection between Bayes sets and conditional performance is very strong, as shown by Pierce (1973) and Robinson (1979a). If $\pi(\theta)$ is a proper prior, and we define the pair $< C^{\pi}(x),\gamma^{\pi}(x) >$ by

$$\gamma^{\pi}(x) = \int_{C^{\pi}(x)} \pi(\theta \mid x)\, d\theta , \tag{7}$$

where $\pi(\theta \mid x) = f(x \mid \theta)\pi(\theta)/\int f(x \mid \theta)\pi(\theta)\,d\theta$, then no *semirelevant* functions exist for $< C^{\pi}(x),\gamma^{\pi}(x) >$. Thus, proper Bayes procedures have the strongest possible conditional properties.

Although the connection between Bayesianity and conditional performance is very strong, the exact link has not yet been established. That is, necessary and sufficient conditions for elimination of relevant, or semirelevant, functions have not yet been established. Although the work of Pierce and Robinson, and also Bondar (1977), establishes links between (possibly improper)

Bayes procedures and nonexistence of relevant sets, the ultimate theorem, giving a necessary and sufficient condition, is still not known. The answer, although still unproven due to mathematical technicalities, seems to be that elimination of relevant functions will occur if and only if the procedure is a limit of Bayes rules. Another step in establishing this connection was taken by Casella and Robert (1988), but the full answer remains an open question in the conditional inference literature.

**Frequentist Conditional Inference**

Although proper Bayes rules have strong conditional properties they do not, in general, have good frequentist properties. Even Bayes rules based on "flat priors", such as a Cauchy, which may exhibit some acceptable frequentist performance, cannot maintain a frequentist confidence guarantee. This is a property shared by Bayes credible sets based on proper prior distributions (Hwang and Casella, 1988). However, limits of Bayes rules, or generalized Bayes rules, can maintain a frequentist guarantee, and such procedures may also have acceptable conditional properties. It is within this class that we can find procedures that have acceptable frequentist (or pre-data) properties and acceptable conditional (or post-data) properties.

A confidence set, $C(x)$, is a $1-\alpha$ frequentist confidence procedure for a parameter $\theta$ if

$$P_\theta\Big(\theta \in C(X)\Big) \geq 1-\alpha \qquad \text{for all } \theta \,, \tag{8}$$

that is, the unconditional coverage probability of $C(x)$ is at least $1-\alpha$. Of course, this pre-data guarantee says nothing of the conditional performance of the procedure $< C(x), 1-\alpha >$. Robinson was able to establish conditional properties for several frequentist procedures by using the fact that they are limits of Bayes rules. In particular, his results for the $t$-interval (Robinson, 1976) rely on this fact. Other results (Robinson, 1979b) for frequentist intervals for location or scale families also use arguments based on limiting Bayesianity. Most conditional properties of limits of Bayes rules deal with relevant, rather than semirelevant, functions, and the existence of $\epsilon > 0$ becomes important in the limit. However, for certain procedures from location families, Robinson (1979b) established the nonexistence of semirelevant functions. In particular, if $X \sim f(x-\theta)$, then the procedure

$$< [x-c, x+c],\ 1-\alpha > \,,$$

$$1-\alpha = \int_{-c}^{c} f(t)\,dt \,, \tag{9}$$

is a $1-\alpha$ frequentist confidence procedure that allows no semirelevant functions.

Using different arguments based on invariance, Bondar (1977) established conditional properties of invariant frequentist sets.

The issue that is at the heart of the frequentist/conditional dichotomy is the assignment of a confidence function to a set $C(x)$. For example, for any set $C(x)$, where $X \sim f(x \mid \theta)$, if we define $\gamma(x)$ by

$$\gamma(x) = \frac{\int_{C(x)} f(x \mid \theta)\pi(\theta)\, d\theta}{\int_{\theta} f(x \mid \theta)\pi(\theta)\, d\theta}, \tag{10}$$

where $\pi(\theta)$ is a proper prior, then the procedure $< C(x), \gamma(x) >$ is free of semirelevant sets. However, if $C(x)$ is also a $1-\alpha$ frequentist confidence procedure, this argument does not imply any conditional properties of $< C(x), 1-\alpha >$. Thus, this type of consideration leads to two questions:

$$\begin{array}{ll} \text{i)} & \text{Is } < C(x),\gamma(x) > \text{ a reasonable frequentist procedure?} \\ \text{ii)} & \text{Is } < C(x),1-\alpha > \text{ a reasonable conditional procedure?} \end{array} \tag{11}$$

Since the work of Robinson, and the others, in the 1970s there has been some progress made on the questions in (11). In Casella (1987) it was argued that, with some regularity conditions, a sufficient condition for the frequentist procedure $< C(x), 1-\alpha >$ to be conditionally acceptable is the existence of a (possibly improper) prior $\pi(\theta)$ such that

$$\gamma(x) = \frac{\int_{C(x)} f(x \mid \theta)\pi(\theta)\, d\theta}{\int_{\theta} f(x \mid \theta)\pi(\theta)\, d\theta} \geq 1-\alpha \text{ for all } x. \tag{12}$$

If (12) is satisfied, then the procedure $< C(x), 1-\alpha >$ allows no negatively biased relevant sets, which is acceptable conditional performance. Furthermore, it was demonstrated that such a property held for the multivariate normal confidence set centered at the positive–part James-Stein estimator. Specifically, if $X \sim N(\theta, I)$, a $p$-variate normal random variable ($p \geq 3$), then the confidence procedure $< C_{\delta}(x), 1-\alpha >$ allows no negatively-biased relevant sets, where

$$C_{\delta}(x) = \{\theta\colon |\theta-\delta(x)| \leq c\},\ \delta(x) = \left(1 - \frac{p-2}{|x|^2}\right)^{+} x\,,$$

$$P\left(\chi_p^2 \leq c\right) = 1-\alpha\,.$$

Such a conditional inference strategy was also promoted in Casella (1988), and some other procedures were also examined. In discussing this paper, a number of alternate strategies were put forth. For example, Berger (1988) advocates an "estimated confidence" approach, where the procedure $< C(x),\gamma(x) >$ would be considered *frequency valid* if

$$E_\theta\gamma(x) \leq P_\theta\Big(\theta \in C(x)\Big), \qquad \text{for all } \theta \; , \tag{13}$$

i.e., on the average, the confidence assertion is conservative. Lu and Berger (1989a, b) have applied these ideas to Stein-type problems. Most recently, Brown and Hwang (1989) have shown that for the confidence set $[x-c,x+c]$, where $X = x$ is an observation from $f(x-\theta)$, the confidence procedure $< [x-c,x+c],1-\alpha >$ is admissible, where $1-\alpha = \int_{-c}^{c} f(t)dt$. The admissibility is with respect to the class of confidence procedures $< [x-c,x+c],\gamma(x) >$ (fixed $c$), where $\gamma(x)$ satisfies $E_\theta\gamma(x) \leq 1-\alpha$ (frequentist validity) and the loss function is $L_c(\theta,\gamma(x)) = (\gamma(x)-I(\theta \in [x-c,x+c]))^2$.

Another alternate strategy was described by Lindsay (1988), who suggested attaching both a frequentist and conditional confidence to a given set $C(x)$. Although this is a sensible approach, it is probably the case that practitioners are more comfortable with one number for a confidence assertion. Thus, this reasonable solution might not find acceptability in practice.

Returning to the questions posed in (11), we might now ask what is the reasonable requirement for the confidence assertion to be attached to $C(x)$. Considering the theories of relevant sets, and how confidence sets are used by practitioners, the following strategy seems most reasonable. For a set $C(x)$, assert confidence $\gamma(x)$ where $\gamma(x)$ satisfies (10) for some (possibly improper) prior $\pi(\theta)$. This strategy assures us that $< C(x),\gamma(x) >$ is conditionally acceptable. Moreover, we require that $\gamma(x)$ be valid as a measure of frequentist confidence. Ideally, we would require that $\gamma(x)$ satisfy (12), which not only renders $< C(x),\gamma(x) >$ frequency valid, but also yields the conditional acceptability of $< C(x),1-\alpha >$. However, condition (12) may not always be attainable and, in such a case, we would settle for $\gamma(x)$ satisfying a condition such as (13). This would give some frequentist acceptability to the procedure $< C(x),\gamma(x) >$.

If neither condition (12) nor condition (13) can be attained by a $\gamma(x)$ satisfying (10), then frequentist acceptability may have to be compromised. The frequentist guarantee of the procedure $< C(x),\gamma(x) >$ may then be based on quantities such as $E_\theta\gamma(X)$, $min_\theta E_\theta\gamma(X)$, or $min_x\gamma(x)$ (as long as these last two quantities are positive). The point should be clear. A guaranteed legitimate conditional inference is of primary importance. After that, the frequentist guarantee should be arrived at in *some* reasonable manner.

These ideas have been investigated, in different forms, by Maatta and Casella (1987), Goutis, Casella and Maatta (1989), Goutis and Casella (1989) for estimating a normal variance, and Hwang and Casella (1988) for estimation of a normal mean.

**Discussion**

The ideas behind conditional inference are deep, and here we have superficially sketched one line of work stemming from the developments of Fisher and Basu. There are many ideas in their work, both implicit and explicit, that have not been mentioned. (For example, Basu is an advocate of the Likelihood Principle; and recent work by Casella and Robert, 1988, suggest that violation of this principle immediately leads to the existence of relevant sets.) However, the ideas of conditional inference play an important role in statistics.

Although it might be argued that searching for relevant sets is an occupation only for the theoretical statistician, we must remember that practitioners are going to make conditional (post-data) inferences. Thus, we must be able to assure the user that any inference made, either pre-data or post-data, possesses some definite measure of validity.

**References**

Basu, D. (1964): Recovery of ancillary information, in *Statistics*, Pergammon Press, Oxford, 7-20.

Basu, D. (1981): On ancillary statistics, pivotal quantities, and confidence statements, in *Topics in Applied Statistics*, Y.P. Chaubey and T.D. Dwivedi (eds.), Concordia University, Montreal, 1-29.

Berger, J. O. (1988): An alternative: The estimated confidence approach, in *Statistical Decision Theory and Related Topics IV*, Volume I, S.S. Gupta and J.O. Berger (eds.), 85-90.

Bondar, J. V. (1977): A conditional confidence principle, *Ann. Statist.* 5, 881-891.

Brown, L. D. (1967): The conditional level of Student's t-test, *Ann. Math. Statist.* 38, 1068-1071.

Brown, L. D. and Hwang, J. T. (1989): Estimated confidence approach and the validity admissibility criterion, technical report, Statistics Center, Cornell University.

Buehler, R. J. (1959): Some validity criteria for statistical inference, *Ann. Math. Statist.* 30, 845-863.

Buehler, R. J. and Feddersen, A. P. (1963): Note on a conditional property of Student's *t*, *Ann. Math. Statist.* 34, 1098-1100.

Casella, G. (1987): Conditionally acceptable recentered set estimators, *Ann. Statist.* 15, 1363-1371.

Casella, G. (1988): Conditionally acceptable frequentist solutions, in *Statistical Decision Theory and Related Topics IV*, Volume I, S.S. Gupta and J.O. Berger (eds.), 73-117.

Casella, G. and Robert, C. (1988): Nonoptimality of randomized confidence sets, Biometrics Unit technical report, Cornell University.

Cox, D.R. and Reid, N. (1987): Parameter orthogonality and approximate conditional inference (with discussion), *J. Roy. Statist Soc. Ser. B* 49, 1-39.

Fisher, R. A. (1956): On a test of significance in Pearson's Biometrika tables (No. 11), *J. Roy. Statist. Soc. Ser. B* 18, 56-60.

Fisher, R. A. (1959): *Statistical Methods and Scientific Inference*, Second edition, Hafner, New York.

Goutis, C. and Casella, G. (1989): Improved invariant confidence intervals for a normal variance, Biometrics Unit technical report BU-1038-M, Cornell University.

Goutis, C., Casella, G. and Maatta, J. M. (1989): Conditional confidence of improved variance estimators, Biometrics Unit technical report BU-1039-M, Cornell University.

Hinkley, D.V. (1980): Likelihood, *Can. J. Statist.* 8, 151-164.

Hwang, J. T. and Casella, G. (1988): Frequentist priors, Biometrics Unit technical report, Cornell University.

Kiefer, J.C. (1977): Conditional confidence statements and confidence estimators, *J. Amer. Statist. Assoc.* 72, 789-827.

Lindsay, B. G. (1988): Discussion, in *Statistical Decision Theory and Related Topics IV*, Volume I, S.S. Gupta and J.O. Berger (eds.), 94-98.

Lu, K. and Berger, J. O. (1989a): Estimated confidence procedures for multivariate normal means, *J. Stat. Plann. Inf.* 23, 1-19.

Lu, K. and Berger, J. O. (1989b): Estimation of normal means: The estimated loss frequentist approach, *Ann. Statist.* 17, 890-906.

Maatta, J. M. and Casella, G. (1987): Conditional properties of interval estimators of a normal variance, *Ann. Statist.* 15, 1372-1388.

Olshen, R. A. (1973): The conditional level of the *F*-test, *J. Amer. Statist. Assoc.* 68, 692-698.

Olshen, R. S. (1977): Comment, *J. Amer. Statist. Assoc.* 72, 144-146.

Pierce, D. A. (1973): On some difficulties with a frequency theory of inference, *Ann. Statist.* 1, 241-250.

Robinson, G. K. (1976): Properties of Student's t and the Behrens-Fisher solution to the two means problem, *Ann. Statist.* 4, 963-971.

Robinson, G. K. (1979a): Conditional properties of statistical procedures, *Ann. Statist.* 7, 742-755.

Robinson, G. K. (1979b): Conditional properties of statistical procedures for location and scale parameters, *Ann. Statist.* 7, 756-771.

Scheffé, H. (1977): A note on a reformulation of the S-method of multiple comparisons, *J. Amer. Statist. Assoc.* 72, 143-146.

Stein, C. (1961): Estimation of many parameters, Inst. Math. Statist. Wald Lectures, unpublished.

Tukey, J. W. (1958): Fiducial inference, Inst. Math. Statist. Wald Lectures, unpublished.

## INTERVENTION EXPERIMENTS, RANDOMIZATION AND INFERENCE

Oscar Kempthorne, Department of Statistics, Iowa
State University, Ames, Iowa

**Abstract**

This essay gives a discussion of processes of design and analysis of a study of the effect of two or more interventions or treatments on a set of experimental material (e.g., an agricultural area, or a set of mice, or a human). The problems of design, which includes, critically, the plan by which treatments are conjoined to experimental units, and of analysis are discussed. The author suggests that everything be based on randomization, both design and analysis by randomization tests and inversion thereof. The problem that usual conventional randomization gives bad plans is discussed and suggestion made to overcome it. Parametric models are not used, so defects in conventional parametric inference do not arise. Discussion is given on subjectivity and objectivity.

**Introduction**

The term *experiment* is commonly interpreted to mean a variety of activities. It can mean nothing more than observation of a piece of space-time; e.g., observing the moon by sending a moon shot. It can mean making a piece of material and measuring attributes of this piece. It can mean doing a study to attempt to determine the effects of a treatment protocol on a disease in humans. It is not entirely unusual to refer to a study estimating an attribute of a defined population such as the human population of the United States as an experiment, though most statisticians would say that such a study is a survey. Then we have the writings of theoretical statisticians that an experiment is a triple $(X, A, P(\theta))$ where $X$ is a sample space, $A$ is an algebra of subsets of $X$ and $P(\theta)$ is a set of probability measures indexed by a parameter $\theta$.

I have taken the position that there is a case for distinguishing three types of *experiment* with associated types of inference that I named sampling, observation and experimental (Kempthorne, 1979).

In the sampling problem, there is a real existent population, say, the totality of human beings of the United States. Each individual has unambiguously defined attributes, such as age, height, weight, amount of education and so on. The problem is very simple to state and to understand; namely, what is the frequency distribution of an attribute in this real population?

---

Preprint Number 88-30, Statistical Laboratory, ISU, November 1988, Department of Statistics, Iowa State University, Ames, Iowa 50011. Journal paper no. J-13287 of the Iowa Agriculture and Home Economics Experimental Station, Ames, Iowa, Project 890.

It is easy to imagine having a huge army of enumerators – measurers, so that every human is located, enumerated and measured. The inference problem in this case is also obvious: as a simple example, there is a population of ages, and this population has a mean. An inference problem is then to obtain data and then to make useful statements about the unknown mean.

In the observation problem, we observe a whole population, but we hope and wish that this population that we observe is *representative of* a much larger population. Our explorers on the moon observed a portion of the surface of the moon over a very brief period (hours, I imagine), but the hope is that the observations are more or less typical of what would be observed over an extensive time. Similarly, we hope that our observations of planet Earth relate to its status over a significant period, e.g., years, decades, or centuries, etc. We are currently concerned about the ozone layer and wonder what its status will be in, say, 20 ot 50 years. Obviously, to speculate about this, we must have observations at a few times and validated dynamic model of how the status changes. So, then, in the observation problem, we must have a model that represents what we hypothesize about the unobserved world, unobserved because it is in the past or in the future, or at present and not looked at.

In the present essay, I wish to address solely the third class of problem, which is easily exemplified. Let me give some examples. Atherosclerosis of the heart is a common enough problem: rather worrying, I am sure, and I know. How should this be treated? There are treatments by drugs, by diet, etc., and there is one treatment that is rather *heavy* – heart bypass surgery. There is then an obvious question. Is it a good idea to treat the sick person with bypass surgery? Other *heavy* questions arise with the disease of cancer in humans. What treatments are effective, which treatments are better than other treatments? The nature of situations of this sort is that we have a problem developing under its own dynamic, and the question is of what intervention will help.

## The Intervention Experiment

A rather generally accepted, and, I imagine, not to be challenged, partial model is that we have materiel and a set of interventions. The partial design of the experiment is to partition the experimental materiel into *pieces* and then place one of the interventions on each piece of materiel. The branch of statistics called the design of experiments was started by R. A. Fisher at the Rothamsted Agricultural Experiment Station. The materiel was agricultural land, planted with certain crops such as wheat, or mangolds, or grass, etc., which was partitioned into pieces called plots, and the treatments were various agricultural interventions such as nutritional supplements. An example that seems superficially quite different is a psychological experiment in which the materiel is part of the life of a human subject for example, the 6 days of a week, and the pieces are human-days. The treatments could be various drug regimes. The aim of the experiment might be to palliate depression, for instance.

The performance of the experiment consists of the following steps:

(i) defining the problem, which will consist of specifying the experimental material and specifying the interventions (treatments) that are to be compared;

(ii) *dividing* the experimental material into *plots*, each of which is to receive a treatment;

(iii) deciding how to conjoin the set of *plots* and the set of treatments, taking into account the totally obvious fact that a *plot* can *receive* only one of the treatments;

(iv) letting the experiment proceed to the prechosen termination point; e.g., the point of harvest of an agricultural crop, or recovery or judged failure of a medical treatment;

(v) taking measurements that are thought to be relevant to the problem;

(vi) *analyzing* the resultant data: I put the word analyzing in quotation marks because this is by no means a well-defined operation; and the *drawing of conclusions*, with the same obscurity;

(vii) discussing usefully how the conclusions can be extended to what is often called the *target population*.

**The "Design" of the Experiment**

It is commonplace among statisticians who actually work with real investigators (not individuals who only write about the design of experiments) to consider *all three* of steps (i), (ii) and (iii) as critical components of the design of the comparative intervention experiment. Both adjectives comparative and intervention are essential.

It is useful, I think, to mention for comparison, the type of study in which the outcome is *thought* or *modelled* to be a realization of a random variable, $X$ say, which is distributed according to a distribution determined by some control variables, say $z$, and indexed by some parameter $\theta$, where $z$ and $\theta$ may be vectors. Such a study is purely mathematical.

It is rather obvious, at least by hindsight, that a natural field for thinking about the comparative intervention experiment is farm agriculture or garden agriculture. Suburbia consists mostly of houses on individual lots with associated grassed areas – commonly called lawns. Almost all suburbanites experience problems with their lawns. The grass is thin, is dying or has died. What should be done to obtain a lawn that is *good looking*? What interventions should be made? In trying to teach the design of experiments I have often used this problem as an example. It is not at all surprising that the formulation of a set of procedures for the experiment was done at the Rothamsted Agricultural Experiment Station. The beginning of *experimental agriculture* was made by Lawes and Gilbert in, say, 1843. The most famous Rothamsted experiment is, surely, the Broadbalk field experiment on wheat which was started in 1852 and has continued to present time. The field, Broadbalk, was divided into 13 plots for different nutritional treatments. The yields of wheat were analyzed in a certain way by Fisher (1921). Later Fisher (1924) gave a data analysis of the yields (or years 1852 to 1918) attempting to determine the influence of rainfall on yield.

The use of intervention studies obviously goes back for centuries or millennia – humans found that eating certain plants was harmful or even fatal. It was only in this century that a partial logic was developed.

That the design and analysis of intervention experiments did not originate in connection with human nutrition or human medical problems is not surprising, perhaps, because the comparative intervention experiment requires conjoining one of several treatments to each experimental unit, e.g., human. There were obviously no ethical problems in treating a plot of land with one of several treatments.

There was the recognition that there was variability between experimental units that received the same treatment, and it was obvious that this variability was not the result of measurement error. The existence of such variability was exhibited completely by the various uniformity trials that were conducted, after agricultural scientists recognized that there were problems of design and of analysis.

**The Field Plot Experiment**

Suppose that our initial problem is that of Lawes and Gilbert in 1843. We wish to determine the effectiveness of several nutritional treatments for wheat. We realize that the yield of wheat grown under the same regime varies over England. Obviously, the yields at Rothamsted will not be the same as the yields in Cornwall or even on a farm 5 miles from Rothamsted. We are able to perform the experiment at Rothamsted and have the field Broadbalk to use. Then, obviously, we can hope only to determine somewhat the effectiveness of the treatments on Broadbalk field of Rothamsted. We realize that we can only, at best, determine the differences among treatments as measured on Broadbalk field in year, say, 1852. Suppose that we can determine these differences exactly. Then to apply the results to what will happen elsewhere and in different years (e.g., 1990), the only process we can use is to assume that the treatment differences will be the same or that the differences are related to some variables that are known for the other circumstances.

This thinking leads me to a view of the fundamental problem of what we might (but should not necessarily) call experimental inference. I state this in very simple form:

We have a collection, a set, of experimental material. We have a set of interventions or treatments. Our task is to form judgments on the effects of the treatments on this collection of material.

The extension of conclusions to some larger set of material is a problem I shall not address. I merely make the comment that making the assumption that the material used in the experiment that is performed is a random sample from some large population of material is unjustifiable, though perhaps the only way to make even a guess.

I shall discuss agronomic field experiments later, but I first wish to consider what I call experimentation on a line.

**Experimentation "On A Line"**

Suppose we have an oil processing plant with an inflowing pipeline of feed stock. We wish to examine the differential effects of some treatment processes; e.g., the use of different catalysts. Then our procedure will be to take time slugs of the input and treat each slug with one or other of the treatments. We shall use time slugs that are separated by intervals necessary to make the alterations in the processing and to allow the processing to reach equilibrium status under each given treatment.

As a result of such considerations we shall have experiment time slugs that can be indexed by 1, 2,..., the integers. Suppose now that we have 4 treatments, say $A$, $B$, $C$ and $D$, and we have decided to use 20 successive time slugs. Then the question must be faced of how we are to assign $A$, $B$, $C$, $D$ to the slugs. An obvious suggestion is to use the sequence $ABCDABCD...$ but only a fool would do this. Why do I say this? There will be undoubtedly a *time trend* in the nature of the feed stock and one would expect there to be variation around the time trend. I put the words *time trend* in bold because I find it difficult to find another term. One would expect that if one made a uniformity trial, thereby using only one treatment – say $A$, that the difference squared between observations on different time slugs would depend on the *distance* between the time slugs. In the particular example I am using, the uniformity trial will have been given by preexperiment records.

There would be no computational difficulty with any treatment assignment in using a linear model,

$$y_2 = \tau_{(i)} + e_i,$$

where $\tau_{(i)}$ is the effect of treatment in slug $i$ and $e_i$ is the error, and then to *assume* that the set $\{e_i\}$ is a realization of 20 independent random Gaussian variables that have mean 0 and variance $\sigma^2$ (unknown). From even an elementary first course in statistics one can set this up as a Gauss Markov Normal Linear (GMNL) model, do the ANOVA, make the usual tests of significance, set up the usual *confidence* intervals, etc.

The experimental scientist with even minuscule understanding of variability should object to the plan – the treatment assignment above and the ensuing analysis as given by the usual elementary procedures in the attempt statement of precision of estimation of the differences between treatments – for the simple reason that treatments $A$ and $B$ are contiguous, treatments $A$ and $C$ occur at *points* that are apart by 2 units, and $A$ and $D$ are contiguous half the time and apart by 3 units the other half. So one would expect the difference between treatments $A$ and $B$ to have lower variance than that between $A$ and $C$.

What then should be done? It is a standard cliche of *the design of experiments* that one has to contemplate analysis to evaluate designs. It is less standard (and even not accepted by some) that the proper analysis (*if there is one*, and this is by no means sure) is determined to a considerable extent by the design.

Suppose that one has used the treatment assignment stated above; i.e., *ABCDABCD...ABCD*. At the end of the experiment, one has observations $y_1$, $y_2, \ldots, y_{20}$. How should one "analyze" the data? I imagine that 10 statisticians would produce perhaps 5 different analyses. There is the *obvious* one mentioned above. A second one would be to note that the whole sequence is made up of 5 *blocks* each containing the 4 treatments $A$, $B$, $C$ and $D$. Then to compound the naiveté, the statistician could say that he is doing a randomized block analysis, though this can reasonably be characterized only as a *block analysis*. But why do this? Such an analysis ignores almost completely that the units are *on a line*.

Why not consider the model

$$y_i = \beta_0 + \beta_1 i + \tau_{(i)} + e_i,$$

or

$$y_i = f_i + \tau_{(i)} + e_i,$$

where $f_i$ is some function of $i$ (e.g., a quadratic or higher degree polynomial) and $e_i$ is a term that is called error?

The range of possible models with regard to the systematic part – the non-*error* part of the model is huge. In our little case, it is just the number of functions definable on the set of 20 values of $i$. It is perhaps of interest to mention that I remember with vividness being given a set of data of an experiment like this and the task of *analyzing* the data when I had completed a bachelor degree in mathematics at Cambridge. I was scared stiff – petrified, then. After many decades of being comfortable with the standard programs of *statistical methods*, I find I am again scared, except when randomization is used.

An aspect of standard statistical methods that should cause questioning, but seems not to, is the nature of *error*. What is this error that statisticians talk and write about? One *part* of error is error of measurement, and this is very easy to understand. We have a process of measurement and often, or always in our imagination, we can measure without affecting the object or entity being observed. We assume without questioning, it seems, that individual unknown errors of measurements are independent realization of a scalar random variable. With this mode of thinking, it is natural to think of a large number of measurements of the entity being measured, and that the error of a particular measurement is the deviation of the result from the average. Curiously then, this error is conceptualized by means of what would be observed with repetition, with what might have happened – a notion objectionable, it seems, to *Bayesians.*

In a real experiment with the usual nature of experimental units, there are, in fact, differences between the units, and there will be differences between units in the absence of measurement error, with the same treatment, as we would observe in a uniformity trial. These are called *plot errors* or *experimental unit errors.* Is it proper to use the term *error* for such variability?

Suppose for definiteness that I wish to quantify the result of applying a treatment to 2 units: I do the *experiment* and I obtain 2 numbers $y_1$ and $y_2$. Is the difference between $y_1$ and $y_2$ an indication of error in this little study? We learned in our elementary statistics the role and importance of replication. I suggest, however, that we, including our founding fathers, have not thought out and told us what replication is. It seems easy and unquestionable that replication consists of repetition under constant circumstances. But we never have constant circumstances. Perhaps nearly so in a chemical or physical laboratory but not in, say, interventional research on humans. Fisher (1937, Sections 25 and 26) gives an interesting and relevant but not totally convincing discussion. In the case of the agronomic field experiment, he says that the problem of the impossibility of testing two or more treatments in the same year and on identically the same land can be overcome by testing the treatments on random samples of the same experimental area. Perhaps this will make my doubts seem reasonable. In the case of a field experimental area that is divided into parts, 2 plots are *the same* only if we agree to say this, and if we look at them sufficiently carefully, they will be found to be different. So it seems that we never have what may be called *real replication* in any sort of intervention experiment. This seems almost an absurd line of thought. We can have replication only in the sense of repeating a set of operations (e.g., of baking a cake).

A *natural* model to characterize the variability of the observations is to assume that the errors, $e_i$, $i = 1(1)20$, are a realization of a short section of a time series; e.g., a moving average process or an autoregressive process. However, one can surmise that the choice of a parametric class of models and subsequent fitting will be difficult. Finally, the assessment of uncertainty in treatment effects will be difficult. It is curious that methods based on such ideas have not been well developed and used.

### The Fundamental Problem of the Intervention Experiment

I take the basic common structure of the intervention experiment to be that we have units of material that we index by $i$, and we have interventions that we index by $j$. If we conjoin unit i and intervention $j$, we obtain an observation

$y_{ij}$. The fundamental problem is that we cannot determine how the observation $y_{ij}$ is caused. We cannot conjoin more than one treatment with unit $i$. If, for instance, we could observe $y_{11}$ and $y_{12}$, we could conclude that the effect of treatment 2 minus the effect of treatment 1 on *unit* 1 is $y_{12} - y_{11}$. We shall observe, say, $y_{11}$ and $y_{22}$. Then the difference $y_{22} - y_{11}$ can be attributed equally well (and equally badly) to this difference being the effect of treatment 2 minus the effect of treatment 1 or the effect of unit 2 minus that of unit 1. It is obvious that we have to deal with a set of units, some of which receive treatment 1 and some treatment 2. Suppose then we observe in a small experiment

$$y_{11} = 10,\ y_{22} = 15,\ y_{31} = 13,\ y_{42} = 20.$$

We are inclined to view that the effect of treatment 2 minus the effect of treatment 1 is

$$\frac{1}{2}(15 + 20) - \frac{1}{2}(10 + 13) = 6$$

But we can equally well conclude that this difference should be attributed to

(unit 2 + unit 4) minus (unit 1 + unit 3)

In fact, the size of the experiment is irrelevant to the difficulty. If we have treatment 1 on 1,000 units and treatment 2 on a different set of 1,000 units, whatever mean difference we observed can be equally well attributed to difference of effects of treatment 1 and treatment 2 or the difference between the 2 sets of units.

This leads to the absurd conclusion that we cannot determine whether any intervention produces some effect. Obviously the conclusion is false. What has often enabled the conclusion that an intervention is, e.g., successful, is a sort of empirical Bayesian reasoning. If, for instance, in the past all humans who have contracted a disease subsequently died, and one individual who contracted the disease and received an intervention survived, then one concludes that the intervention was successful. It may be, of course, that there is something unique about the individual and, thus, the intervention has not produced the successful outcome. One guesses that most so-called quack remedies have come about by this route.

This procedure is, of course, the method of historical controls, which has been very successful in many contexts. The method has been successful when the result produced after intervention is hugely different from the historical record.

The first act that must be considered in thinking about an intervention is to ask what the historical record is without intervention and with intervention. Such questioning is usual, of course, in the case of treatment for illness, especially when the intervention is not reversible or removable; e.g., in a partial gastrectomy. In many situations with a new intervention, there is no historical record of the outcome from it. In many cases, the outcome without intervention and with intervention is very variable. Insofar as there is a historical record, it is imprecise and exhibits variability. It would then be very difficult to determine a historical control.

Even though the idea of a historical control is very appealing, there is a very difficult problem of deciding whether a proposed historical control is appropriate. What indeed makes a historical record relevant to evaluation of proposed intervention? In raising this question, I am thinking about interventional studies in connection with human illness and disease. We are told frequently that an attempt to determine if an intervention helps must incorporate its own controls. An exemplar case in which controls must be included in the experiment is that of agricultural research; for example, evaluation of a nutritional treatment on farm animals or farm crops.

Holland (1986) has written very informatively on the general problem I am discussing.

**Design and Analysis**

These are surely interrelated. The quality of a design can be determined only by means of the method of analysis and the quality of the conclusions. So the first step in considering design must revolve around the method of analysis.

The first step in *standard* theory of data analysis is to assume that the data $D$ are a realization of a random variable $X$ that has a distribution function $F_X$, which depends on a parameter $\theta$. The next step is to determine if the data are in agreement with a particular value $\theta_0$.

This step in Neyman-Pearson-Wald theory is to construct a rule for rejecting the hypothesis that $\theta = \theta_0$. This rule is to have the property that the probability under the model that it rejects $\theta = \theta_0$ when $\theta$ is in fact $\theta_0$ is some pre-chosen $\alpha$. Then, with this done for every $\theta_0$, the values of $\theta$ that are not rejected by this rule are said to constitute a $(1 - \alpha)$ confidence set for the unknown $\theta$.

Related to this process, but different from it, is the use of significance levels, often called $P$ values. Inversion of the whole family of related significance tests of $\theta = \theta_0$ for a set of values of $\theta_0$ gives a region of values of $\theta$ that *agree with the data* to a designated extent.

My preference is to regard the regions so obtained as consonance regions, regions that specify values of $\theta$ that are consonant with the data at chosen levels.

These procedures, however characterized by particular words, do not give probabilities of hypotheses such as probability that $\theta$ belongs to any chosen region of the parameter space.

If, then, the aim of the whole exercise, design, performance and analysis of the experiment is the obtaining of such probabilities, the procedures are totally unsuccessful.

The group of statisticians known as Bayesians take the position that the aim of all investigation must be the obtaining of such probabilities. Then it is obvious that one can reach the result with the introduction of a prior distribution. Unfortunately there is no logic that forces choice of a prior. It is the conclusion of this line of development that the probability outcome is a belief probability that depends critically, obviously, on the prior belief probability.

My opinion is that the processes of science and technology do not require belief probabilities. The processes of science and technology require the obtaining of data under circumstances chosen by the investigator, and analysis of the data, which consists of making judgment of whether the data are consonant with particular models suggested by previous investigations or of determining new models from the data that are obtained. The idea that one has a realization from the *holy trinity* (to use a phrase of Basu) is simply ludicrous, so ludicrous that I can only suggest that those who base their ideas of learning about the real world, its present position and its dynamics have no experience of the nature of the processes one must use. One never knows the model! Did Newton know of the inverse square gravitation law? I say, "Obviously not". He and other scientists knew that motion of the planets was elliptic – they knew this by observation and data analysis. The *Bayesians* write as though the past workers *knew* that the law of force was $d^{\gamma}$, where $d$ is the distance and $\gamma$ is a parameter, and that they also had a belief distribution or a prior distribution on $\gamma$.

Another example that comes to my mind, though I have no depth of understanding, is the nature of the universe. It is expanding it seems, but will it continue to do so, or will it stop expanding or stay as it is, or start contracting and reach the size of a golf ball, or something even smaller? The idea that analysis of astronomical data should use a parametric model determined by some $\theta$ with a prior belief distribution on $\theta$ seems to me to be an antithesis of scientific method.

I therefore take the view that the Bayesian prescription, which is being heavily touted as the prescription by which all the uncertainty about this world in which we have to live can be handled, is not worth considering. The prescription is very beautiful in its simplicity and its power. There are many nice theorems in its theory. But it is based on assumptions and ideas that cannot be validated. It is true, of course, that any *reasonable* prior will be overcome by data eventually if the data come from an unvarying stochastic process. This, however, is essentially useless in that (a) any individual has a finite life and (b) the models that are consonant with past data change with new data. A critical process of science is the determination of a model that is consonant with all data accumulated in the past and then challenging that model, which is done only by new *experiments* and determining if predictions from the *old* model are realized in the *new* experiment. The lesson of science of the past century is surely that the models of yesteryear, while having predictive value for circumstances under which they were developed, are found to fail. It follows then that evaluations of goodness of fit (e.g., of the question of whether a prediction and the actual realization agree) is an essential element of science. It is, of course, an essential element of decision making. Where does the particular $(\mathfrak{X}, A, P, \theta)$ come from? The very neat presentations start off with the assumption that this is known. How silly this is! I think I have said enough.

**Randomization "Inference"**

I have tried to communicate my opinion that the usual frequentist theory and Bayesian theory, which purport to address the problems of inference and decision making, are failures. The failure of frequentist theory is not as deep, because it does recognize, though not at all adequately, that a stochastic model for a particular situation is a *pure invention*, which must be discovered, checked out and validated by means of real world data.

It is useful, perhaps, to discuss the matter of subjectivity and objectivity, which seems to require discussion forever (see, for example, Berger and Berry, 1988). The background seems to be the perception that Neyman-Pearson-Wald theory claims to be objective in contradistinction to Bayesian theory, which is subjective. The described polarity is partly *fake*. The real story is that both theories qua theories are theories, and neither is subjective or objective, just as a theory is not heavy or light in the sense of weight avoirdupois.

The only question is whether the practical use of either of the two *rival theories* is subjective or objective. My answer to this is that the NPW theory is partly objective in that the statistical models it uses must be confronted by the associated data, even though theory books say nothing about this. Practitioners of Bayesian theory (*if there really are any*) seem to *pull* their models and their prior distributions *out of thin air* but obviously do not. They do, however, make beliefs an absolutely essential component of their procedures, and any reasonable use of language must characterize the introduction of beliefs as subjective.

In the Bayesian framework, the conditional distribution of the supposed random variable given the parameter value is checkable. If for instance $X|\theta$ is $N(\mu, \sigma^2)$, we can check this by looking at a normal plot. If, however, we wish to adjoin to this the assumption that $\theta$ is $N(\nu, \phi^2)$, how are we to check the appropriateness of this assumption? Someone else could declare that he would like to assume that $\theta$ is distributed Cauchy (or whatever). Even more simply, where do $\nu$ and $\phi^2$ come from? The fact that the values seem not to matter (but, of course, they do!) gives me no comfort, and I think I am not alone.

The obvious conclusions that should be drawn from the objective-subjective polarity that seems to be necessary are twofold:

(a) use of NPW theory requires data confrontation, which is *not* discussed in any theory book but uses use portions of general distribution theory and, obviously, significance tests to make such confrontation;

and

(b) use of Bayesian theory requires data confrontation, but this is not discussed in any exposition of the theory – I include exposition by *any* of the purported founding fathers. I shall not give references. Let any reader of this essay pull his (her) favorite exposition and examine it with respect to what I am discussing; in fact, I think, Bayesian users use the distribution theory and significance tests that NPW users use; finally, Bayesian theory is subjective in that a prior is *plucked out of thin air* or quasi-derived by theory which itself is not validated for use even though based on axioms that seem (but are not) unchallengeable.

The story really is that NPW theory is the half-clothed emperor while Bayesian theory is the emperor without any clothes.

My discussion does not include empirical Bayes procedures which depend on data analysis in the choice of constituent distributions and face the same difficulties as NPW theory.

**Where Do I Come Out?**

I have given my views about the general mix of NPW decision theory and Bayesian theory. It seems to me that there are huge lacunae or gaps between the currently available theory and needed applications.

I now turn to the intervention experiment problem. My perception of the history is that our founding father, Fisher, recognized almost all the problems that I have mentioned, but was not as explicit as he could have been.

I am of the opinion that the assumption in a comparative intervention experiment that the outcome is a random variable from a probability distribution of a family of distributions indexed by some parameter of interest is not supportable.

So the question then is: Can anything be done? An answer is that something can be done; namely, use randomization in the conjoining of units and treatments and then use tests of significance (= tests of consonance) that are based on the frame of reference induced by the randomization process used.

Obviously, I am of the opinion that tests of significance are useful. If one regards them as useless, one is, it seems, in the position of being unable to determine objectively that a data set is not consonant with a particular model.

The value of randomization and the randomization test of significance in the randomized intervention experiment is that the probabilities that arise in the justification are not *belief probabilities* but are *frequency-in-repetition* probabilities determined by the randomization process used.

I think that most of the criticism of use of $P$ values comes from a literal interpretation of Neyman-Pearson theory with its accept-reject rules and its type I error. Such a test of $\theta = 0$, say, carries with it the idea that $\theta$ may really be 0. In the significance testing outlook the achieved significance level is a measure of strength of evidence against the hypothesis $\theta = 0$. Use of the significance test of $\theta = 0$ carries no implication that $\theta$ may be exactly zero. Also, no one should have a strongly different outlook if $P$ were 0.049 rather than 0.051 as Neyman-Pearson theory suggests.

The determination of confidence intervals or regions or, as I prefer, consonance intervals or regions for a parameter $\theta$ requires a formulation of how results resulting from $\theta_2$ will differ from results from $\theta_1$. In the case of intervention experiments, the idea is used that if a unit with intervention j gives a result of $y$ then with that same unit intervention $j'$ would give the result $y + (\tau_{j'} - \tau_j)$.

**What Randomization Process to Use?**

This is, I judge, the basic question to be addressed. I think it was not addressed properly in past years. In experimentation on a line with say 8 units and 2 treatments denoted by $A$ and $B$, the plan

$$AAAABBBB$$

is obviously a bad one.

What makes a plan *bad*? It is obvious that, with $n$ units and $t$ treatments, there are $n^t$ possible treatment assignments, so in the case of 8 units and 2 treatments, there are 64 possible assignments, 2 of which are completely useless. The first plan is bad because the units are on a line. The 4 units that receive $B$ occur later than the 4 units that receive $A$. In the second plan, $B$ occurs after $A$ in step. The third one that I give is a *sandwich* plan, which was discussed by Yates (1939). A plan is bad if the treatment assignment *favors* or *seems* to *favor* the treatments unequally. If one *knows nothing* about the units and they are labelled 1, 2 to 8, the first plan is not "bad". What one wants is that the plan be *balanced* with respect to the variability among the units that *one thinks may be present.* Choice of randomization process is then a matter of informal Bayesian thinking. A plan is *bad* is the investigator thinks so.

With 8 units on a line, *one may have the opinion* that the position of the units on the line tells one nothing about the variability among the units. One may think that most of the variability is expressed by a difference between the first 4 units and the second 4 units. One would then use this partition as a block partition. But, obviously, doing this is only part of the problem. One would still have to decide how to place $A$ and $B$ within each block. The sandwich plan seems not unreasonable. Another plan would be to partition the 8 units segmentally in *blocks of 2.* Then one would have to decide how to place the treatments within the resultant blocks. It is *reasonable* to surmise that

$$| AB | AB | AB | AB |$$

is a *bad* plan.

Suppose we wish to compare 2 treatments on a piece of land. We could partition the land into 2 pieces, one of which would receive $A$ and the other $B$. This would be an appallingly bad choice. Why? What informative model can one use? How could one obtain an idea of error of conclusions? We could divide into 4 pieces of land, into 8 pieces, into 16 pieces, and then decide on a partitioning of the pieces into blocks. We could partition the land into a $2 \times 2$ array and assign the treatments according to a $2 \times 2$ Latin square. We could partition the land into a $4 \times 4$ array and then use a plan in which $A$ and $B$ each occur twice in each row and in each column. There are undoubtedly many other possibilities.

How should one choose among all the possibilities? Why does the problem of choice arise? It arises because we have to decide how to partition the experimental material into *pieces* such that all subpieces of a piece receive the same treatment and then, of course, assign the treatments to the pieces. In the case of the agronomical field plot trial, the *pieces* are called *plots*, and the choice of plots is a matter that is discussed under the rubric *Field plot technique*. I shall not discuss this.

It is obvious intuitively that the *pieces*, the *plots* or the *experimental units* should be partitioned into subsets that are as *alike* as possible, with a subset for each treatment. But one can only guess about the *alikeness* of the units. One's guesses about *alikeness* may prove to be very poor. The actual experiment must be such that one can form a judgment about the *alikeness* of the units and then apply that judgment to form objective judgment about the *alikeness* of units receiving different treatments.

I am saying nothing new in these remarks. The ideas are all in Fisher's *The Design of Experiments*. Fisher discussed only two designs, the randomized block design and the Latin square design in that book. Various other designs are discussed by Cochran and Cox (1957). Later, Yates initiated the ideas of incomplete block designs and designs for two-way elimination of heterogeneity.

The ideas used for making analysis of the resultant data were those of linear models and analysis of variance. Fisher proved (insofar as Fisher proved anything!) that, if one used the customary randomization of the randomized block design and of the randomized Latin square design, then treatment comparisons were unbiased (meaning that the comparisons estimated by the use of the ordinary linear models and the method of least squares were unbiased for what one would observe if one could assign every treatment to every unit). Also the variance over randomizations of estimated treatment comparisons could be estimated by analysis of variance, if unit-treatment additivity holds, though Fisher was not aware of this requirement. Later, Yates gave the idea that the design should be unbiased in the sense that the expectation of the treatment mean square should equal the expectation of the residual (error) mean square in the absence of treatment effects.

The properties indicated in the previous paragraph hold in the case of the randomized design only if each block comprises a completely randomized design. The requirement for the Latin square design is unclear, except that the properties are realized if one chooses a Latin square plan from the totality of Latin squares of the given size.

It was realized, for example, by Grundy and Healy (1950), that the use of such randomization gave some realizations that were *bad*. For example, on a piece of land with 3 blocks of 4 plots, the blocks being aligned, one might obtain the plan:

*BCAD*
*BCAD*
*BCAD*

This is obviously a *bad* plan. An $8 \times 8$ Latin square design involving several factors each at 2 levels could result in the levels of one of the 2 level factors occurring in the 4 quarters of the square. Grundy and Healy made a suggestion of a restricted randomization plan. Youden (1956) discussed the problem, as did Sutter, Zyskind and Kempthorne (1963). There has been extensive work in recent years by Bailey and others.

The whole line of development with regard to restricted randomization appears to have been dominated by analysis of variance unbiasedness.

It is worthwhile to note that the randomized block design is a restriction of the completely randomized design and that the Latin square design is a restriction of a particular randomized block design, so the idea of restricted randomization goes back to the beginnings of the subject of design.

In recent years, I (Kempthorne, 1986a, b) have reached the opinion that the whole matter of randomization, and associated estimation and tests of significance, needs to be rethought in what is, conceptually, a very simple way. We realize, or should do so, that use of the *classical* designs is based on a sort of informal Bayesian process, in which one guesses or judges, or suspects or surmises (*but does not believe*) that the pattern of variability among the experimental units is such and such; for example, units within blocks are very much alike, while the units in different blocks differ appreciably.

The suggested procedure is that the experimenter specifies a set of plans, which he *surmises* will give *fair* comparisons among the treatments. He (she) then uses this set as a randomization frame for choice of plan that is used and for the randomization test of the null hypothesis of no treatment differences and for the randomization test of any shift alternative by adjusting the data to the null hypothesis.

In the case of *experimentation on a line*, the only attribute of a unit that is known is $x$, equal to its position. One can then pick out of the totality of plans, those for which $\Sigma x$ is nearly the same for the various treatments and $\Sigma x^2$ is nearly the same: Any one plan in which this occurs can be regarded as a *systematic design*, of course. Indeed, any plan produced by randomizations looks to be systematic if one looks at it long enough.

I use the case of experimentation on a line because the implications are obvious. The extension of the basic idea to experimentation on a plane or a set of units in $R^k$ is intuitively clear but not easy to implement.

My discussion brings to mind the argumentation in the '30s and '40s about the value of systematic designs. Fisher (1937) gives a discussion on this that is useful but not forcing. He assumed, without even mentioning so, that the *proper* way to analyze a systematic design was by means of the same linear model as that he used for a randomized design. He then gave compelling reasons in that framework for his view that the systematic Latin square designs were variance biased in the sense that the expectation under the null hypothesis of the treatment mean square would be less than the expectation of the *error* mean square. I say that Fisher's discussion is not forcing because it is not at all clear that the analysis of the data set resulting from any plan should be based on the obvious Gauss-Markov-Normal-Linear Model (GMNLN) theory. Considerations of expectations, variances and covariances under randomization does suggest that GMNLN theory can be used as approximating randomization distribution theory if the classical randomization procedures are followed.

### The Work of R. A. Bailey

Bailey (1983, 1985) has written very informatively on restricted randomization versus blocking and cites much literature that is strongly relevant. I suggest that these papers be read. She made (Bailey, 1983, p. 17) critical remarks about blocking that are very similar to those I have made in this essay and in Kempthorne (1986b, c), where I failed badly in not knowing and recognizing her work.

It appears that, if the plots (or units) lie be in a regular configuration with *nice dimensions* (e.g., a $2 \times 4$ or $8 \times 8$ array), one can bring ideas of permutation groups to bear.

I have three comments on this line of work. First, it seems that it is only in very special cases that the conditions demanded can be met. What is a good thing to do, for instance, with 12 units on a line and 3 treatments? Second, the requirement is imposed that the design has to be valid in the sense that the analysis of variance based on a linear model gives a treatment mean square and error mean square that have equal expectations under the randomization in the absence of treatment effects. Third, along with the use of analysis of variance, which I have just questioned, there is the problem of how to make tests of significance and how to make interval statements about treatment effects. This is where Kempthorne *came in* some decades ago. In his book (Kempthorne, 1952), he took the viewpoint that GMNLN theory can be used as an approximation to randomization theory, with respect to estimation of effects, estimation of error and statistical tests (and, hence, intervals on parameters). It is rather obvious, I think, that with restricted randomization this will not happen. It does not happen with small *classical* restricted randomized experiments; e.g., the $3 \times 3$ Latin square design (Kempthorne, 1952, pp. 193-195). It is on the basis of such thinking that I advocate the construction of a list of acceptable plans and using this list for design and statistical testing. The point is that an estimate and standard error of estimate are useless except for the construction of a pivotal, and a pivotal with distribution over 2 or 6 points is rather useless.

### Some Closing Remarks on D. Basu

I have found myself in an anomalous position with respect to the writings of Basu (to whom I have referred at times as *my beloved enemy*). It is obvious that Basu is highly expert in *mathematical statistics* at an advanced measure theory level. I can only admire this aspect. It is also obvious that Basu is deeply interested in inference. I have found that I agree rather strongly with some of his criticisms of NPW theory. However, I judge that Basu is *a sort of Bayesian*, and it is clear from the present essay, I imagine, that I am strongly averse to Bayesian writing that I have seen.

I am particularly averse to the introduction of formal Bayesian processes in the design and analysis of the comparative intervention experiment. I would like to read an account by a dedicated Bayesian of a real experimental situation, with the real outcome and with the statement of conclusions. In the absence of such, I suggest that Bayesian writings be ignored.

I am not at all clear on whether Basu has written on the problems I discuss. I hope that I have not committed any injustices.

I attempted (Kempthorne, 1980) to give my reactions to Basu's writing on the Fisher randomization (Basu, 1980) and decided that repetition of this would serve no useful purpose. The aspect that I did not emphasize then is the matter of design. The obviously Bayesian nature of design surely needs consideration. The discussion or argumentation of 1980 had little, if any, relevance to the problems of experimental method.

**References**

Bailey, R. A. (1983): Restricted randomization, *Biometrika* 70, 183-198.

Bailey, R. A. (1985): Restricted randomization and blocking, *International Statistical Review* 53, 171-182.

Bailey, R. A. (1986): Randomization constrained, *Encyclopedia of Statistical Sciences* 7, 519-524.

Basu, D. (1980). Randomization analysis of experimental data: The Fisher randomization test, *J.A.S.A.* 75, 575-582.

Berger, J. O. and D. A. Berry (1988): Statistical analysis and the illusion of objectivity, *American Scientist* 76, 159-165.

Cochran, W. G. and G. Cox (1957): *Experimental Designs*, Wiley, New York.

Finch, P. D. (1986): Randomization I, *Encyclopedia of Statistical Sciences* 7, 516-519.

Fisher, R. A. (1921): Studies in crop variation. I. An examination of the yield of dressed grain from Broadbalk, *J. Agric. Sci.* 11, 107-135.

Fisher, R. A. (1924): The influence of rainfall on the yield of wheat at Rothamsted, *Phil. Trans. Roy. Soc.* (B) 213, 89-142.

Fisher, R. A. (1935): *The Design of Experiments*, 2nd ed., 1937, Oliver and Boyd, Edinburgh.

Grundy, P. M. and M. J. R. Healy (1950): Restricted randomization and quasi-Latin squares, *J.R.S.S.* (B) 12, 286-2981.

Holland, P. W. (1986): Statistics and causal inference, *J.A.S.A.* 81, 945-970.

Kempthorne, O. (1952): *The Design and Analysis of Experiments*, Wiley, New York (reprinted Krieger).

Kempthorne, O. (1955): The randomization theory of experimental inference, *J.A.S.A.* 50, 946-967.

Kempthorne, O. (1975): Inference from experiments and randomization, in: *A Survey of Statistical Design and Linear Models*, ed. J. N. Srivastava, North Holland Publishing Co., Amsterdam, pp. 303-331.

Kempthorne, O. (1977): Why randomize?, *J. Statistical Planning & Inference* 1, 1-25.

Kempthorne, O. (1979): Sampling inference, experimental inference and observation inference, *Sankhya* 40, 115-145.

Kempthorne, O. (1984): Statistical methods and science, in: W. G. Cochran's *Impact on Statistics*, ed. P. S. R. S. Rao, Wiley, New York, pp. 287-308.

Kempthorne, O. (1986a): Comment on paper by D. Basu, *J.A.S.A.* 75, 584-587.

Kempthorne, O. (1986b): Randomization II, *Encyclopedia of Statistical Sciences* 7, 519-524.

Kempthorne, O. (1986c): Comparative experiments and randomization, in: *Statistical Design: Theory and Practice*, eds. C. E. McCulloch, S. J. Schwager, G. Casella, S. R. Searle, Cornell University, Ithaca, New York, pp. 43-48.

Kempthorne, O. and L. Folks (1971): *Probability, Statistics and Data Analysis*, Iowa State University Press, Ames.

Sutter, G. J., G. Zyskind, and O. Kempthorne (1963): Some aspects of constrained randomization, ARL 63-18 of Aeronautical Research Laboratories.

Yates, F. (1933): The formation of Latin squares for use in agricultural experiments, *Emp. J. Exp. Agric.* 1, 235-244.

Yates, F. (1939): The comparative advantages of systematic and randomized arrangements in the design of agricultural and biological experiments, *Biometrika* 30, 440-469.

Yates, F. (1970): A fresh look at the basic principles of the design and analysis of experiments, in: *Experimental Designs: Selected Papers of Frank Yates,* C. B. E., F. R. S., Hafner, Darien, Connecticut.

Youden, W. J. (1956): Randomization and experimentation, *Ann. Math. Stat.* 27, 1185-1186.

# ANCILLARITY

E. L. Lehmann,[1] Department of Statistics, University of California, Berkeley, California

and

F. W. Scholz, Boeing Computer Services, Seattle, Washington

## Introduction

A statistic is ancillary if its distribution does not depend on the parameters of the model. It might appear at first sight as if ancillary statistics could make no contribution to inference about these parameters. However, as was pointed out by Fisher who first defined and named the concept (1925, 1934, 1935, 1936), this appearance is deceptive. By themselves ancillaries of course carry no information about the parameters, but they may be very useful in conjunction with other parts of the data.

Ancillarity has connections with many other statistical concepts, among them sufficiency, group families, conditionality, completeness, information, prerandomization, and mixtures. Its most important impact on statistical methodology comes from the suggestion that inference should be carried out conditionally given an ancillary statistic rather than unconditionally. For small samples, the resulting conditional procedures can be less efficient than their unconditional counterparts; however, they have the advantage of greater relevance to the situation at hand and frequently are simpler. Typically, the efficiency difference tends to disappear as the sample size becomes large (see for example Barndorff-Nielsen, 1983, and Liang, 1984).

Since ancillaries typically are not unique, the recommendation to condition on an ancillary is not sufficiently specific. Conditioning comes closest to its purpose of making the inference relevant to the situation at hand if the ancillary is maximal, i.e. if there exists no other (nonequivalent) ancillary of which it is a function. The concept of maximal ancillary, which is basic to the theories of ancillarity and conditioning, was introduced by Basu (1959) who showed that maximal ancillaries always exist,[2] but noted that even they may not be unique. In the same paper he also pointed out some measure theoretic complications which require the slightly weaker definition of essential maximality for their resolution. Further results and some basic examples were given in Basu (1964) and some additional generalizations in Basu (1967).

---

[1] Research partially supported by NSF grant DMF-8908670.

[2] For a more precise statement see Theorem 3.

Ancillarity is in a certain sense the dual of sufficiency. If $T$ is a sufficient statistic, then any inference can be based solely on $T$, and the conditional distribution of the full data set $X$ given $T$ is independent of the parameters. Conversely, if $V$ is ancillary, inference may be based entirely on the conditional distribution of $X$ given $V$, while the distribution of $V$ is independent of the parameters. In this duality, a maximal ancillary corresponds to a minimal sufficient statistic. They differ however in that a minimal sufficient statistic is essentially unique and that explicit methods for its construction are available, neither of which is the case for maximal ancillaries.

Systems including sufficient and ancillary statistics as special cases are discussed in Basu (1967). Another common generalization of both sufficiency and ancillarity are the corresponding concepts (partial sufficiency and partial ancillarity) in the presence of nuisance parameters. Discussions of these concepts can be found, for example, in Dawid (1975), Basu (1977), and Barndorff-Nielsen (1978).

General discussions of various aspects of ancillarity are given by Cox and Hinkley (1974), Hinkley (1980b), Buehler (1982), Kalbfleisch (1982), and Lehmann (1986). A recent important development is the extension to asymptotic ancillarity, i.e. statistics with limit distribution independent of the parameters, and from that to higher order and local ancillaries. In the present paper, we shall restrict attention to exact ancillaries with respect to all unknown parameters, i.e. in theoriginal sense considered by Fisher and Basu. However, work on both partial and approximate ancillaries is included in the references.

## Relation to Other Concepts

### 1. Group families

A *group family* or *transformation model* is obtained by subjecting a random variable with a fixed distribution to a group $\mathcal{G}$ of transformations. Any statistic $V(X)$ that is invariant under $\mathcal{G}$ is ancillary. Thus in particular a maximal invariant with respect to $\mathcal{G}$ is ancillary.

**Example 1. Location family.**

Let $X = (X_1,\ldots,X_n)$ be distributed according to a location family with density

$$f(x_1 - \theta) \cdots f(x_n - \theta).$$

This is a group family obtained by subjecting a random variable $X = (X_1,\ldots,X_n)$ with density $f(x_1,\ldots,x_n)$ to the group of transformations

$$X_i' = X_i + c, \qquad i = 1,\ldots,n, \qquad -\infty < c < \infty.$$

A maximal invariant is the set of differences

$$Y = (X_1 - X_n, \ldots, X_{n-1} - X_n).$$

This is the example with which Fisher introduced the concept of ancillarity.

For some general results for the case of group families see Barndorff-Nielsen (1980).

## 2. Mixture experiments

Suppose a family of experiment $\mathcal{E}_z$, $z \in \mathcal{Z}$ is available, each experiment consisting of a family of distributions $\mathcal{P}_z = \{P_{z,\theta}, \theta \in \Omega\}$, labeled by the same parameter $\theta$, i.e. corresponding to the same states of nature. A value of $z$ is selected according to a known distribution $\Pi$ and the experiment $\mathcal{E}_z$ is performed, resulting in the observation of a random quantity $X$ with distribution $P_{z,\theta}$. For the final result $X$ of such a mixture experiment, $Z$ is ancillary since its distribution $\Pi$ is known.

**Example 2. Two workers.**

Let

$$\mathcal{E}_0 = (X, \mathcal{P}), \quad \mathcal{P} = \{P_\theta, \theta \in \Omega\}$$

$$\mathcal{E}_1 = (Y, \mathcal{Q}), \quad \mathcal{Q} = \{\mathcal{Q}_\theta, \theta \in \Omega\}$$

be two experiments, corresponding for example to two different workers $A$ and $B$ performing a needed experimental task. One of the workers is chosen at random (with probability 1/2 each) and is assigned to perform the experiment. Here a random variable taking on the values of 0 and 1 as worker $A$ or $B$ is chosen plays the role of $Z$. The example, which was first discussed in this context by Cox (1958), makes clear the appeal of conditioning on the experiment actually performed.

Mixture models appear to represent a rather special case of models admitting ancillaries but in fact, unlike group families, they cover all cases. To see this, suppose that $X$ is distributed according to one of the distributions $P_\theta$, $\theta \in \Omega$ and that $V$ is ancillary for $X$. For each value $v$, let $\mathcal{E}_v$ be the experiment consisting in observing a random quantity $X'$, distributed according to the conditional distribution of $X$ given $v$. Then $X'$ is the outcome of a mixture experiment and its distribution is the same as that of $X$.

Some authors have introduced distinctions between real and conceptual (Basu, 1964) or experimental and mathematical (Kalbfleisch, 1975, 1982) ancillaries. However, these distinctions require going outside the postulated models and are based on considerations involving other models.

## 3. Conditionality; pre-randomization

Fisher's suggestion that inference should be conditional on an ancillary is called the *principle of conditionality*. As was discovered by A. Birnbaum (1962), conditionality has surprisingly strong consequences for the foundations of statistics since in conjunction with sufficiency it implies the likelihood principle. For discussions of this result and its consequences see Rao (1971), Basu (1975), Joshi (1983), Berger and Wolpert (1984), and Evans, Fraser and Monette (1986).

Typically, conditioning on ancillaries seems reasonable. However, it runs into difficulty when the design involves deliberate randomization (e.g. random selection of a sample, random assignment of subjects, or random choice of a Latin square). Since the random selection process with known probabilities is ancillary, the conditionality principle would require conditioning on the selected arrangement, thus largely vitiating the purposes of randomization. This difficulty is discussed, for example, in Basu (1969, 1978, 1980), Berger and Wolpert (1984), and Finch (1986).

## 4. Sufficiency

Sufficient statistics provide data reduction without loss of information. The amount of reduction that can be achieved in this way depends on the situation.

**Example 1. Location family (continued).**

If the density $f$ in Example 1 is the standard normal density, sufficiency reduces the full $n$-dimensional sample $X_1,\ldots,X_n$ to the single statistic $\bar{X} = \Sigma_{i=1}^{n} X_i/n$, regardless of the size of $n$. On the other hand, if $f$ is, for example, the logistic, Cauchy, or double exponential density, the minimal sufficient statistic is the set of order statistics $X_{(1)} \leq \ldots \leq X_{(n)}$, so that there is hardly any reduction. As discussed in Lehmann (1981), the amount of reduction depends essentially on how much of the ancillary information the minimal sufficient statistic retains.

## 5. Completeness

The most favorable situation for reduction by means of a sufficient statistic $T$ is that in which all ancillaries are independent of $T$. A sufficient condition for this to occur is given by the following result which (together with a converse) is known as Basu's theorem (Basu, 1955, 1958, 1982 and Koehn and Thomas, 1975).

**Theorem 1. (Basu).**

If $T$ is boundedly complete, then every ancillary is independent of $T$.

That bounded completeness is not necessary for every ancillary to be independent of $T$ can be seen for instance from examples in which the constants are the only ancillaries. A condition that is necessary, but not sufficient, is provided by the concept of *weak completeness*, introduced by Basu and Ghosh (1968), and independently in the present context by Lehmann (1981) under the term $\mathcal{F}_0$-completeness.

**Definition 1.**

A statistic $T$ is weakly complete with respect to a family $\mathcal{P}^T = \{P_\theta, \theta \in \Omega\}$ of distributions of $T$ if

$$E_\theta f(T) = 0 \text{ for all } \theta \in \Omega \Rightarrow f(t) = 0 \text{ (a.e. } \mathcal{P}^T)$$

for all two-valued functions $f$.

As we shall see later, this concept is central to the study of maximal ancillaries.

**Note.** A (not very useful) completeness condition that is both necessary and sufficient for every ancillary to be independent of $T$ is given by Lehmann (1981).

## 6. Conditionality and sufficiency in conflict

The principles of conditionality and sufficiency may conflict, as in the following example of Becker and Gordon (1983), which is essentially equivalent to one considered in a different content by Fisher (1956, p. 47).

**Example 3. Quadrinomial.**

Consider $n$ quadrinomial trials with the probabilities of the four outcomes being

$$p_1 = \frac{1+\theta}{5}, \quad p_2 = \frac{1-\theta}{5}, \quad p_3 = \frac{1-\theta}{5}, \quad p_4 = \frac{2+\theta}{5}, \quad -1 < \theta < 1,$$

and with $N_1,\dots,N_4$ denoting the numbers of the trials resulting in these outcomes. Then $T = (N_1, N_2+N_3, N_4)$ is minimal sufficient and it appears that there are no ancillaries based on $T$. On the other hand, $A = (N_1+N_2, N_3+N_4)$ is clearly ancillary, and so is $B = (N_1+N_3, N_2+N_4)$.

It seems clear to the present authors that here sufficiency should be given priority over ancillarity, and inference should be based on $T$. For otherwise, given a trinomial situation with probabilities $((1+\theta)/5, (1-2\theta)/5, (2+\theta)/5))$, (the distribution of $T$), we would prefer a procedure that would require dividing the trials in the middle category, each with probability 1/2 between two artificial subcategories. This seems very unappealing.

## 7. Similar regions and regions of Neyman structure

A set $S$ in the sample space is a similar region with respect to a family $\mathcal{P} = \{P_\theta, \theta \in \Omega\}$ if $P_\theta(X \in S)$ does not depend on $\theta$, i.e. if its indicator is ancillary. The set $S$ is said to have Neyman structure with respect to a sufficient statistic $T$ if the conditional probability

$P(X \in S|t)$ is independent of $t$ a.e.

Suppose now that $T$ is boundedly complete. Then by Theorem 1 every ancillary – and therefore the indicator $I_S$ of any similar region – is independent of $T$ and therefore has Neyman structure. The characterization of all similar regions as having Neyman structure in the presence of a complete sufficient statistic is therefore mathematically (although not in its interpretation) equivalent to Theorem 1.

## 8. Information

Fisher's primary interest in introducing ancillary statistics was the *recovery of information*. If $I_X(\theta)$ and $I_{\hat{\theta}}(\theta)$ denote the amount of Fisher information in the sample $X$ and the maximum likelihood estimator $\hat{\theta}$ respectively, then it will often happen that[1] $I_{\hat{\theta}}(\theta) < I_X(\theta)$, so that $\hat{\theta}$ is not fully informative. Fisher discovered that the lost information can be recovered if there exists an ancillary statistic $V$ such that $(\hat{\theta}, V)$ is sufficient, in the following sense. If $I_{\hat{\theta}|v}(\theta)$ is the information carried by $\hat{\theta}$ in the conditional distribution given $V = v$, then

$$EI_{\hat{\theta}|V}(\theta) = I_X(\theta). \tag{1}$$

For a discussion of the implementation of this program in two important classes of models, see Barndorff-Nielsen (1980). When (1) holds, the average conditional information equals the whole information in the sample; for particular values of $v$, the conditional information of $\hat{\theta}$ given $v$ may be smaller or larger than $I_X(\theta)$.

Recall now the other motive for conditioning on ancillaries: to make the inference more relevant to the situation at hand. Cox (1971) points out that ancillaries are therefore most useful when the amount $I_v(\theta)$ of information in the conditional distribution of $X$ given $v$ varies widely with $v$, so that some values of $v$ are much more informative than others. This point is nicely illustrated by Example 2, where conditioning on the chosen worker seems particularly important when there is a big difference in the quality of their work.

In the light of this remark, Cox suggests that when the maximal ancillary is not unique, that ancillary should be preferred for which $I_v(\theta)$ is most variable, e.g. for which the variance $var[I_V(\theta)]$ is the largest.

## Weak Completeness

The central concept for the characterization of maximal ancillaries is weak completeness. It is easy to see that the definition of weak completeness given in the preceding section is equivalent to the following statement.

---

[1]We have here assumed for the sake of simplicity that $\theta$ is real valued.

The family $\mathcal{P} = \{P_\theta, \theta \in \Omega\}$ is weakly complete if any measurable set $A$ with probability independent of $\theta$ has probability 0 or 1. (2)

This is the form in which the definition was given by Basu and Ghosh (1969). A simple restatement of (2) yields Theorem 2.

**Theorem 2.**

A family $\mathcal{P}$ admits no nontrivial ancillaries (i.e. any ancillary statistic is almost surely constant) if and only if $\mathcal{P}$ is weakly complete.

To illustrate the situation of no ancillaries consider the following examples.

**Example 3. No ancillaries.**

Let $X_i$ be independent $N(\theta_i, 1)$, $i = 1,\ldots,n$. Then $X = (X_1,\ldots,X_n)$ is complete, hence weakly complete, and so there are no ancillaries.

**Example 4. Sequential binomial sampling.**

Consider a sequence of binomial trials, with success probability $p$ and a stopping rule (with probability 1 of eventually stopping). This can be represented by a random walk in the plane starting at the origin, with a unit step to the right for a success and a unit step up for a failure. The stopping rule is represented by a set of stopping points. The observation is a path starting at (0, 0) and ending at some stopping point $(a, b)$. Since every path ending at $(a, b)$ has probability $p^a(1 - p)^b$, it follows that the coordinates $(a, b)$ of the stopping point constitute a sufficient statistic, which may or may not be complete (necessary and sufficient conditions for completeness are given in Lehmann and Stein, 1950). The path itself is of course not complete except in the rare cases in which there is only one path to each stopping point.

(i) In light of this it is very surprising that not only the endpoint but also the path itself is weakly complete, provided the stopping rule has a finite boundary point on the $x$ or the $y$ axis. To see this let $S$ be a set of paths with $P_p(S) = c \; \forall \; p \in (0, 1)$. Suppose the stopping rule has a finite boundary point $(0, k)$ for some $k \geq 1$. Then the path $\pi_0$ from (0, 0) to $(0, k)$ is either contained in $S$ or in its complementary set of paths $S^c$. It follows that either $c = P_p(S) \geq P_p(\pi_0) = (1 - p)^k \rightarrow 1$ or $1 - c = P_p(S^c) \geq P_p(\pi_0) = (1 - p)^k \rightarrow 1$ as $p \rightarrow 0$ so that either $c = 1$ or $c = 0$. The case of finite boundary point $(k, 0)$ is treated similarly. Hence there are no ancillaries.

(ii) If there is no bound on the stopping rule along the $x$- or $y$-axis, then weak completeness may not obtain as the following example shows. Perform the binomial trials in pairs until the first time that either (*success*, *failure*) or (*failure*, *success*) is observed. Then the set $S$ of paths that end in (*failure*, *success*) has probability 1/2 for all $p \in (0, 1)$.

**Note.** Exactly the same result as in Example 4 with the same proof applies to sequential sampling from trinomial (or any multinomial) trials.

The following example is due to Basu and Ghosh (1969) where many additional examples can be found.

**Example 5. Two-point location families.**

Let $X$ take on the two values $\theta$ and $\theta + c$ with probabilities

$$P(X = \theta) = \pi,$$

$$P(X = \theta + c) = 1 - \pi, \quad -\infty < \theta < \infty,$$

$\pi$ and $c$ known. Then $X$ is weakly complete provided $\pi \neq 1/2$, but not when $\pi = 1/2$. In the latter case any set $A$ whose complement is $A + c$ has probability $1/2$, independent of $\theta$.

It turns out that Theorem 2 is a special case of a general characterization of maximality for an ancillary statistic $V$, given in its proper setting in Theorem 4. Loosely, this characterization finds $V$ to be maximal if and only if the family of conditional distributions of $X$ given $V$ is weakly complete. In the situation of Theorem 2, where $V$ is constant, this family of conditional distributions coincides with the family $\mathcal{P}$ of distributions for $X$.

In the case when the only ancillary statistics are the a.s. constant functions there (usually) does not exist a maximal ancillary (due to null set problems) but a maximal ancillary $\sigma$-field $\mathcal{A}_m$ does exists, see Theorem 2. The reason is that not every $\sigma$-field is induced by a statistic. Since the $\sigma$-field induced by an a.s. constant function is essentially equivalent to $\mathcal{A}_m$ (to be made precise below) it makes sense to call such an a.s. constant function essentially maximal ancillary; the alternative would be to admit that there are no maximal ancillary statistics due to null set problems. This state of affairs carries over to the general case and the above loosely stated characterization is that of essential maximal ancillarity. Bearing this in mind one may want to accept that characterization and skip or skim the next two sections.

## Notation and Definitions

Let $(\mathcal{X}, \mathcal{B})$ be an arbitrary measurable space and $\{P_\theta, \theta \in \Omega\}$ be a family of probability measures on $\mathcal{B}$. Considering $\mathcal{X}$ as the sample space we denote the random element in $\mathcal{X}$ by $X$ and write $P_\theta(X \in B) = P_\theta(B) \ \forall\ B \in \mathcal{B}$. We now give some definitions and a theorem taken from Basu (1959).

**Definition 2.**

A $\sigma$-field $\mathcal{A} \subset \mathcal{B}$ is said to be *ancillary* if $P_\theta(A)$ is constant in $\theta \in \Omega \ \forall\ A \in \mathcal{A}$.

**Comment.** One easily sees that $\mathcal{A}$ is ancillary iff $\int f(x) dP_\theta(x)$ is constant in $\theta \in \Omega$ for all integrable and $\mathcal{A}$-measurable functions $f$: $\mathfrak{X} \to R$.

**Definition 3.**

If $V$: $(\mathfrak{X}, \mathfrak{B}) \to (\mathfrak{Y}, \mathfrak{C})$ is a statistic $(\mathcal{A}_V := V^{-1}(\mathfrak{C}) \subset \mathfrak{B})$ then $V$ is said to be ancillary if $\mathcal{A}_V$ is ancillary.

**Comment.** Rather than dealing with (ancillary) statistics we follow Basu's example and continue the following theoretical exposition in terms of (ancillary) $\sigma$-fields. When dealing with concrete examples we will use the more intuitive term *statistic* in place of $\sigma$-field. Hence it is understood that the following definitions in terms of $\sigma$-fields have analogous counterparts in terms of *statistics*.

**Definition 4.**

An ancillary $\sigma$-field $\mathcal{A} \subset \mathfrak{B}$ is said to be *maximal ancillary* if there exists no other ancillary $\sigma$-field $\mathcal{A}^* \subset \mathfrak{B}$ such that $\mathcal{A} \subset \mathcal{A}^*$.

**Theorem 3. (Basu, 1959).**

Given an ancillary $\sigma$-field $\mathcal{A} \subset \mathfrak{B}$ there exists a maximal ancillary $\sigma$-field $\mathcal{A}_m \subset \mathfrak{B}$ such that $\mathcal{A} \subset \mathcal{A}_m$.

**Definition 5.**

Two $\sigma$-fields $\mathcal{A}_1, \mathcal{A}_2 \subset \mathfrak{B}$ are said to be *essentially equivalent* if for any $A_1 \in \mathcal{A}_1$ $(A_2 \in \mathcal{A}_2)$ there exists an $A_2 \in \mathcal{A}_2$ $(A_1 \in \mathcal{A}_1)$ such that

$$P_\theta(A_1 \Delta A_2) = 0 \quad \forall\, \theta \in \Omega.$$

**Definition 6.**

Any ancillary $\sigma$-field that is essentially equivalent to a maximal ancillary $\sigma$-field is called *essentially maximal ancillary.*

**Comment.** Although Theorem 3 guarantees the existence of a maximal ancillary $\sigma$-field $\mathcal{A}_m$ containing any given ancillary $\sigma$-field $\mathcal{A}$ the same does not necessarily hold for statistics. The reason is that $\mathcal{A}_m$ is usually too rich to be generated by any statistic $V$.

The following definition of conditional weak completeness is a direct adaptation of the concept of weak completeness to the conditioned case.

**Definition 7.**

$X$ given $\mathcal{A}$ is said to be *conditionally weakly complete* if for any given function

$$g(x) = a(x) I_B(x) + b(x) I_{B^c}(x)$$

with $B \in \mathfrak{B}$, $a(\cdot)$ and $b(\cdot)$ $\mathcal{A}$-measurable and such that

$$\forall\, \theta \in \Omega \;\; E_\theta(g(X)|\mathcal{A}) = 0 \;\text{ a.s. } (P_\theta)$$

we have

$$\forall\, \theta \in \Omega \;\; P_\theta(g(X) = 0|\mathcal{A}) = 1 \;\;\text{ a.s. } (P_\theta),$$

i.e. $P_\theta(g(X) = 0) = 1 \;\forall\, \theta \in \Omega$.

An equivalent formulation of Definition 7, without the *a.s.* qualifiers, is Definition 7′.

**Definition 7′.**

$X$ given $\mathcal{A}$ is said to be *conditionally weakly complete* if for any given function

$$g(x) = a(x)I_B(x) + b(x)I_{B^c}(x)$$

with $B \in \mathfrak{B}$, $a(\cdot)$ and $b(\cdot)$ $\mathcal{A}$-measurable and such that

$$E_\theta(I_A(X)g(X)) = 0 \quad \forall\, \theta \in \Omega \quad \text{and } \forall\, A \in \mathcal{A}$$

we have $P_\theta(g(X) = 0) = 1 \quad \forall\, \theta \in \Omega$.

Note that Definitions 7 and 7′ are not contingent on the existence of regular conditional distributions. However, if $X$ admits regular conditional distributions given $\mathcal{A}$ a natural question is: how does weak completeness of a family of regular conditional probability distributions relate to the conditional weak completeness of $X$ given $\mathcal{A}$ defined above? Lemma 1 will provide a partial answer under certain regularity conditions. These conditions are as follows:

i) $\Omega$ is a separable topological space,

ii) $\mathcal{A}$ is generated by the ancillary statistic $V$: $(\mathfrak{X}, \mathfrak{B}) \to (\mathfrak{Y}, \mathfrak{C})$,

iii) $\forall$ v $\in \mathfrak{Y}$: $\{f_\theta(\cdot|v),\ \theta \in \Omega\}$ is a family of conditional densities for $X$ given $V = v$ with respect to a $\sigma$-finite dominating measure $\mu$ on $(\mathfrak{X}, \mathfrak{B})$,

iv) $\exists\, N \in \mathfrak{C}$ with $P(V \in N) = 0$ so that $\forall$ v $\in N^c$ we have $f_\theta(x|v) \to f_{\theta_0}(x|v)$ a.s. in $x[\mu]$ whenever $\theta \to \theta_0$,

**Lemma 1.**

Under conditions i) - iv) the weak completeness of the families $\{f_\theta(\cdot|v),\ \theta \in \Omega\}$ $\forall$ v $\in N_1^c$ with $P(N_1) = 0$ implies the conditional weak completeness (Definition 7) of $X$ given $\mathcal{A}$.

**Proof:** Let g be as in Definition 7, then for any $\theta \in \Omega$ we have

$$0 = E_\theta(g(X)|V) = \int g(x) f_\theta(x|V)\, d\mu(x) \quad \text{a.s. } P. \tag{3}$$

Since the exceptional null set may depend on $\theta$ (through $f_\theta$) we invoke (3) for all $\theta$ in a countably dense subset of $\Omega$. Using Scheffé's theorem in conjunction with iv) it follows that there exists a set $N_0 \in \mathcal{C}$ with $P(V \in N_0) = 0$ such that for $v \in N_0^c$ we have

$$0 = \int g(x) f_\theta(x|v)\, d\mu(x) \quad \forall\, \theta \in \Omega$$

which by weak completeness of the conditional densities entails for all $v \in N_0^c$

$$\int I_{\{g(X)=0\}} f_\theta(x|v)\, d\mu(x) = 1 \quad \forall\, \theta \in \Omega,$$

hence $P_\theta(g(X) = 0) = 1\ \forall\, \theta \in \Omega$.

It is not clear whether the converse of Lemma 1 is true under the stated conditions.

**Characterization of Maximal Ancillarity**

The following theorem will give necessary and sufficient conditions for an ancillary $\sigma$-field $\mathcal{A} \subset \mathcal{B}$ to be essentially maximal ancillary. A special case of Theorem 4 was proved by Basu and Ghosh (1969) for the case of a dominated location family.

**Theorem 4.**

If $\mathcal{A} \subset \mathcal{B}$ is ancillary, then the following statements are equivalent:

i) $\mathcal{A}$ is essentially maximal ancillary.

ii) $\nexists\, B \in \mathcal{B}$ such that $P_\theta(B|\mathcal{A})$ admits a version $\psi_B$ ($\mathcal{A}$-measurable) independent of $\theta \in \Omega$ with $P(0 < \psi_B(X) < 1) > 0$.

iii) $X$ given $\mathcal{A}$ is conditionally weakly complete.

**Proof. i) $\Rightarrow$ ii):** $\mathcal{A}$ be ancillary and let $B \in \mathcal{B}$ be such that $P_\theta(B|\mathcal{A})$ admits a version $\psi_B$ ($\mathcal{A}$-measurable) independent of $\theta \in \Omega$. First note that the smallest $\sigma$-field $\mathcal{A}_B$ containing both $\mathcal{A}$ and $B$ is ancillary, since

$$P_\theta(A \cap B) = \int_A \psi_B(x)\, dP_\theta(x) \quad A \in \mathcal{A}$$

is independent of $\theta \in \Omega$ ($\mathcal{A}$ is ancillary and $\psi_B$ is $\mathcal{A}$-measurable and independent of $\theta \in \Omega$) and since this property extends to all of $\mathcal{A}_B$ by the usual unique measure extension.

Next let $A_0 = \{x \in \mathfrak{X} : 0 < \psi_B(x) < 1\} \in \mathcal{A}$. Assuming $\mathcal{A}$ to be essentially maximal ancillary we can find $A_1 \in \mathcal{A}$ such that $N = A_1 \Delta (A_0 \cap B) \in \mathcal{A}_B$ has probability zero for all $\theta \in \Omega$. Then

$$I_{A_1}(X) = I_{A_0}(X) I_B(X) \quad \forall\, X \in N^c$$

and taking conditional expectation given $\mathcal{A}$ we have

$$I_{A_1}(X) = I_{A_0}(X) \psi_B(X) \quad \text{a.s. } P_\theta\ \theta \in \Omega$$

which implies $P(0 < \psi_B < 1) = 0$, thus i) $\Rightarrow$ ii).

**ii) $\Rightarrow$ iii):** Let

$$\begin{aligned} g(x) &= a(x) I_B(x) + b(x) I_{B^c}(x) \\ &= (a(x) - b(x)) I_B(x) + b(x) \end{aligned}$$

$B \in \mathfrak{B}$, $a(\cdot)$ and $b(\cdot)$ $\mathcal{A}$-measurable such that

$$\forall\, \theta \in \Omega \quad E_\theta(g(X)|\mathcal{A}) = 0 \quad \text{a.s. } P_\theta. \tag{4}$$

Let $C_0 = \{x \in \mathfrak{X} : a(x) \neq b(x)\}$, $B_0 = C_0 \cap B$ and

$$\begin{aligned} \psi_{B_0}(x) &= b(x)/(b(x) - a(x)) && \in C_0 \\ &= 0 && x \in C_0^c \end{aligned}$$

The condition (4) on $g$ implies that $\psi_{B_0}$ may serve as a $\theta$-independent version of $P_\theta(B_0|\mathcal{A})$ $\forall\, \theta \in \Omega$, since

$$\begin{aligned} 0 &= E_\theta(I_{C_0}(X) g(X)|\mathcal{A}) \\ &= (a(X) - b(X))\ P_\theta(B_0|\mathcal{A}) + b(X) I_{C_0}(X) \quad \text{a.s. } P_\theta, \end{aligned}$$

i.e.

$$\begin{aligned} X \in C_0 \ \Rightarrow P_\theta(B_0|\mathcal{A}) &= b(X)/(b(X) - a(X)) \\ &= \psi_{B_0}(X) \quad \text{a.s. } P_\theta \end{aligned}$$

and for $X \in C_0^c \Rightarrow P_\theta(B_0|\mathcal{A}) = 0 = \psi_{B_0}(X)$ a.s. $P_\theta$.

Condition (4) also implies

$$0 = E_\theta(g(X) I_{C_0^c}(X)|\mathcal{A}) = g(X) I_{C_0^c}(X) \qquad \text{a.s. } P_\theta\ \forall\, \theta \in \Omega. \tag{5}$$

Since $P(0 \leq \psi_{B_0} \leq 1) = 1$

$$\text{ii)} \Rightarrow P(\psi_{B_0} \in \{0, 1\}) = 1$$

$$\Rightarrow I_{B_0}(X) = \psi_{B_0}(X) \quad \text{a.s.} \quad P_\theta \ \forall\ \theta \in \Omega$$

$$\Rightarrow g(X)I_{C_0}(X) = 0 \ \text{ a.s.} \quad P_\theta \ \forall\ \theta \in \Omega$$

since

$$\begin{aligned} 0 &= E_\theta(g(X)I_{C_0}(X)|\mathcal{A}) \\ &= (a(X) - b(X))\psi_{B_0}(X) + b(X)I_{C_0}(X) \\ &= (a(X) - b(X))I_{B_0}(X) + b(X)I_{C_0}(X) \\ &= I_{C_0}(X)g(X) \quad \text{a.s.} \ P_\theta \ \forall\ \theta \in \Omega. \end{aligned}$$

This together with (5) implies $P_\theta(g(X) = 0) = 1 \ \forall\ \theta \in \Omega$, i.e. ii) $\Rightarrow$ iii).

**iii) $\Rightarrow$ i):** By theorem 4.1 there exists a maximal ancillary $\sigma$-field $\mathcal{A}_m \supset \mathcal{A}$. Let $D_0 \in \mathcal{A}_m$ and for some fixed $\theta_0 \in \Omega$ and some version $P_{\theta_0}(D_0|\mathcal{A})$ let $\psi_{D_0}(x) := P_{\theta_0}(D_0|\mathcal{A})$, then for $A \in \mathcal{A}$:

$$\begin{aligned} \int_A P_\theta(D_0|\mathcal{A})dP_\theta &= P_\theta(A \cap D_0) = P_{\theta_0}(A \cap D_0) \\ &= \int_A \psi_{D_0}(x)dP_{\theta_0}(x) = \int_A \psi_{D_0}(x)dP_\theta(x), \end{aligned}$$

i.e. $\psi_{D_0}$ may serve as a $\theta$-independent version of $\mathrm{P}_\theta(D_0|\mathcal{A}) \ \forall\ \theta \in \Omega$. Let

$$g(x) = I_{D_0}(x) - \psi_{D_0}(x)$$

then $\forall\ \theta \in \Omega \quad E_\theta(g(X)|\mathcal{A}) = 0$ a.s. $P_\theta$, which under iii) implies $P_\theta(g(X) = 0) = 1 \ \forall\ \theta \in \Omega$, i.e.

$$\psi_{D_0}(X) = I_{D_0}(X) \quad \text{a.s.} \ P_\theta \quad \forall\ \theta \in \Omega$$

which shows $\mathcal{A}$ and $\mathcal{A}_m$ to be essentially equivalent, i.e. iii) $\Rightarrow$ i) q.e.d.

**Examples**

In the examples that follow it is understood that when claiming maximal ancillarity what is really meant is essential maximal ancillarity. However, these two concepts coincide when the null set issues do not arise, as in situations when that ancillary is discrete.

**Example 6.**

With probability 1/2 let $X_1,\ldots,X_n$ be i.i.d. from $N(\theta, 1)$, and with probability 1/2 from $N(\theta, 2)$. Let $I = 1$ or 0 as the first or the second case obtains. Then $V = (I, X_1-X_n,\ldots,X_{n-1}-X_n)$ is maximal ancillary since $(I, X_1,\ldots,X_n)$ is equivalent to $(\bar{X}, V)$ and the conditional distribution of $\bar{X}$ given $V$ is complete.

**Example 7.**

Let $X_1,\ldots,X_n$ be i.i.d. with continuous and strictly increasing c.d.f. $F$. This model is invariant under the group $G$ of common, continuous, strictly increasing transformations $X_i' = g(X_i)$, $i = 1,\ldots,n$. Maximal invariant is the vector of ranks $(R_1,\ldots,R_n)$ of the $n$ $X$'s. Since the group $G$ is transitive, the maximal invariant is ancillary. Is it maximal ancillary? Since the conditional distribution of the $X$'s given the ranks is the same as the joint distribution of the rank permuted order statistics and since the distribution of the latter is complete, hence weakly complete, it follows that the ranks are maximal ancillary.

**Example 8.**

In Example 6.2, suppose attention is restricted to $F$ with median 0. Now the ranks are no longer maximal ancillary since the ranks together with the number of positive observations are ancillary. This latter ancillary is maximal since the order statistic given the number of positive and negative observations are complete. (We are dealing with $n_+$ and $n_-$ functions on $(0, \infty)$ and $(-\infty, 0)$, respectively. Note: This maximal ancillary is a maximal invariant under a smaller group than in Example 6.2, namely the group $G$ of transformations $g$ which are continuous, strictly increasing and satisfy $g(0) = 0$.)

**Example 9.**

Let $X_1,\ldots,X_n$ be i.i.d. $N(\theta, 1)$. Here of course the vector of differences $(X_1-X_n,\ldots,X_{n-1}-X_n)$ is maximal ancillary since the distribution of $\bar{X}$ is complete.

As has been pointed out by Basu (1959) and others, maximal ancillarity does not mean that there are no other maximal ancillaries. As a well known example, in the present case with $n = 2$, we have that $V = (X_2-X_1)sign(\bar{X})$ is ancillary. To see that it is also maximal note that $(X_1, X_2)$ is equivalent to $(\bar{X}, V)$ and that $\bar{X}$ and $V$ are independent. Now the completeness of $\bar{X}$ entails the conditional weak completeness of $(\bar{X}, V)$ given $V$.

Another maximal ancillary is $V' = X_2-X_1$. Which of these two ancillaries is preferable? The Cox criterion discussed in part 4 of the section "Relations to Other Concepts" does not distinguish between them; however, a criterion advanced by Barnard and Sprott (1971) applies and gives preference to $X_2-X_1$ since it is invariant under translations (see Padmanabhan, 1977).

**Example 10.**

Let $X_1,\ldots,X_n$ be i.i.d. uniform on $[\theta, \theta+1)$. Here (denoting by $[x]$ the integer part of $x$) $(X_1-X_n,\ldots,X_{n-1}-X_n)$ together with $X_n - [X_n]$ are ancillary and are easily seen to be maximal ancillary since the conditional distribution of $[X_n]$ (all that is left of the data for any fixed $\theta$) given that ancillary is just a one point distribution which is complete.

Basu (1964) treats this example in the case $n = 1$; Basu and Ghosh (1969) treat the same example for the case of arbitrary $n$ for which they determine the maximal ancillary $\sigma$-field.

Basu and Ghosh (1969) show that a sufficient condition for weak completeness of the location family of densities $\{f(x-\theta)\colon \theta \in R\}$ is that the characteristic function $\hat{f}(t) = \int exp(-itx)f(x)dx$ of $f$ has at most a finite number of roots on the real line.

**Example 11. (Basu and Ghosh).**

Let $X$ have density $f(x-\theta)$ with $f(x) = x^2 exp(-x^2/2)/\sqrt{2\pi}$. Since $\hat{f}(t) = (1-t^2)exp(-t^2/2)$ which has only two roots it follows that $X$ is weakly complete and hence admits only the *a.s.* constant functions as ancillaries.

**Example 12. The general location family.**

Let $X_1,\ldots,X_n$ be i.i.d. $\sim f(x-\theta)$ where $f(x)$ is a density with respect to Lebesgue measure on $R$. The differences $V = (V_2,\ldots,V_n) = (X_1-X_2,\ldots,X_1-X_n)$ are ancillary and the question is for which $f$ may one claim also maximal ancillarity? Examples 6.4 and 6.5 show that the answer depends on $f$. The conditional density of $U = X_1$ given $V = v = (v_2,\ldots,v_n)$ is $h_\theta(u|v) = cf(u-\theta)f(u-\theta-v_2)\cdots f(u-\theta-v_n)$ with $c$ being the appropriate normalizing constant. Since this yields a univariate location family $\{h_\theta(u|v) = h_v(u-\theta)\colon \theta \in R\}$ with $h_v(x) = cf(x)f(x-v_2)\cdots f(x-v_n)$ one could appeal to the above sufficient criterion of Basu and Ghosh to establish weak completeness for this family by showing that $\hat{h}_v(t)$ has only a finite number of roots.

Unfortunately, the Basu-Ghosh criterion of a finite number of roots frequently is not satisfied and then does not provide an answer concerning maximality. Examples for which this is the case are the Cauchy and double exponential distributions with $n = 2$.

**Example 13.**

Cox and Hinkley (p. 33, 1974) give the following simplified version of an example due to Basu (1964) which points out the dilemma of multiple ancillaries. Consider $N$ quadrinomial trials with probabilities

$$\tfrac{1}{6}(1-\theta),\quad \tfrac{1}{6}(1+\theta),\quad \tfrac{1}{6}(2-\theta),\quad \tfrac{1}{6}(2+\theta).$$

If the number of outcomes in the four categories are $X$, $U$, $Y$, $V$, respectively, then $X + U$ is ancillary, as is $X + V$. The question is whether either is maximal ancillary. The answer is somewhat surprising and still mostly a conjecture.

(i) First consider the case of $X + U$. The conditional distribution of $(X, Y)$ given $X + U = m$, $Y + V = n$, $(n + m = N)$ is that of two independent binomial random variables, distributed respectively as $b(p_1, m)$ and $b(p_2, n)$ with $p_1 = (1 - \theta)/2$ and $p_2 = (2 - \theta)/4$. Since the conditional expectation of $X/m - 2Y/n + 1/2$ vanishes for all $\theta$ we do not have conditional bounded completeness, whenever $m \geq 1$ and $n \geq 1$. If $m = 0$ or $n = 0$ completeness follows easily.

To establish weak completeness (conditionally) one needs to show that for any indicator function $f(x, y)$ with constant conditional expectation for all $\theta$ it follows that $f$ is either identically one or zero with conditional probability one. For $0 \leq \alpha \leq 1$ consider therefore the following identify for all $\theta$:

$$\sum_{x=0}^{m} \sum_{y=0}^{n} f(x, y)\binom{m}{x}\binom{n}{y}\left(\frac{1-\theta}{2}\right)^{x}\left(\frac{1+\theta}{2}\right)^{m-x}$$

$$\times \left(\frac{2-\theta}{2}\right)^{y}\left(\frac{2+\theta}{2}\right)^{n-y} \equiv \alpha.$$

Show that $f \equiv 0$ and $f \equiv 1$, or equivalently that $\alpha = 0$ and $\alpha = 1$, are the only solution. Reparametrizing $\lambda = (1 - \theta)/(1 + \theta)$ the identity becomes

$$\sum_{x=0}^{m} \sum_{y=0}^{n} f(x, y)\binom{m}{x}\binom{n}{y}\lambda^{x}(1 + 3\lambda)^{y}(3 + \lambda)^{n-y}$$

$$\equiv \alpha 4^{n}(1 + \lambda)^{m+n}.$$

Comparing the coefficients of $\lambda^i$ and $\lambda^{m+n-i}$ for $i = 0, 1, 2$ on both sides of the identity and exploiting the binary nature of $f$ it is easy yet tedious to show weak completeness for the following cases: 1) $n = 1$ and $m = 1$, $m \geq 3$ and 2) $n = 2$ and $m \geq 1$. For the case $(m, n) = (2, 1)$ we don't have weak completeness as can easily be seen by using $f(0, 1) = f(2, 0) = 1$ and $f(x, y) = 0$ otherwise.

Using the reparametrization $\lambda = (2-\theta)/(2+\theta)$ one can show weak completeness for all $(n, m)$ with 3) $m = 1$ and $n \geq 1$ and 4) $m = 2$ and $n \geq 1$ (no counter example here). The above approach does not appear promising for the situations $n \geq 3$ and $m \geq 3$.

(ii) Similar results can be obtained when considering the other ancillary, $X + V$, except that the above counter example does not obtain, i.e. the conditional distribution of $(X, Y)$ given $X + V = m$ is weakly complete for $(m, n)$ in the following cases 1) $n = 1$ and $m \geq 1$, 2) $n = 2$ and $m \geq 1$, 3) $m = 1$ and $n \geq 1$, 4) $m = 2$ and $n \geq 1$.

What does this mean with respect to maximal ancillarity of $X + U$ and $X + V$? For $N = m + n \leq 5$ the latter is maximal ancillary whereas the former is maximal ancillary for $N = 1, 2, 4, 5$ but not for $N = 3$. Maximality in the cases $N > 5$ at this point can only be conjectured.

**References**

Amari, S. (1982): Geometrical theory of asymptotic ancillarity and conditional inference, *Biometrika* 69, 1-17.

Andersen, E. B. (1967): On partial sufficiency and partial ancillarity, *Skand. Aktuar. Tidskr.* 50, 137-152.

Barnard, G. A. and Sprott, D. A. (1971): A note on Basu's examples of anomalous ancillary statistics, in *Foundations of Statistics*, eds. Godambe and Sprott, Holt, Rinehart and Winston, Toronto.

Barndorff-Nielsen, O. (1978): *Information and Exponential Families in Statistical Theory*, John Wiley, New York.

Barndorff-Nielsen, O. (1980): Conditionality resolutions, *Biometrika* 67, 293-310.

Barndorff-Nielsen, O. (1983): On a formula for the distribution of the maximum likelihood estimator, *Biometrika* 70, 343-365.

Basu, D. (1955): On statistics independent of a complete sufficient statistic, *Sankhyā* 15, 377-380.

Basu, D. (1958): On statistics independent of a sufficient statistic, *Sankhyā* 20, 223-226.

Basu, D. (1959): The family of ancillary statistics, *Sankhyā* (A) 21, 247-256.

Basu, D. (1964): Recovery of ancillary information, *Sankhyā* (A) 26, 3-16.

Basu, D. (1967): Problems relating to the existence of maximal and minimal elements in some families of statistics (subfields), in *Proc. Fifth Berkeley Symp. on Math. Statistics and Probability*, Vol. 1, 41-50, Univ. of Calif. Press.

Basu, D. (1969): Role of the sufficiency and likelihood principles in sample survey theory, *Sankhyā* 31, 441-454.

Basu, D. (1971). An essay on the logical foundations of survey sampling, in *Foundations of Statistics*, eds. Godambe and Sprott, Holt, Rinehard and Winston, Toronto.

Basu, D. (1975): Statistical information and likelihood (with discussion), *Sankhyā* 37, 1-71.

Basu, D. (1977): On the elimination of nuisance parameters, *J. Amer. Statist. Assoc.* 72, 355-366.

Basu, D. (1978): Relevance of randomization in data analysis, in *Survey Sampling and Measurement*, ed. Namboodiri, 267-292.

Basu, D. (1980): Randomization analysis of experimental data: The Fisher randomization test (with discussion), *J. Amer. Statist. Assoc.* 75, 575-595.

Basu, D. (1982): Basu theorems, *Encycl. Statist. Sci.* 1, 193-196.

Basu, D. and Ghosh, J. K. (1969): Invariant sets for translation-parameter families of measures, *Ann. Math. Statist.* 40, 162-174.

Becker, N. and Gordon, I. (1983): On Cox's criterion for discriminating between alternative ancillary statistics, *Int. Statist. Rev.* 51, 89-92.

Berger, J. O. and Wolpert, R. L. (1984): The likelihood principle, IMS Lecture Notes-Monograph Series, Vol. 6.

Birnbaum, A. (1962): On the foundations of statistical inference (with discussion), *J. Amer. Statist. Assoc.* 57, 269-306.

Buehler, R. J. (1982): Some ancillary statistics and their properties, *J. Amer. Statist. Assoc.* 77, 581-589.

Cox, D. R. (1958): Some problems connected with statistical inference, *Ann. Math. Statist.* 29, 357-372.

Cox, D. R. (1971): The choice between alternative ancillary statistics, *J. Roy. Statist. Soc.* (B) 33, 251-255.

Cox, D. R. (1980): Local ancillarity, *Biometrika* 67, 279-286.

Cox, D. R. (1984): Discussion of a paper by F. Yates, *J. Roy. Statist. Soc.* (A) 147, 451.

Cox, D. R. and Hinkley, D. V. (1974): *Theoretical Statistics*, Chapman and Hall, London.

Dawid, A. P. (1975): On the concepts of sufficiency and ancillarity in the presence of nuisance parameters, *J. Roy. Statist. Soc.* B 37, 248-258.

Efron, B. and Hinkley, D. V. (1978): Assessing the accuracy of the maximum likelihood estimators: observed versus expected Fisher information, *Biometrika* 65, 457-487.

Evans, M., Fraser, D. A. S. and Monette, G. (1986): On principles and arguments to likelihood, *Can. J. Statist.* 14, 181-200.

Finch, P D. (1986): Randomization I, *Encycl. Statist. Sci.* 7, 516-519.

Fisher, R. A. (1925): Theory of statistical estimation, *Proc. Camb. Philos. Soc.* 22, 700-725.

Fisher, R. A. (1934): Two new properties of mathematical likelihood, *Proc. Roy. Soc.* (A) 144, 285-307.

Fisher, R. A. (1935): The logic of inductive inference, *J. Roy. Statist. Soc.* 98, 39-54.

Fisher, R. A. (1936): Uncertain inference, *Proc. Am. Acad. Arts and Sci.* 71, 245-258.

Fisher, R. A. (1948): Conclusions fiduciaries, *Ann. Inst. Henri Poincaré* 10, 191-213.

Fisher, R.A. (1956): *Statistical Methods and Scientific Inference*, Oliver and Boyd, Edinburgh.

Fraser, D. A. S. (1973): The elusive ancillary, in *Multivariate Statistical Inference*, ed. D. G. Kabe and R. P. Gupta, North Holland, American Elsevier.

Fraser, D. A. S. and Reid, N.. (1988): On conditional inference for a real parameter: A differential approach on the sample space, *Biometrika* 75, 251-264.

Gordon, I. (1983): Ancillarity and minimal sufficiency, *Austral. J. Statist.* 25, 273-277.

Hinkley, D. V. (1980a): Likelihood as approximate pivotal distribution, *Biometrika* 67, 287-292.

Hinkley, D. V. (1980b): Fisher's development of conditional inference, in *R. A. Fisher: An Appreciation*, 101-108, Springer, New York.

Hosoya, Y. (1988). The second order Fisher information, *Biometrika* 75, 265-274.

Joshi, V. M. (1983): Likelihood principle, *Encycl. Statist. Sci.* 4, 644-647.

Kalbfleisch, J. D. (1975): Sufficiency and conditionality (with discussion), *Biometrika* 62, 251-268.

Kalbfleisch, J. D. (1982): Ancillary statistics, *Encycl. Statist. Sci.* 1, 77-81.

Kempthorne, O. (1986): Randomization II, *Encycl. Statist. Sci.* 7, 519-524.

Koehn, U. and Thomas, D. L. (1975): On statistics independent of a sufficient statistic: Basu's lemma, *Amer. Statist.* 29, 40-42.

Lehmann, E. L. (1981): An interpretation of completeness and Basu's theorem, *J. Amer. Statist. Assoc.* 76, 335-340.

Lehmann, E. L. (1986): *Testing Statistical Hypotheses*, 2nd ed., John Wiley, New York.

Lehmann, E. L. and Stein, C. (1950): Completeness in the sequential case, *Ann. Math. Statist.* 21, 376-385.

Liang, K. Y. (1984): The asymptotic efficiency of conditional likelihood methods, *Biometrika* 71, 305-313.

Padmanabhan, A. R. (1977): Ancillary statistics which are not invariant, *Amer. Statistician* 31, 124.

Rao, C. R. (1971): Some aspects of statistical inference in problems of sampling from finite populations (with discussion), in *Foundations of Statistics*, eds. Godambe and Sprott, Holt, Rinehart and Winston, Toronto.

Sandved, E. (1967): A principle for conditioning on an ancillary statistic, *Skand. Aktuar. Tidskr.* 50, 39-47.

Sandved, E. (1972): Ancillary statistics in models without and with nuisance parameters, *Skand. Aktuar. Tidskr.* 55, 81-91.

Skovgaard, I. (1986): Successive improvement of the order of ancillarity, *Biometrika* 73, 516-519.

Sverdrup, E. (1966): The present state of the decision theory and the Neyman-Pearson theory, *Rev. Int. Statist. Inst.* 34, 309-333.

Young, A. (1986): Conditioned data-based simulations: Some examples from geometrical statistics, *Rev. Int. Statist. Inst.* 54, 1-13.

# THE PITMAN CLOSENESS OF STATISTICAL ESTIMATORS: LATENT YEARS AND THE RENAISSANCE

Pranab Kumar Sen, University of North Carolina at Chapel Hill

**Abstract**

The Pitman closeness criterion is an intrinsic measure of the comparative behavior of two estimators (of a common parameter) based solely on their joint distribution. It generally entails less stringent regularity conditions than in other measures. Although there are some undesirable features of this measure, the past few years have witnessed some significant developments on Pitman-closeness in its tributaries, and a critical account of the same is provided here. Some emphasis is placed on nonparametric and robust estimators covering fixed-sample size as well as sequential sampling schemes.

## Introduction

In those days prior to the formulation of statistical decision theory (Wald, 1949), the reciprocal of variance [or mean square error (MSE)] of an estimator ($T$) used to be generally accepted as an universal measure of its precision (or efficiency). The celebrated Cramér-Rao inequality (Rao, 1945) was not known that precisely although Fisher (1938) had a fair idea about such a lower bound to the variance of an estimator. The use of mean absolute deviation (MAD) criterion as an alternative to the MSE was not that popular (mainly because its exact evaluation often proved to be cumbersome), while other loss functions (convex or not) were yet to be formulated in a proper perspective. In this setup, Pitman (1937) proposed a novel measure of closeness (or nearness) of statistical estimators, quite different in character from the MSE, MAD and other criteria. Let $T_1$ and $T_2$ be two rival estimators of a parameter $\theta$ belonging to a parameter space $\Theta \subseteq R$. Then $T_1$ is said to be closer to $\theta$ than $T_2$, in the Pitman sense, if

$$P_\theta\{|T_1 - \theta| \leq |T_2 - \theta|\} \geq 1/2, \forall \theta \in \Theta, \tag{1}$$

with strict inequality holding for some $\theta$. Thus, the Pitman-closeness criterion (PCC) is an intrinsic measure of the comparative behavior of two estimators. Note that in terms of the MSE, $T_1$ is better than $T_2$, if

$$E_\theta(T_1 - \theta)^2 \leq E_\theta(T_2 - \theta)^2, \forall \theta \in \Theta, \tag{2}$$

with strict inequality holding for some $\theta$; for the MAD criterion, we need to replace $E_\theta(T - \theta)^2$ by $E_\theta|T - \theta|$. In general, for a suitable nonnegative loss function $L(a, \theta) : R \times R \to R^+$, $T_1$ dominates $T_2$ if

$$E_\theta[L(T_1, \theta)] \le E_\theta[L(T_2, \theta)], \forall \theta \in \Theta, \tag{3}$$

with strict inequality holding for some $\theta$. We represent (1), (2) and (3) respectively as

$$T_1 \succ_{PC} T_2, T_1 \succ_{MSE} T_2 \text{ and } T_1 \succ_L T_2. \tag{4}$$

It is clear from the above definitions that for (2) or (3), one needs to operate the expectations (or moments), while (1) involves a distributional operation only. Thus, in general, (2) or (3) may entail more stringent regularity conditions (pertaining to the existence of such expectations) than needed for (1). In this sense, the PCC is solely a distributional measure while the others are mostly moment based ones, and hence, from this perspective, the PCC has a greater scope of applicability (and some other advantages too). On the other hand, other conventional measures, such as (2) or (3), may have some natural properties which may not be shared by the PCC. To illustrate this point, note that if there are three estimators, say, $T_1$, $T_2$ and $T_3$, of a common parameter $\theta$, such that

$$E_\theta(T_1 - \theta)^2 \le E_\theta(T_2 - \theta)^2$$

and

$$E_\theta(T_2 - \theta)^2 \le E_\theta(T_3 - \theta)^2, \forall \theta \in \Theta, \tag{5}$$

then, evidently, $E_\theta(T_1 - \theta)^2 \le E_\theta(T_3 - \theta)^2$, $\forall \theta \in \Theta$. Or, in other words, the MSE criterion has the transitivity property, and this is generally the case with (3). However, this transitivity property may not always hold for the PCC. That is, $T_1$ may be closer to $\theta$ than $T_2$, and $T_2$ may be closer to $\theta$ than $T_3$ (in the Pitman sense), but $T_1$ may not be closer to $\theta$ than $T_3$ in the same sense!. Although a little artificial, it is not difficult to construct suitable examples testifying the intransitivity of the PCC (Blyth, 1972). Secondly, the measure in (2) or (3) involves the marginal distributions of $T_1$ and $T_2$, while (1) involves the joint distribution of $(T_1, T_2)$. Hence, the task of verifying the dominance in (1) may require more elaborate analysis. This was perhaps the main reason why in spite of a good start and notable contributions by Geary (1944) and Johnson (1950), the use of PCC remained somewhat skeptical for more than thirty years! In fat, the lack of transitivity of the PCC in (1) caused some difficulties in extending the pairwise dominance in (1) to that within a suitable class of estimators. Only recently, such results have been obtained by Ghosh and Sen (1989) and Nayak (1990) for suitable *families of equivariant estimators.* We shall comment on them in a later section. Thirdly, in (1), when both $T_1$ and $T_2$ have continuous distributions and $T_2 - T_1$ has a non-atomic distribution, the $\le$ sign may as well be replaced by $<$ sign, without affecting the probability inequality. However, if $T_2 - T_1$ has an atomic distribution, the two probability statements involving $\le$ and $<$ signs, respectively, may not agree, and somewhat different conclusions may crop up in the two cases. Although this anomaly can be eliminated by attaching suitable probability (viz., 1/2) for the tie

($|T_1 - \theta| = |T_2 - \theta|$), the process can be somewhat arbitrary and less convincing in general. Fourthly, the definitions in (1) through (3) need some modifications in the case where $\underset{\sim}{\theta}$ (and $\underset{\sim}{T}$) are $p$-vectors, for some $p > 1$. The MSE criterion lends itself naturally to an appropriate *quadratic error loss*, where for some chosen positive definite (p.d.) matrix $\underset{\sim}{Q}$, the distance function is taken as $\| \underset{\sim}{T} - \underset{\sim}{\theta} \|_{\underset{\sim}{Q}}$, and given by

$$\| \underset{\sim}{T} - \underset{\sim}{\theta} \|^2_{\underset{\sim}{Q}} = (\underset{\sim}{T} - \underset{\sim}{\theta})' \underset{\sim}{Q} (\underset{\sim}{T} - \underset{\sim}{\theta}) \tag{6}$$

The use of the Fisher Information matrix $\underset{\sim}{\mathfrak{I}}_{\underset{\sim}{\theta}}$ as $\underset{\sim}{Q}$ leads to the so-called Mahalanobis distance. Recall that

$$E_{\underset{\sim}{\theta}} \| \underset{\sim}{T} - \underset{\sim}{\theta} \|^2_{\underset{\sim}{Q}} = Trace\left[ \underset{\sim}{Q} E_{\underset{\sim}{\theta}} \{ (\underset{\sim}{T} - \underset{\sim}{\theta})(\underset{\sim}{T} - \underset{\sim}{\theta})' \} \right], \tag{7}$$

so that (2) entails only the computation of the mean product error (or dispersion) matrix of $\underset{\sim}{T}_1$ and $\underset{\sim}{T}_2$. On the other hand, if instead of $|T_1 - \theta|$ and $|T_2 - \theta|$, in (1), we use $\| \underset{\sim}{T}_1 - \underset{\sim}{\theta} \|_{\underset{\sim}{Q}}$ and $\| \underset{\sim}{T}_2 - \underset{\sim}{\theta} \|_{\underset{\sim}{Q}}$, the probability statement may be a more involved function of the actual distribution of $(\underset{\sim}{T}_1, \underset{\sim}{T}_2)$ and of the $\underset{\sim}{Q}$. Although in some special cases this can be handled without too much of complications (see for example, Sen, 1989a), in general, we may require more stringent regularity conditions to verify (1) in the vector case. In the asymptotic case, however, an equivalence of BAN estimators and Pitman-closest ones may be established under very general regularity conditions (viz., Sen, 1986), so that (1) and (2) may have asymptotic equivalence. But, in the multiparameter case, best estimators, in the sense of having a minimum value of (7) may not be BAN. A natural reference is the so-called Stein paradox (viz, Stein, 1956) for the estimation of the mean vector of a multivariate normal distribution. For $p$, the dimension of the multivariate normal law, greater than 2, Stein (1956) showed that the sample mean vector [although being the *maximum likelihood estimator* (MLE)] is not *admissible*, and later on, James and Stein (1962) constructed some other estimators which dominate the MLE in the light of (2) [as amended in (7)]. Such *Stein-rule* or *shrinkage estimators* are typically non-linear and are non-normal, even asymptotically. Thus, they are not BAN. So, a natural question arose: Do the Stein-rule estimators dominate their classical counterparts in the light of the PCC? An affirmative answer to this question has recently been provided by Sen, Kubokawa and Saleh (1989), and we shall discuss this in a later section. Fifthly, we have tacitly assumed so far that we have a conventional *fixed-sample size case*. There are, however, some natural situations calling for suitable *sequential schemes*, so that one may also like to inquire how far the PCC remains adoptable in such a sequential scheme. Some studies in this direction have been made very recently by Sen (1989a), and we shall discuss some of these results in a later section. Another direction in which the PCC has proven to be a very useful avenue for comparing estimators is the employment of more general loss functions (instead of the Euclidean norm or the usual quadratic norm) in the definition in (1). In the context of estimation of the dispersion matrix of a multivariate normal distribution and parameters in some other distributions belonging to the exponential family, one may adopt the entropy (or some related) loss functions which when incorporated in (1) lead to a more general formulation. This has been termed the generalized Pitman nearness criterion (GPNC) (viz., Khattree, 1987, for the dispersion matrix estimation problem). We shall review some of the developments in this area in the last section.

As has been mentioned earlier, for nearly four decades, there were not much activities in this general arena, while the past ten years have witnessed a remarkable growth of the literature on the PCC. This renaissance is partly due to the work of C.R. Rao (1981) who clearly pointed out the shortcomings of the MSE or the quadratic error loss and explained the rationality of the PCC (which attaches less importance to *large deviations*). The work of Efron (1975) also deserves a special mention: the feasibility of an estimator dominating the classical MLE of the univariate normal mean in the light of the PCC clearly points out the adaptability of the PCC in a more general situation where other forms of admissibility criteria may not work out well. A somewhat comparable picture in both the works of Efron (1975) and Rao (1981) might have been based on the MAD criterion which attaches less importance to large deviations than the MSE criterion. However, in the general multiparameter case, the MAD criterion may lose its appeal to a greater extent. This is mainly due to the following factors: (i) lack of invariance under suitable groups of transformations usually employed in multiparameter estimation problems, (ii) complexity of the definitions and (iii) need for the estimation of nuisance parameters (such as the reciprocal of the density functions) in the definition of the norm itself which usually requires really large sample sizes! One might also argue in favor of some other criteria. Hwang (1985) has considered the *stochastic dominance* criterion based on the marginal distributions with an arbitrarily chosen cut-off point, and this in turn introduces some arbitrariness in the adaptation of his measure; the dominance may not hold uniformly in the choice of such a cut-off point. Brown, Cohen and Strawderman (1976) advocated the use of some non-convex loss functions. We have no definite prescription in favor of the PCC, MAD, such non-convex loss functions or the stochastic dominance criterion, although the PCC may have some natural appeal. In passing, we may remark that some controversies have been reported in Roberts and Hwang (1988), although it is very hard to endorse fully the views expressed in this report. We would like to bypass these by adding that *let the cliff fall where it belongs to*! In our opinion, in spite of some of the shortcomings of the PCC, as have been mentioned earlier, the developments in the past decade have, by far, been much more encouraging to advocate in favor of the use of PCC (or the GPNC) in a variety of statistical models which will be considered here in the subsequent sections. We also refer to a recent Panel Discussion on Pitman nearness of statistical estimators at the International Conference on Recent Developments in Statistical Data Analysis and Inference (in honor of C.R. Rao) at Neuchatel, Switzerland (August 24, 1989), where some of these issues have been discussed critically, and a report of these findings is accounted in Mason, Keating, Sen and Rao (1990). As with any other measure, there are pathological examples where the PCC may not appear to be that rational, but in real applications, we will rarely be confronted with such artificial cases. On the other hand, in the conventional linear models and in multivariate analyses, some theoretical studies (supplemented by numerical investigations) made by Mason, Keating, Sen and Blaylock (1990) justify the appropriateness of the PCC, even when a dominance may not hold for the entire parameter space. Finally, in the asymptotic case where the sample size is large enough to justify the usual regularity conditions needed to use simplified distribution theory for the estimators, for a wider class of nonparametric and robust estimators, we may justify the adaptation of the PCC on a very broad ground. We shall stress this point in the subsequent sections. All in all, we welcome the renaissance of the PCC and look forward to further developments in this fruitful area of statistical research.

**PCC in the Single Parameter Case**

In this section, we stick to the basic definition in (1) and examine the Pitman-closeness of a general class of statistical estimators. According to (1), rival estimators are compared two at a time, while (2) or (3) lends itself readily to suitable classes of estimators. This prompted Ghosh and Sen (1989) to consider Pitman closest estimators within reasonable classes of estimators. In this context, we may remark that under (2), the celebrated Rao-Blackwell theorem depicts the role of *unbiased, sufficient statistics* in the construction of such optimal estimators. Ghosh and Sen (1989) have shown that under appropriate regularity conditions, a *median unbiased* (MU) estimator is Pitman-closest within an appropriate class of estimators. Recall that an estimator $T$ of $\theta$ is MU if

$$P_{\theta}\{T \leq \theta\} = P_{\theta}\{T \geq \theta\}, \forall\, \theta \in \Theta, \tag{8}$$

and $T_0$ is Pitman-closest within a class of estimators ($\mathfrak{C}$), if (1) holds for $T_1 = T_0$ and every $T_2 \in \mathfrak{C}$. In many applications, $T_0$ is a function of a (*complete*) *sufficient statistic* and $T_2 = T_0 + Z$, where $Z$ is *ancillary*. Then, note that

$$\begin{aligned}\left[|T_0 - \theta| \leq |T_2 - \theta|\right] &\Leftrightarrow \left[|T_0 - \theta|^2 \leq (T_0 - \theta + Z)^2\right] \\ &\Leftrightarrow \left[2Z(T_0 - \theta) + Z^2 \geq 0\right],\end{aligned} \tag{9}$$

while by Basu's (1955) theorem, $T_0$ and $Z$ are independently distributed. Since $Z^2$ is a nonnegative random variable, the MU character of $T_0$ ensure that the right hand side of (9) has probability $\geq 1/2$, $\forall\, \theta \in \Theta$. This explains the role of MU sufficient statistics in the characterization of the Pitman-closest estimators.

However, the following theorem due to Ghosh and Sen (1989) presents a broader characterization.

**Theorem 1.**

Let $T$ be MU-estimator of $\theta$ and let $\mathcal{C}$ be the class of all estimators of the form $U = T + Z$, where $T$ and $Z$ are independently distributed. Then $P_\theta\{|T-\theta| \leq |U-\theta|\} \geq 1/2$, for all $\theta \in \Theta$ and $U \in \mathcal{C}$.

Theorem 1 typically relates to the estimation of location parameter ($\theta$) in the usual location-scale model where the class $\mathcal{C}$ relates to suitable *equivariant estimators* (relative to appropriate groups of transformation). Various examples of this type have been considered by Ghosh and Sen (1989). In the context of the estimation of the scale parameter, the PCC has been studied in a relatively more detailed manner. Keating (1985) considered a general scale family of distributions, and confined himself to the class ($\mathcal{C}^0$) of all estimators which are scalar multiple of the usual MLE; however, he did not enforce any equivariance considerations to clinch the desired Pitman-closest property. Keating and Gupta (1984) considered various estimators of the scale parameter of a normal distribution, and compared them in the light of the PCC. Again in the absence of any equivariance considerations, their result did not lead to the desired Pitman-closest characterization. The following theorem due to Ghosh and Sen (1989) provides the desired result.

**Theorem 2.**

Let $\mathcal{C}^*$ be the class of all estimators of the form $U = T(1 + Z)$, where $T$ is MU for $\theta$ and is nonnegative, while $T$ and $Z$ are independently distributed. Then, $P_\theta\{|T-\theta| \leq |U-\theta|\} \geq 1/2$, $\forall\, \theta \in \Theta$, $U \in \mathcal{C}^*$.

Both these theorems have been incorporated in the PC characterization of BLUE (*best linear unbiased estimators*) of location and scale parameters in the complete sample as well as censored cases (Sen, 1989b); equivariance plays a basic role in this context too. Further note that if $T$ has a distribution symmetric about $\theta$, then $T$ is MU for $\Theta$. This sufficient condition for $T$ is easy to verify in many practical applications. Similarly, if the conditional distribution of $T$, given $Z$, is symmetric about $\theta$, then in Theorem 1, we may not need the independence of $T$ and $Z$. The uniform distribution on $[\theta - \frac{1}{2}\delta, \theta + \frac{1}{2}\delta]$, $\delta > 0$, provides a simple example of the latter (Ghosh and Sen, 1989).

We shall now discuss some further results on PCC in the single parameter case pertaining to the asymptotic case and to sequential sampling plans. The current literature on theory of estimation is flooded with asymptotics. Asymptotic normality, asymptotic efficiency and other asymptotic considerations play a vital role in this context. An estimator ($T_n$) based on a sample of size $n$ is termed a BAN (*best asymptotically normal*) estimator of $\theta$ if the following two conditions hold:

$$n^{\frac{1}{2}}(T_n - \theta) \text{ is asymptotically normal } (0, \sigma_T^2) \tag{10}$$

[which is the AN (*asymptotically normal*) criterion], and

$$\sigma_T^2 = \frac{1}{\mathfrak{J}_\theta}, \text{ where } \mathfrak{J}_\theta \text{ is the Fisher information of } \theta \tag{11}$$

[which is the B (*bestness*) criterion]. Let us now consider the class $\mathfrak{C}_A$ of estimation $\{U_n\}$ which admit an asymptotic representation of the form:

$$U_n - \theta = n^{-1}\sum_{i=1}^{n} \psi_\theta(x_i) + o_p\left(n^{-\frac{1}{2}}\right), \text{ as } n \to \infty, \tag{12}$$

where the *score function* $\psi_\theta(\cdot)$ may depend on the method of estimation and the model; $E_\theta\psi_\theta(X_i) = 0$ and $E_\theta\psi_\theta^2(x_i) = \sigma_U^2 < \infty$. Recall that for a BAN estimator of $\theta$, we would have a representation of the form (12) where $\mathfrak{J}_\theta \cdot \psi_\theta(x_i) = f_\theta'(x_i, \theta)/f(x_i, \theta)$, $f(\cdot)$ is the probability density function and $f_\theta'$ is its first order derivative w.r.t. to $\theta$. Note further that $E_\theta\{[f_\theta'(x_1; \theta)/f(x_1; \theta)]^2\} = \mathfrak{J}_\theta$, so that for a BAN estimator, $E_\theta\{\psi_\theta(x_1)f_\theta'(x_1)/f(x_1; \theta)\} = 1$, $\forall\ \theta$. Thus, if we let

$$\xi_n = n^{-\frac{1}{2}}\sum_{i=1}^{n} (\partial/\partial\theta) \log f(X_i; \theta), \tag{13}$$

then for a BAN estimator $T_n$, we have under the usual regularity conditions that as $n \to \infty$,

$$\left(n^{\frac{1}{2}}(T_n - \theta), \xi_n\right) \underset{\mathfrak{D}}{\to} \mathcal{N}_2\left((0, 0), \begin{bmatrix} \mathfrak{J}_\theta^{-1} & 1 \\ 1 & \mathfrak{J}_\theta \end{bmatrix}\right). \tag{14a}$$

Consider now the class $\mathfrak{C}^0$ of all estimators $\{U_n\}$, such that as $n \to \infty$,

$$\left(n^{\frac{1}{2}}(U_n - \theta), \xi_n\right) \underset{\mathfrak{D}}{\to} \mathcal{N}_2\left((0, 0), \begin{bmatrix} \sigma_U^2 & 1 \\ 1 & \mathfrak{J}_\theta \end{bmatrix}\right), \tag{14b}$$

where $\sigma_U^2 \geq \mathfrak{J}_\theta^{-1}$, and the equality sign holds whenever $U_n$ is a BAN estimator of $\theta$. Note that the $\sqrt{n}$-consistency of $U_n$ entails the unit covariance term. As such, by an appeal to Theorem 1 of Sen (1986) we conclude that the BAN estimator satisfying (14a) is asymptotically (as $n \to \infty$) a Pitman-closest estimator of $\theta$ (within the class $\mathfrak{C}^0$).

Note that this characterization is localized to the class of asymptotically normal estimators. In the context of estimation of location (or simple regression) parameter, incorporating robustness considerations (either on a local or global basis), various other estimators have been considered by a host of workers.

Among those, the *M-*, *L-* and *R-estimators* deserve special mention. The *M-estimators* are especially advocated for plausible local departures from the assumed model, and they retain high efficiency for the assumed model and at the same time possess good local robustness properties. The *R*-estimators are based on appropriate rank statistics and possess good global robustness properties. *L*-estimators are based on linear functions of order statistics with a similar robustness consideration in mind. In general, these *M*-, *L*- and *R*-estimators satisfy the AN condition in (10) through appropriate representations of the type (12), where $\psi_\theta(x) = \psi(x-\theta)$; see for example, Sen (1981, Ch. 8). From considerations of bestness based on the minimum (asymptotic) MSE, the optimal *M*-, *L*- and *R*-estimators all satisfy the bestness condition in (11). Hence, we conclude that an *M*-, *L*- or *R*- estimator of $\theta$ having the BAN character in the usual sense is also asymptotically Pitman-closest. This places the PCC in a very comparable stand in the asymptotic case. Note that being a completely distributional measure, the PCC does not entail the computation or convergence of the actual MSE of the estimators, and hence (14a) requiring the usual conditions needed for the BAN property, also leads to the desired PCC property.

We consider now some recent results on PCC in the sequential case (Sen, 1989a). Note that for the estimation of the mean of a normal distribution with unknown variance $\sigma^2$, generally a sequential sampling plan is advocated to ensure some control on the performance characteristics (which can not be done in a fixed sample procedure). In this setup, the *stopping number* $N$ is a positive integer valued random variable such that for every $n \geq 2$, the event $[N = n]$ depends only on $\{s_k^2,\ k \leq n\}$, where $s_k^2$ is the sample variance for the sample size $k$, $k \geq 2$. It is known that $\{\bar{X}_k,\ k \geq 1\}$ and $\{s_k^2,\ k \leq n\}$ are mutually independent, and hence, given $N = n$ (i.e., the $s_k^2$, $k \leq n$), $T_n = \bar{X}_n$ satisfies the conditions of Theorem 1, so that $\bar{X}_N$ has the Pitman-closest character. This simple observation can be incorporated in a formulation of the PC characterization of sequential estimators. Let $\{X_i,\ i \geq 1\}$ be a sequence of independent and identically distributed (i.i.d.) random variables (r.v.) with a distribution function (d.f.) $F_\theta(x)$, $x \in R$, $\theta \in \Theta \subset R$. For every $n \geq 1$, consider the transformation:

$$(X_1,\ldots,X_n) \rightarrow (T_n,\ \underset{\sim}{V}_n,\ \underset{\sim}{W}_n) \qquad (\underset{\sim}{V}_n \text{ could be vacuous}) \tag{15}$$

$\underset{\sim}{W}_n$ is a $(n-k-1)$-vector and $\underset{\sim}{V}_n$ is a $k$-vector, where $k$ is a nonnegative integer. Let $\mathfrak{B}_T^{(n)}$ and $\mathfrak{B}_W^{(n)}$ be the sigma sub-fields generated by $T_n$ and $\underset{\sim}{W}_n$, respectively, for $n \geq 1$,

$$[N = n] \text{ is } \mathfrak{B}_W^{(n)}\text{-measurable,} \tag{16}$$

$$T_n \text{ is MU for } \theta, \tag{17}$$

$$\begin{aligned} Z_n = \ & v_n(\underset{\sim}{W}_n) \text{ is } \mathfrak{B}_W^{(n)}\text{-measurable and} \\ & T_n \text{ and } \underset{\sim}{W}_n \text{ are independently distributed.} \end{aligned} \tag{18}$$

As in Theorem 1, let $\mathcal{C}^0$ be the class of all (sequential) estimators of the form $U_N = T_N + Z_N$. Then, under (16), (17) and (18),

$$P_\theta\{|\,T_N - \theta| \leq |\,U_N - \theta|\} \geq 1/2, \quad \forall\ U_N \in \mathcal{C}^0 \text{ and } \theta \in \Theta. \tag{19}$$

A similar extension of Theorem 2 to the sequential case works out under (16) - (18).

The characterization of PC of sequential estimators made above is an exact one, in the sense that it holds for an arbitrary stopping number ($N$) so long as $N$ satisfies (16). In the context of bounded-width confidence intervals for $\theta$ or minimum risk (point) estimation of $\theta$ (and in some other problems too), the stopping number $N$ is indexed by a positive real number $d$ (i.e., $N = N_d$), such that $N_d$ is well defined for every $d > 0$ (and $N_d$ is usually $\downarrow$ in $d$). In this setup, one considers an asymptotic model where $d \downarrow 0$. Often, there exists a sequence $\{n_d^0\}$ of positive integers ($n_d^0$ is $\downarrow$ in $d$), such that $n_d^0 \to \infty$, as $d \downarrow 0$, and further,

$$(n_d^0)^{-1}\ N_d \xrightarrow{p} 1, \text{ as } d \downarrow 0.$$

In such a case, we may extend the PC characterization to the class of BAN (sequential) estimators, without necessarily requiring (16). Consider the BAN estimators treated in (10) through (14), but now adapted to the stopping number $\{N_d\}$. Suppose that the $U_n$ [in (12)] satisfy an Anscombe-type condition [Anscombe (1952)] that for every $\epsilon > 0$ and $\eta > 0$, there exist a $\delta > 0$ and an integer $n_0$, such that

$$P\left\{\max_{m:|m/n-1| \leq \delta} n^{\frac{1}{2}}\,|\,U_m - U_n| > \epsilon\right\} < \eta,\ \forall\ n \geq n_0. \tag{20}$$

This Anscombe-condition holds for the $\xi_n$ in (13) under no extra regularity conditions. On the other hand, (20) is also a byproduct of (weak) invariance principles for the $U_n$, which have been studied extensively in the literature [viz., Sen (1981), Ch. 3-8]. Thus, we may replace $\{U_{N_d}, T_{N_d}\}$ by $\{U_{n_d^0}, T_{n_d^0}\}$, as $d \downarrow 0$, and then make use of (14) to characterize the desired PC property of the sequential BAN estimators. Note that, in general, M-estimators of locations are not scale-equivariant (so as to qualify for the class $\mathcal{C}$ in Theorem 1), and $L$- and $R$-estimators of location may not also belong to this class. Thus, in finite sample case, the PC characterization may not apply to these estimators. But, in the asymptotic case (sequential or fixed-sample size setup), the PC characterization holds in spite of the fact that these estimators may not belong to the class $\mathcal{C}$ or that (16) may not hold.

To sum up the main findings on PCC in the uniparameter case, we observe that the MU property (along with ancillarity and sufficiency) provide us with the desired tool for finding the Pitman-closest estimators in the fixed-sample as well as sequential cases. In the asymptotic case, BAN estimators enjoy the PC-property, and this naturally raises the question: What is the relationship of the PCC and the (asymptotic) variance of an estimator? Following the lead of Rao et al. (1986) and Keating and Mason (1985), Peddada and Khattree (1986) studied this problem; however, their main results pertain to two estimators, say, $T_1$ and $T_2$, which are distributed independently of each other, and hence, the conclusions derived from these results may not apply to an usual situation where the two rival estimators of a common parameter $\theta$ are not independently distributed. Moreover, as they were assuming normality in most of the cases (treated by them), more general results for such models can be obtained from Sen (1986).

**PCC in the Multiparameter Case**

There has been a lot of research work on the PCC in the multiparameter case, including *shrinkage* and sequential estimators. Let us consider the case of a vector $\underset{\sim}{\theta} = (\theta_1,\ldots,\theta_p)'$ of parameters, where $\underset{\sim}{\theta} \in \Theta \subset R^p$, for some $p \geq 1$. Let $\underset{\sim}{T} = (T_1,\ldots,T_p)'$ be an estimator of $\theta$. First, we need to extend the definition of the distance $|T - \theta|$ in (1) to the multiparameter case. Although the Euclidean norm is a possibility, since the different components of $\underset{\sim}{T}$ may have different importance (and they are generally not independent), a more general quadratic norm is usually adopted. We may define

$$\| \underset{\sim}{d} \|^2_{\underset{\sim}{Q}} = \underset{\sim}{d}' \underset{\sim}{Q} \underset{\sim}{d}, \ \underset{\sim}{d} \in R^p, \tag{21}$$

where $\underset{\sim}{Q}$ is a given p.d. matrix. It is not uncommon to use some other metric (viz., entropy, etc.), so that we may as well take a general

$$L(\underset{\sim}{T}, \underset{\sim}{\theta}), \text{ satisfying the usual properties of a 'norm'.} \tag{22}$$

In the last section, in the context of estimation of dispersion matrices of multivariate normal distributions, we shall use such norms. As an extension of (1) and following the lead of Peddada (1985), we consider the following *generalized Pitman nearness criterion* (GPNC): An estimator $\underset{\sim}{T}_1$ is GPN closer than $\underset{\sim}{T}_2$ if

$$P_{\underset{\sim}{\theta}}\left\{L(\underset{\sim}{T}_1, \underset{\sim}{\theta}) \leq L(\underset{\sim}{T}_2, \underset{\sim}{\theta})\right\} \geq 1/2, \forall \underset{\sim}{\theta} \in \Theta. \tag{23}$$

In the context of multivariate location models and in other situations too, it is quite possible to identify a class of estimators similar to that in Theorem 1. However, this would rest on plausible extensions of the notion of median unbiasedness in the multiparameter case. Since the components of $\underset{\sim}{T}$ may not be all independent and $\underset{\sim}{Q}$ in (21) may not be a diagonal matrix, the MU property for each coordinate of $\underset{\sim}{T}$ may not suffice. For our purpose, under (21), it seems that the following definition of multivariate MU property may suffice.

We say that $\underset{\sim}{T}$ is MU for $\underset{\sim}{\theta}$, if

$$\underset{\sim}{\ell}'(\underset{\sim}{T} - \underset{\sim}{\theta}) \text{ is MU for } 0, \text{ for every } \underset{\sim}{\ell} \in R^p, \underset{\sim}{\theta} \in \Theta. \tag{24}$$

In passing, we may remark that if $\underset{\sim}{T}$ has a distribution *diagonally symmetric* about $\underset{\sim}{\theta}$, then (24) holds, although the converse is not necessarily true. Recall that $\underset{\sim}{T}$ has a diagonally symmetric d.f. around $\underset{\sim}{\theta}$ if $\underset{\sim}{T} - \underset{\sim}{\theta}$ and $\underset{\sim}{\theta} - \underset{\sim}{T}$ both have the same d.f.

**Theorem 3.**

Let $\underset{\sim}{T}$ be a MU-estimator of $\underset{\sim}{\theta}$ [in the sense of (24)], and let $\mathfrak{C}$ be the class of all estimators of the form $\underset{\sim}{U} = \underset{\sim}{T} + \underset{\sim}{Z}$, where $\underset{\sim}{T}$ and $\underset{\sim}{Z}$ are independently distributed. Then for any arbitrary p.d. $\underset{\sim}{Q}$,

$$P_\theta\Big\{\| \underset{\sim}{T} - \underset{\sim}{\theta} \|_{\underset{\sim}{Q}} \leq \| \underset{\sim}{U} - \underset{\sim}{\theta} \|_{\underset{\sim}{Q}}\Big\} \geq 1/2, \quad \forall\, \underset{\sim}{\theta} \in \underset{\sim}{\Theta}, \underset{\sim}{U} \in \mathfrak{C}. \tag{25}$$

The proof is simple (Sen, 1989a) and is omitted. As a simple example illustrating (25), consider the case where $\underset{\sim}{X}_1, \ldots, \underset{\sim}{X}_n$ are i.i.d. r.v.'s having the multinormal distribution with mean vector $\underset{\sim}{\theta}$ and dispersion matrix $\underset{\sim}{\Sigma}$. Then $\underset{\sim}{T}_n = N^{-1}\Sigma_{i=1}^n \underset{\sim}{X}_i$ is MU in the sense of (24). Further, for known $\underset{\sim}{\Sigma}$, $\underset{\sim}{T}_n$ is sufficient for $\underset{\sim}{\theta}$, and the class $\mathfrak{C}$ consists here of all estimators of the form $\underset{\sim}{T}_n + \underset{\sim}{Z}_n$ where $\underset{\sim}{Z}_n$ is ancillary; this rests on the group of affine transformations $\underset{\sim}{X}_i \to \underset{\sim}{a} + \underset{\sim}{B}\underset{\sim}{X}_i$, $\underset{\sim}{B}$ nonsingular and $\underset{\sim}{a}$ arbitrary. Thus, by Theorem 3, within the class of such equivariant estimators of $\underset{\sim}{\theta}$, the sample mean $\underset{\sim}{T}_n$ (MLE) is the Pitman-closest one. By using the classical Helmert transformation for the multivariate normal vectors, it can be shown that the conclusion remains true in the case of unknown (but nonsingular) $\underset{\sim}{\Sigma}$. Moreover, the interesting feature of this example [or (25)] is that the construction of $\underset{\sim}{T}$ or the class $\mathfrak{C}$ does not depend on $\underset{\sim}{Q}$ in (21). In the multiparameter case, we shall study the GPNC for the *Stein-rule* or *shrinkage estimators*, and in that context, it will be seen that neither these estimators belong to the class $\mathfrak{C}$ nor their dominance may hold for all $\underset{\sim}{Q}$ (i.e., for a given $\underset{\sim}{Q}$, the construction of PC $\underset{\sim}{T}_n$ may generally depend on $\underset{\sim}{Q}$, and this $\underset{\sim}{T}_n$ may not retain its optimality simultaneously for all $\underset{\sim}{Q}$, possibly different from the adapted one). For the time being, we refrain ourselves from generalizing Theorem 2 to the vector-case; we shall make comments on it in the last section. Perhaps, it will be to our advantage to discuss the sequential analogue of Theorem 3, i.e., a multi-parameter extension of (19). Let us consider the same model as in (14) – (18) with the exception that in (15), $\underset{\sim}{T}_n$ is a vector and in (18), $\underset{\sim}{Z}_n$ is a vector too. Then the following result is proved in Sen (1989a):

Under (16), (18) and (24), for the class $\mathfrak{C}^0$ of (sequential) estimators of the form $\underset{\sim}{U}_N = \underset{\sim}{T}_N + \underset{\sim}{Z}_N$, we have

$$P_\theta\left\{\| T_N - \theta \|_Q \leq \| U_N - \theta \|_Q\right\} \geq 1/2, \quad \forall\, \theta \in \Theta,\ U_N \in \mathcal{C}^0, \tag{26}$$

for any arbitrary (p.d.) $Q$.

Again as an illustration, we may consider the multinormal mean vector ($\theta$) estimation problem when the covariance matrix ($\Sigma$) is arbitrary and unknown. Ghosh, Sinha and Mukhopadhyay (1976) and others have considered suitable stopping numbers ($N$) which are based solely on the sample covariance matrices $\{S_n;\ n > p\}$, so that (16) and (18) hold (for $T_n = \bar{X}_n$, $n \geq 1$). Further, (24) follows from the diagonal symmetry of the d.f. of $\bar{X}_n$ (around $\theta$), $\forall$ $n \geq 1$. Hence, (26) holds.

Let us next consider the asymptotic case parallel to that in the previous section. As in (10) – (11), a BAN estimator $T_n$ is characterized by its asymptotic (multi-) normality along with the fact that the dispersion matrix of this asymp totic distribution is equal to $\mathfrak{J}_\theta^{-1}$, where $\mathfrak{J}_\theta$ is the Fisher information matrix. The representation in (12) also extends readily to this multiparameter case, and (13) relates to a stochastic p-vector which has the dispersion matrix $\mathfrak{J}_\theta$. Consider then the class $\mathcal{C}^0$ of all estimators $\{U_n\}$ for which

$$\begin{bmatrix} n^{\frac{1}{2}}(U_n - \theta) \\ \xi_n \end{bmatrix} \xrightarrow[\mathfrak{D}]{} \mathcal{N}_{2p}\left[\begin{bmatrix} 0 \\ 0 \end{bmatrix}, \begin{bmatrix} \nu, & I \\ I, & \mathfrak{J}_\theta \end{bmatrix}\right], \tag{27}$$

where $\nu - \mathfrak{J}_\theta^{-1}$ is positive semi-definite, and the $\sqrt{n}$-consistency of $U_n$ entails the identity matrix $I$ in (27); for a BAN estimator $T_n$, $\nu = \mathfrak{J}_\theta^{-1}$. Finally, in (21), it seems quite appropriate to let $Q = \mathfrak{J}_\theta$. Then, by Theorem 1 of Sen (1986) we conclude that within the class $\mathcal{C}^0$ of estimators which are asymptotically multinormal and for which (27) holds [with $\mathfrak{J}_\theta^{-1}$, being replaced by the asymptotic dispersion matrix of $n^{1/2}(U_n - \theta)$], the BAN estimators are Pitman-closest with respect to the norm in (21), where $Q = \mathfrak{J}_\theta$.

The interesting feature is that we are no longer restricting ourselves to the class $\mathcal{C}$ of estimators (which are generally equivariant), but the Pitman-closest property depends on the adaptation of $Q = \mathfrak{J}_\theta$. For an arbitrary $Q$, this property may not hold. The asymptotic theory of Pitman-closeness of sequential estimators runs parallel to that in the concluding part of last section, and hence, we do not repeat these details.

In multiparameter estimation problems, the usual MLE may not be *admissible* (in the light of quadratic error loss functions). Stein (1956) considered the simple model that $X$ has a multi-normal distribution with mean vector $\theta$ and dispersion matrix, say, $I_p$, for some $p \geq 1$. He showed that though $X$ is the

MLE of $\underset{\sim}{\theta}$ for all $p \geq 1$, it is inadmissible for $p \geq 3$. James and Stein (1962) constructed a shrinkage version which dominates $\underset{\sim}{X}$ in quadratic error loss. Sparked by this *Stein-phenomenon*, during the past twenty-five years, a vast amount of work has been done in improving the classical estimators in various multiparameter estimation problems by suitable shrinkage versions; these improvements being judged by the smallness of appropriate quadratic error loss function based *risks*. Coming back to the multivariate normal law, such shrinkage or Stein-rule estimators do not belong to the class $\mathfrak{C}$ considered in Theorem 3! Thus, the characterization of PC made in Theorem 3 is not applicable to such shrinkage estimators. This raises the question: Does the usual Stein-rule estimator have the PC property too? The answer is affirmative in a variety of situations, and moreover, this PC dominance may hold even under less restrictive regularity conditions.

Rao (1981) initiated renewed interest in the PCC by showing that some simple shrinkage estimators may not be the Pitman closest ones! He actually argued that the usual quadratic error loss function places undue emphasis on large deviations which may occur with small probability, and hence, minimizing the mean square error may insure against large errors in estimation occurring more frequently rather than providing greater concentration of an estimator in neighborhoods of the true value. Since, typically, a Stein-rule estimator is non-linear and may not have (even asymptotically) multi-normal law, Rao's criticism is more appropriate in this context. Actually, Rao, Keating and Mason (1986) and Keating and Mason (1988) have shown by extensive numerical studies that for the p-variate normal distribution, for $p \geq 2$, the James-Stein estimator is closer (in the Pitman sense) than the MLE $\underset{\sim}{X}$. The quadratic error loss criterion may also cause some difficulties in the usual linear models when the incidence (design) matrix is nearly singular; in such a case, a *ridge regression* estimator is generally preferred. In this context too, one may enquire whether such ridge regression estimators have the Pitman closeness property. This issue has been taken up by Mason, Keating, Sen and Blaylock (1990), and both theoretical and numerical studies are made. So long as the incidence matrix is non-singular, a ridge estimator may not dominate the classical least square estimator in the PCC, although it fares well over a greater part of $\Theta$. The lack of dominance arises mainly due to the fact that as $\underset{\sim}{\theta}$ moves away from the pivot, the performance of a ridge estimator may deteriorate, so that the inequality in (23) may not hold for all $\theta$ belonging to $\Theta$, although it generally holds for all $\underset{\sim}{\theta} : \|\underset{\sim}{\theta}\| < C$, where $C$ is related to the factor $k$ ( $> 0$) arising in the construction of a ridge estimator. Their study also covers the comparison of two arbitrary linear estimators in the light of the PCC.

The interesting fact is that the PCC may not even need that $p$ is $\geq 2$ (comparable to $p \geq 3$ for the quadratic error loss)! Even for $p = 1$, $X \sim \mathcal{N}(\theta, 1)$, Efron (1975) showed that for

$$\delta = X - \Delta(X); \quad \Delta(x) = \tfrac{1}{2}\left[min\{x, \Phi(-x)\}\right], \; x \geq 0, \tag{28}$$

$[\Delta(-x) = -\Delta(x),\ x \geq 0$ and $\Phi(\cdot)$ is the standard normal d.f.], (1) holds for $T_1 = \delta$ and $T_2 = X$. He made some conjectures for $p \geq 2$. For the multivariate normal mean estimation problem, a systematic account of the PC dominance of Stein-rule estimators is given by Sen, Kubokawa and Saleh (1989). Consider first the model that for some positive integer $p$, $\underset{\sim}{X}$ has a $p$-variate normal distribution with mean vector $\underset{\sim}{\theta}$ and dispersion matrix $\sigma^2 \underset{\sim}{V}$, where $\underset{\sim}{V}$ is known (and p.d.), while $\underset{\sim}{\theta}$ and $\sigma^2$ are unknown. Also assume that $s^2$ is an estimator of $\sigma^2$, such that (i) $ms^2/\sigma^2 \underset{\mathfrak{D}}{=} \chi^2_m$, a r.v. having the central chi square distribution with $m$ ($\geq 1$) degrees of freedom (DF), and (ii) $s^2$ is distributed independently of $\underset{\sim}{X}$. [In actual application, $\underset{\sim}{X}$ may be the sample mean vector or a suitable linear estimator (of regression parameters, for example) and $s^2$ is the residual mean square (with $m = n - q$, for some $q \geq 1$]. Keeping in mind the loss function in (21), we may consider a Stein-rule estimator of the form

$$\underset{\sim}{\delta}_\phi = \left[\underset{\sim}{I} - \phi(\underset{\sim}{X}, s^2)s^2 \| \underset{\sim}{X} \|^{-2}_{\underset{\sim}{Q}, \underset{\sim}{V}} \underset{\sim}{Q}^{-1} \underset{\sim}{V}^{-1}\right]\underset{\sim}{X}, \tag{29}$$

where $\phi(\underset{\sim}{x}, s^2)$ is a nonnegative r.v. bounded from above by a constant $c_p$ (depending on $p$) (with probability one), and $\| \underset{\sim}{X} \|^{2}_{\underset{\sim}{Q}, \underset{\sim}{V}} = \underset{\sim}{X}' \underset{\sim}{V}^{-1} \underset{\sim}{Q}^{-1} \underset{\sim}{V}^{-1} \underset{\sim}{X}$. Note that estimators of this type with a different bound for $\phi(\cdot)$ (and for $p \geq 3$) were considered by Stein (1981), and hence, we regard them as Stein-rule estimators. Then, we have the following result due to Sen et al. (1989).

**Theorem 4.**

Assume that $p \geq 2$, and

$$0 \leq \phi(\underset{\sim}{X}, s^2) \leq (p-1)(3p+1)/(2p), \text{ for every } (\underset{\sim}{X}, s^2) \text{ a.e.} \tag{30}$$

Then $\underset{\sim}{\delta}_\phi$, given by (29), is closer than $\underset{\sim}{X}$ in the Pitman sense [i.e., (23) holds for $\underset{\sim}{T}_1 = \underset{\sim}{\delta}_\phi$, $\underset{\sim}{T}_2 = \underset{\sim}{X}$ and $L(\underset{\sim}{T}, \underset{\sim}{\theta}) = \| \underset{\sim}{T} - \underset{\sim}{\theta} \|^2_{\underset{\sim}{Q}}$].

If $\sigma^2$ were known, then in (29) and (30), we would have taken $\phi(\underset{\sim}{X}, \sigma^2)$ instead of $\phi(\underset{\sim}{X}, s^2)$. In this sense, the classical James-Stein (1962) estimator is a special case of (29). We may take $\phi(\underset{\sim}{X}, s^2) = a : 0 < a < (p-1)(3p+1)/2p$, and consider the following versions:

$$\underset{\sim}{\delta}_a = \underset{\sim}{X} - as^2 \| \underset{\sim}{X} \|^{-2}_{\underset{\sim}{Q}, \underset{\sim}{V}} \underset{\sim}{Q}^{-1} \underset{\sim}{V}^{-1} \underset{\sim}{X}, \tag{31}$$

$$\underset{\sim}{\delta}^+_a = \underset{\sim}{X} - \min\left\{as^2 \| \underset{\sim}{X} \|^{-2}_{\underset{\sim}{Q}, \underset{\sim}{V}},\ \underset{\sim}{X}' \underset{\sim}{V}^{-1} \underset{\sim}{X} \| \underset{\sim}{X} \|^{-2}_{\underset{\sim}{Q}, \underset{\sim}{V}}\right\} \underset{\sim}{Q}^{-1} \underset{\sim}{V}^{-1} \underset{\sim}{X}, \tag{32}$$

so that $\underset{\sim}{\delta}_a$ is a James-Stein estimator and $\underset{\sim}{\delta}^+_a$ is the so-called *positive-rule* version. Then again (23) holds with $\underset{\sim}{T}_1 = \underset{\sim}{\delta}^+_a$, $\underset{\sim}{T}_2 = \underset{\sim}{\delta}_a$, $L(\underset{\sim}{T}, \underset{\sim}{\theta}) = \| \underset{\sim}{T} - \underset{\sim}{\theta} \|^2_{\underset{\sim}{Q}}$ and

$0 < a < (p-1)(3p+1)/2p$. Thus, the positive rule version dominates the classical James-Stein version in the light of the PCC as well. It may be remarked that for the quadratic error loss dominance, Stein (1981) had $p \geq 3$ and $0 \leq a \leq 2(p-2)$, while here $p \geq 2$ and $0 \leq a \leq (p-1)(3p+1)/2p$. For $p \in [2, 5]$, $(p-1)(3p+1)/2p > 2(p-2)$. For $p \geq 6$, in (30), we may as well replace $(p-1)(3p+1)/2p$ by $2(p-2)$. The main motivation of the upper bound in (30) was to include the case $p = 2$ and to have a larger shrinkage factor for smaller values of $p$.

The proof of Theorem 4 depends on some intricate properties of noncentral chi square densities which may have some interest on their own. Basically, to verify (23) for $\underset{\sim}{T}_1 = \underset{\sim}{\delta}_\phi$ and $\underset{\sim}{T}_2 = \underset{\sim}{X}$, it follows through some standard steps that a sufficient condition is

$$P_\lambda\left\{\chi^2_{p,\lambda} \geq \lambda + c\chi^2_m\right\} \geq 1/2, \ \forall \lambda \geq 0, m \geq 1, p \geq 2, \tag{33}$$

where $c = (p-1)(3p+1)/(4pm)$, $\chi^2_{p,\lambda}$ has the noncentral chi square d.f. with $p$ DF and noncentrality parameter $\lambda$ ($\geq 0$), and $\chi^2_m$ has the central chi square d.f. with $m$ DF, independently of $\chi^2_{p,\lambda}$. The trick was to show that the left hand side of (33) in $\lambda$ ($\geq 0$) and that as $\lambda \to \infty$, it converges to 1/2. Sen et al. (1989) also considered the case of $\underset{\sim}{X} \sim \mathcal{N}_p(\underset{\sim}{\theta}, \underset{\sim}{\Sigma})$, $\underset{\sim}{\Sigma}$ arbitrary (p.d.), $\underset{\sim}{S} \sim Wishart(\underset{\sim}{\Sigma}, p, m)$ independently of $\underset{\sim}{X}$ with $m \geq p$, and considered the usual shrinkage estimator

$$\underset{\sim}{\delta}^*_\phi = X - (m-p+1)^{-1}\phi(\underset{\sim}{X}, \underset{\sim}{S})d_m \| X \|^{-2}_{\underset{\sim}{S}^{-1}} \underset{\sim}{Q}^{-1}\underset{\sim}{S}^{-1}\underset{\sim}{X}, \tag{34}$$

where $d_m = ch_{min}(\underset{\sim}{Q}\,\underset{\sim}{S})$ and $\phi(\underset{\sim}{x}, \underset{\sim}{S})$ has the same bound as in (30). Then, for every $p \geq 2$, (23) holds for $\underset{\sim}{T}_1 = \underset{\sim}{\delta}^*_\phi$ and $\underset{\sim}{T}_2 = \underset{\sim}{X}$.

Let us now consider the asymptotic picture relating to the Stein-rule estimators under the PCC. Generally, we have a sequence $\{\underset{\sim}{T}_n\}$ of estimators, such that as $n \to \infty$,

$$n^{\frac{1}{2}}(\underset{\sim}{T}_n - \underset{\sim}{\theta}) \underset{\mathfrak{D}}{\to} \mathcal{N}_p(\underset{\sim}{0}, \underset{\sim}{\Sigma}), \ \underset{\sim}{\Sigma} \text{ p.d.}, \tag{35}$$

and, also, we have a sequence $\{\underset{\sim}{S}_n\}$ of stochastic matrices, such that

$$\underset{\sim}{S}_n \to \underset{\sim}{\Sigma}, \text{ in probability, as } n \to \infty. \tag{36}$$

Thus, a suitable test statistic for testing the hypothesis of a null pivot is

$$\mathcal{L}_n = n\underset{\sim}{T}'_n\underset{\sim}{S}^{-1}_n\underset{\sim}{T}_n, \tag{37}$$

so that an asymptotic version of (34) is

$$\underset{\sim}{\delta}^{0*}_{\phi,n} = \phi(\underset{\sim}{T}_n, \underset{\sim}{S}_n)d_m\mathcal{L}^{-1}_n\underset{\sim}{Q}^{-1}\underset{\sim}{S}^{-1}_n\underset{\sim}{T}_n. \tag{38}$$

This form is of sufficient generality to cover a large class of $\{\underset{\sim}{T}_n\}$, both of parameter and nonparameter forms. In particular, for $R$- and $M$-estimators, for $\mathcal{L}_n$ in (37), instead of $\underset{\sim}{T}_n$, suitable rank or $M$-statistics may also be used. Also, in (38), a null pivot has been used; the modifications for a general $\underset{\sim}{\theta}_0$ are straightforward. Now, if $\underset{\sim}{\theta} \neq \underset{\sim}{0}$, then $n^{-1}\mathcal{L}_n \xrightarrow{p} \underset{\sim}{\theta}'\underset{\sim}{\Sigma}^{-1}\underset{\sim}{\theta}$, as $n \to \infty$, so that $\mathcal{L}_n^{-1} \xrightarrow{p} 0$, as $n \to \infty$. Thus, for any fixed $\underset{\sim}{\theta} \neq \underset{\sim}{0}$,

$$\sqrt{n} \left\| \underset{\sim}{T}_n - \underset{\sim}{\delta}^{0*}_{\phi,n} \right\|_{\underset{\sim}{Q}} \xrightarrow{p} 0, \text{ as } n \to \infty, \tag{39}$$

so that asymptotically the Stein-rule version becomes stochastically equivalent to the classical version. For this reason, the asymptotic dominance picture has been considered in the case where $\underset{\sim}{\theta}$ belongs to a Pitman-neighborhood of the assumed pivot $(\underset{\sim}{0})$. Thus, we may consider a sequence $\{K_n\}$ of local (Pitman-) alternatives

$$K_n : \underset{\sim}{\theta} = \underset{\sim}{\theta}_{(n)} = n^{-\frac{1}{2}}\underset{\sim}{\lambda},\ \underset{\sim}{\lambda} \in R^p. \tag{40}$$

Further, by virtue of (36), we may replace $\underset{\sim}{S}_n$ by $\underset{\sim}{\Sigma}$, and appeal to Theorem 4 (where $s^2$ is taken as 1 and $\underset{\sim}{V} = \underset{\sim}{\Sigma}$). As such, we obtain that for every $\phi(\cdot)$, satisfying (40),

$$\lim_{n\to\infty} P\left\{\sqrt{n} \left\| \underset{\sim}{\delta}^{0*}_{\phi,n} - \underset{\sim}{\theta} \right\|_{\underset{\sim}{Q}} \leq \sqrt{n} \left\| \underset{\sim}{T}_n - \underset{\sim}{\theta} \right\|_{\underset{\sim}{Q}} \middle| K_n \right\} \geq 1/2. \tag{41}$$

Thus, the usual robust and nonparametric Stein-rule estimators enjoy the Pitman closeness property in the asymptotic case (and for Pitman-alternatives) under less restrictive regularity conditions (than in the conventional case of quadratic error losses).

Let us now consider sequential Stein-rule estimators and discuss their dominance in the light of the PCC. Consider a simple model: $\{\underset{\sim}{X}_i,\ 1 \geq 1\}$ are i.i.d.r.v. with $\mathcal{N}_p(\underset{\sim}{\theta},\ \sigma^2\underset{\sim}{I}_p)$ d.f.; $\underset{\sim}{\theta}$ and $\sigma^2$ are unknown. Let $s_n^2 = (np)^{-1}\Sigma_{i=1}^n(\underset{\sim}{X}_i - \underset{\sim}{\bar{X}}_n)' \times (\underset{\sim}{X}_i - \bar{X}_n)$; $\bar{X}_n = n^{-1}\Sigma_{i=1}^n \underset{\sim}{X}_i$, and consider a stopping number $N$, such that for every $n \geq 2$, $[N = n]$ depends only on $\{s_k^2,\ k \leq n\}$. Let then

$$\underset{\sim}{\delta}_N^b = \left\{1 - bs_N^2\left(N\|\underset{\sim}{\bar{X}}_N\|^2\right)^{-1}\right\} \underset{\sim}{\bar{X}}_N, \tag{42}$$

where

$$0 < b \leq (p-1)(3p+1)/(2p),\ p \geq 2. \tag{43}$$

We may even allow $b$ to be replaced by $\phi(\underset{\sim}{\bar{X}}_N,\ s_N^2)$, where $\phi(\cdot)$ satisfies (40). Again note that $[N = n] \Leftrightarrow [s_k^2,\ k \leq n]$, so that by virtue of the independence of

$\{\underset{\sim}{\bar{X}}_n\}$ and $\{s_n^2\}$, given $[N = n]$, $\underset{\sim}{\bar{X}}_n$ has a multinormal distribution $(\underset{\sim}{\theta}, \frac{1}{n}\underset{\sim}{\Sigma})$, independently of the $s_k^2$, $k \geq 2$. However, the shrinkage factor $\left(bs_N^2\left(N\|\underset{\sim}{\bar{X}}_N\|^2\right)^{-1}\right)$ in (42) depends on all the r.v.'s ($N$, $\underset{\sim}{\bar{X}}_N$ and $s_N^2$). Hence, the simple proof for (26) may not be adaptable in this more complex situation. Nevertheless, it has been shown by Sen (1989a) that by virtue of certain log-concavity property of the noncentral chi square density and the non-sequential results in Sen, Kubokawa and Saleh (1989) the following result holds.

**Theorem 5.**

For the class of Stein-rule estimators in (42), whenever the stopping number $N$ satisfies (16) [with $\underset{\sim}{W}_n = (s_2^2, \ldots, s_n^2)$, $n \geq 2$], for every $b \in (0, (p-1)(3p+1)/2p]$,

$$P_{\underset{\sim}{\theta}}\left\{\left\|\underset{\sim}{\delta}_N^b - \underset{\sim}{\theta}\right\|_{\underset{\sim}{Q}} \leq \left\|\underset{\sim}{\bar{X}}_N - \underset{\sim}{\theta}\right\|_{\underset{\sim}{Q}}\right\} \geq 1/2, \ \forall\, \underset{\sim}{\theta}\ \sigma. \tag{44}$$

In passing, we may remark that a parallel dominance result under a quadratic error loss has been proved by Ghosh, Nickerson and Sen (1987). In the fixed-sample size case, the PC dominance of $\underset{\sim}{\delta}_\phi^*$ in (44) has been established for an arbitrary (p.d.) $\underset{\sim}{\Sigma}$. On the other hand, for arbitrary $\underset{\sim}{\Sigma}$, the sequential case either in terms of the PCC or a quadratic error loss has not yet been resolved.

The asymptotic theory of sequential shrinkage estimation in the light of the PCC has been worked out systematically in Sen (1987a, b; 1989c, d). The basic idea is to incorporate (19) for the proposed stopping rules, verify (20) as amended in the multivariate case, and then by appeal to (35) through (41) completing the proof. Although, in the cited references, suitable quadratic error losses were used, our (35) through (41) ensure that the results remain adaptable in the PCC as well. Further, in this asymptotic setup, the covariance matrix $\underset{\sim}{\Sigma}$ can be quite arbitrary (p.d.). In the case of a quadratic error loss, the actual asymptotic risk functions were replaced by asymptotic distributional risk functions, so that the desired dominance results could be obtained under less restrictive regularity conditions. In the case of PCC, this replacement makes no difference in the asymptotic picture, and therefore, there is no need to assume additional regularity conditions under which the asymptotic limits of the actual quadratic error loss based risks exist. In the case of shrinkage estimation, there is a technical problem in finding an asymptotically optimal stopping time, and this has been discussed in detail in Sen (1989d).

## GPNC and Estimation of a Dispersion Matrix

To motivate, let us consider the problem of estimating the dispersion matrix $\underset{\sim}{\Sigma}$ (p.d. but arbitrary) of a multinormal distribution. An unbiased estimator of $\underset{\sim}{\Sigma}$ is $\underset{\sim}{S} = (n-1)^{-1}\Sigma_{i=1}^n(\underset{\sim}{X}_i - \underset{\sim}{\bar{X}}_n)(\underset{\sim}{X}_i - \underset{\sim}{\bar{X}}_n)'$, where $\underset{\sim}{X}_1, \ldots, \underset{\sim}{X}_n$ are i.i.d.r. vectors and $\bar{X}_n = n^{-1}\Sigma_{i=1}^n X_i$. Note that $A = (n-1)S \sim Wishart(\Sigma, n-1, p)$. One possibility is to take $\theta = vec(\Sigma)$ and the class $\mathcal{C}_1$ of equivariant estimators $T = vec(cA)$, $c > 0$, under the quadratic error loss function $L(T, \theta)$ as in (21) - (23). But the natural appeal for such a quadratic error loss function is not so convincing in this setup, and other forms of loss functions have been considered by various workers (viz., Haff, 1980, Sinha and Ghosh, 1987, and others). A popular choice is the so-called *entropy loss function*:

$$L(S, \Sigma) = tr(S\Sigma^{-1}) - log \mid S\Sigma^{-1}\mid - p; \tag{45}$$

a second one

$$L(S, \Sigma) = tr(S\Sigma^{-1} - I)^2 \tag{46}$$

also deserves mention. [For the estimation of the precision matrix $\Sigma^{-1}$, $S^{-1}$ is a natural choice, and in (45) or (46), we may replace $S$ and $\Sigma^{-1}$ by $S^{-1}$ and $\Sigma$, respectively.] Consider the class of estimation ($\mathcal{C}_1$) of the form

$$\{cA : c > 0 \text{ and } (n-1)S \sim W(\Sigma, n-1, p)\}. \tag{47}$$

Also, consider the GPNC in (23). Then the following result is due to Khattree (1987).

**Theorem 6.**

Let $0 < a_2 < a_1 < 1$ and $a_iA \in \mathcal{C}_1$, $i = 1, 2$. Also, let $c_{p,n} = med\{\chi^2_{p(n-1)}\}$. Then $a_1A >_{GPN} a_2A$ under the loss function in (45) if and only if

$$p\ log(a_1/a_2) > (a_1 - a_2)c_{p,n}. \tag{48}$$

Also, let $c^*_{p,n} = med\{\tau_p\}$ where $\tau_p = [tr(WW')] \div [tr(W)]$ and $W \sim Wishart(I, n-1, p)$. Then, under (46), $a_1A >_{GPN} a_2A$ iff

$$c^*_{p,n} < 2(a_1 + a_2)^{-1}. \tag{49}$$

Thus, if we let $a_0 = p/c_{p,n}$ and $a_0^* = 1/c^*_{p,n}$, then within the class $\mathcal{C}_1$ of estimators of $\Sigma$, $a_0A$ (or $a_0^*A$) is a unique best (in the GPNC sense) estimator of $\Sigma$ under the entropy loss [or (46)], and this can not be improved within this class $\mathcal{C}_1$.

It may be noted that $\mathcal{C}_1$ is the class of equivariant estimators under the (full affine) group of transformations:

$$X \to a + BX,\ A \to BAB',\ B \text{ nonsingular, } a \text{ arbitrary.} \tag{50}$$

Sinha and Ghosh (1987) and Sinha (1988) also considered a larger class $\mathcal{C}_2$ of the form:

$$\mathcal{C}_2 = \Big\{ \underset{\sim}{T}\underset{\sim}{Q}\underset{\sim}{T}' : A = \underset{\sim}{T}\underset{\sim}{T}' \sim W(\underset{\sim}{\Sigma}, n\text{-}1, p);$$

$$\underset{\sim}{Q} = Diag(q_1,\ldots,q_p),\ q_j > 0, \text{ for } j = 1,\ldots,p \Big\}, \tag{51}$$

and established the inadmissibility of the class $\mathcal{C}_1$ relative to the class $\mathcal{C}_2$, under various loss functions. A natural question arises in this context: Are the estimators in the class $\mathcal{C}_2$ admissible in the GPN sense? To address this problem properly, we may note that the entropy loss in (45) was first introduced in the univariate case by James and Stein (1961); in this special case, $\mathcal{C}_1 \equiv \mathcal{C}_2$ contains the class of scalar multiples of the sample variance, and hence, the PC of an estimator can as well be judged by using the usual quadratic error loss. This was accomplished by Ghosh and Sen (1989) (from the PCC point of view). This equivalence result does not, however, hold generally for the multivariate case, and hence, a different approach is needed. The class $\mathcal{C}_2$ is too big, and although for suitable subclasses of $\mathcal{C}_2$ (defined by imposing additional partial ordering), admissibility of estimators in the GPN sense can be established, such a result may not generally hold for the entire class $\mathcal{C}_2$. This is being explored in detail (viz., Sen, Nayak and Khattree, 1990). The following results are worth mentioning in this context:

(i) Within the class $\mathcal{C}_2$, no estimator of $\underset{\sim}{\Sigma}$ is GPN-optimal!

(ii) Let $\underset{\sim}{D}_2 = Diag(d_{21},\ldots,d_{2p})$ with $d_{2j}^{-1} = med(\chi^2_{n+p-2j})$, for $j = 1,\ldots,p$, and let $\hat{\underset{\sim}{\Sigma}}_2 = \underset{\sim}{T}\underset{\sim}{D}_2\underset{\sim}{T}'$. Also, let

$$\mathcal{C}_3 = \{\underset{\sim}{A} \in \mathcal{C}_2 : \underset{\sim}{Q} - \underset{\sim}{D}_2 = \text{positive semi-definite (p.s.d.)}\}; \tag{52}$$

$$\mathcal{C}_4 = \{\underset{\sim}{A} \in \mathcal{C}_2 : \underset{\sim}{D}_2 - \underset{\sim}{Q} = \text{p.s.d.}\}. \tag{53}$$

Then, within the subclass $\mathcal{C}_3$, $\hat{\underset{\sim}{\Sigma}}_2$ is GPN-optimal. Within the subclass $\mathcal{C}_4$, no estimator of $\Sigma$ is GPN-optimal.

(iii) Let $\underset{\sim}{D}_1 = Diag(d_{11},\ldots,d_{1p})$ with $d_{1j} = (n + p - 2j)^{-1}$, for $j = 1,\ldots,p$, and let $\hat{\underset{\sim}{\Sigma}}_1 = \underset{\sim}{T}\underset{\sim}{D}_1\underset{\sim}{T}'$. Then, $\hat{\underset{\sim}{\Sigma}}_1$ is the James-Stein estimator of $\underset{\sim}{\Sigma}$, and its properties have already been studied by Sinha (1988). The usual estimator of $\underset{\sim}{\Sigma}$ is $\hat{\underset{\sim}{\Sigma}}_0 = (n-1)^{-1}\underset{\sim}{A}$. Then, although there is no GPN-optimal estimator of $\underset{\sim}{\Sigma}$ within the class $\mathcal{C}_2$, both $\hat{\underset{\sim}{\Sigma}}_1$ and $\hat{\underset{\sim}{\Sigma}}_2$ dominate the classical estimator $\hat{\underset{\sim}{\Sigma}}_0$ in the GPN-sense.

**Acknowledgements**

Thanks are due to the reviewer for his very critical reading of the manuscript leading to various improvements in the presentation.

**References**

Anscombe, F. J. (1952): Large sample theory of sequential estimation, *Proc. Cambridge Phil. Soc.* 48, 600-607.

Basu, D. (1955): On statistics independent of a complete sufficient statistic, *Sankhya* 15, 377-380.

Blyth, C. R. (1972): Some probability paradoxes in choice from among random alternatives, *J. Amer. Statist. Assoc.* 67, 366-373.

Brown, L. D., Cohen, A. and Strawderman, W. E. (1976): A complete class theorem for strict monotone likelihood ratio with applications, *Ann. Statist.* 4, 712-722.

Efron, B. (1975): Biased versus unbiased estimation, *Adv. Mathem.* 16, 259-277.

Fisher, R. A. (1938): Statistical Theory of Estimation, Calcutta University Readership Lectures.

Geary, R. C. (1944): Comparison of the concepts of efficiency and closeness for consistent estimates of a parameter, *Biometrika* 33, 123-128.

Ghosh, M. and Sen, P. K. (1989): Median unbiasedness and Pitman closeness, *J. Amer. Statist. Assoc.* 84, 1089-1091.

Ghosh, M., Sinha, B. K. and Mukhopadhyay, N. (1976): Multivariate sequential point estimation, *J. Multivar. Anal.* 6, 281-294.

Haff, L. R. (1980): Empirical Bayes estimation of the multivariate normal covariance matrix, *Ann. Statist.* 8, 586-597.

Hwang, J. T. (1985): Universal domination and stochastic domination: Estimation simultaneously under a broad class of loss functions, *Ann. Statist.* 13, 295-314.

James, W. and Stein, C. (1962): Estimation with quadratic loss, *Proc. 4th Berkeley Symp. Math. Statist. Prob.* 1, 361-379.

Johnson, N. L. (1950): On the comparison of estimators, *Biometrika* 37, 281-287.

Keating, J. P. (1985): More on Rao's phenomenon, *Sankhya, Ser. A*, 47, 18-21.

Keating, J. P. and Gupta, R. C. (1984): Simultaneous comparison of scale parameters, *Sankhya, Ser. B*, 46, 275-280.

Keating, J. P. and Mason, R. L. (1985a): Practical relevance of an alternative criterion in estimation, *Amer. Statist.* 39, 203-205.

Keating, J. P. and Mason, R. L. (1985b): Pitman's measure of closeness, *Sankhya, Ser. B.*, 47, 22-32.

Keating, J. P. and Mason, R. L. (1988): James-Stein estimation from an alternative perspective, *Amer. Statist.* 42, 160-164.

Khattree, R. (1987): On comparison of estimators of dispersion using generalized Pitman nearness criterion, *Commun. Statist. Theor. Meth.* 16, 263-274.

Mason, R. L., Keating, J. P., Sen, P. K. and Blaylock, N. W. (1990): Comparison of linear estimators using Pitman's measure of closeness, *J. Amer. Statist. Assoc.* 85, in press.

Mason, R. L., Keating, J. P., Sen, P. K. and Rao, C. R. (1990): Pitman nearness of statistical estimators: A panel discussion on some recent developments, *Comput. Statist. & Data Anal.* 8, in press.

Nayak, T. (1990): Estimation of location and scale parameters using generalized Pitman's measure of closeness, *J. Statist. Plan. Infer.* 24, 259-268.

Peddada, S. D. (1985): A short note on Pitman measure of nearness, *Amer. Statist.* 39, 298-299.

Peddada, S. D. and Khattree, R. (1986): On Pitman nearness and variance of estimators, *Commun. Statist. Theor. Meth.* 15, 3005-3017.

Pitman, E. J. G. (1937): The closest estimates of statistical parameters, *Proc. Cambridge Phil. Soc.* 33, 212-222.

Rao, C. R. (1945): Information and accuracy attainable in the estimation of statistical parameters, *Bull. Calcutta Math. Soc.* 37, 31.

Rao, C. R. (1981): Some comments on the minimum mean square as a criterion of estimation, in *Statistics and Related Topics*, eds. M. Csorgo, D. A. Dawson, J. N. K. Rao and A. K. Md. E. Saleh, North Holland, Amsterdam, pp. 123-143.

Rao, C. R., Keating, J. P. and Mason, R. L. (1986): The Pitman nearness criterion and its determination, *Commun. Statist. Theor. Meth.* 15, 3173-3191.

Robert, C. and Hwang, J. T. (1988): A dubious criterion: Pitman closeness, preprint.

Sen, P. K. (1981): *Sequential Nonparametrics: Invariance Principles and Statistical Inference*, John Wiley & Sons, Inc., New York.

Sen, P. K. (1986): Are BAN estimators the Pitman closest ones too?, *Sankhya, Ser. A.*, 48, 61-68.

Sen, P. K. (1987a): Sequential Stein-rule maximum likelihood estimators: General asymptotics, in *Statistical Decision Theory and Related Topics*, IV, eds. S. Gupta and J. O. Berger, Springer-Verlag, Vol. 2, 195-208.

Sen, P. K. (1987b): Sequential shrinkage U-statistics: General asymptotics, *Rev. Brasileira Prob. Estatst.* 1, 1-21.

Sen, P. K. (1989a): On the Pitman closeness of some sequential estimators, Inst. Statist. Univ. North Carolina Mimeo Rep. No. 1788.

Sen, P. K. (1989b): Optimality of the BLUE and ABLUE in the light of the Pitman closeness of statistical estimators, Inst. Statist. Univ. North Carolina Mimeo Rep. No. 1863. To appear in *Colloq. Limit Theo. Probab. Statist.* (ed. P. Révesz).

Sen, P. K. (1989c): Asymptotic theory of sequential shrunken estimation of statistical functionals, *Proc. 4th Prague Conference Asymp. Meth. Statist.*, eds. P. Mandel and M. Hušková, Charles Univ., Prague, pp. 83-100.

Sen, P. K. (1989d): Statistical functionals, stopping times and asymptotic minimum risk property, Inst. Statist. Univ. North Carolina Mimeo Rep. No. 2002.

Sen, P. K., Kubokawa, T. and Saleh, A. K. M. E. (1989): The Stein paradox in the sense of the Pitman measure of closeness, *Ann. Statist.* 17, 1375-1386.

Sen, P. K., Nayak, T. and Khattree, R. (1990): Comparison of equivariant estimators of a dispersion matrix under generalized Pitman nearness criterion, preprint.

Sinha, B. K. (1988): Inadmissibility of the best equivariant estimators of the variance-covariance matrix, the precision matrix and the generalized variance: A survey, in *Advances in Multivariate Statistical Analysis*, eds. S. Dasgupta and J. K. Ghosh, Statist. Pub. House, Calcutta, pp. 483-497.

Sinha, B. K. and Ghosh, M. (1987): Inadmissibility of the best equivariant estimator of the variance covariance matrix and the generalized variance under entropy loss, *Statist. Dec.* 5, 201-227.

Stein, C. (1956): Inadmissibility of the usual estimator of the mean of a multivariate normal distribution, *Proc. 3rd Berkeley Symp. Math. Statist. Prob.* 1, 197-209.

Stein, C. (1981): Estimation of the mean of a multivariate normal distribution, *Ann. Statist.* 9, 1135-1151.

Wald, A. (1949): Statistical decision functions, *Ann. Math. Statist.* 20, 165-20.

# UNBIASED SEQUENTIAL BINOMIAL ESTIMATION

Bimal K. Sinha, University of Maryland-Baltimore County,

and

Bikas K. Sinha, Indian Statistical Institute

**Abstract**

We review the literature on unbiased estimation of some functions of the Bernoulli parameter $p$ in the sequential case. Connections between the so-called efficient and inefficient sampling plans through the well known concept of sufficiency which have been explored recently are also presented.

## Introduction

Under the set up of independent identical Bernoulli trials with parameter p, various aspects of unbiased estimation of a parametric function g(p) have been studied in the literature. Early works of Girshick, Mosteller and Savage (1946), Wolfowitz (1946, 1947), Lehmann and Stein (1950), De Groot (1959) and Wasan (1964) are devoted to some general results on sequential binomial estimation. Later works by Gupta (1967), Sinha and Sinha (1975), Sinha and Bhattacharya (1982) and Sinha and Bose (1985) deal with problems related to unbiased estimation of $1/p$. Recently Bose and Sinha (1984) studied the connections between the so-called efficient and inefficient Bernoulli sampling plans through the well known concept of sufficiency of statistical experiments.

Our object in this paper is to present a comprehensive review of most of the available results in this area. We omit proofs of all the results. However, detailed and exact references to various results are provided.

The next section is devoted to setting up the notations, nomenclature, and definition of efficient sampling plans. In the third section, we provide results on efficient sampling plans. The problem of unbiased estimation of $1/p$, which has received considerable amount of attention in the literature, is discussed in fourth section. In fifth section, we discuss the connection between efficient and inefficient sampling plans via the concept of sufficiency. Some concluding remarks are made in the last section.

## Notations and Nomenclature

Let ($Z_i$, $i = 1, 2\ldots$) be an i.i.d. sequence of Bernoulli variates with $P(Z_i = 1) = p$ and $P(Z_i = 0) = 1 - p = q$ (say). We assume $p \in \Omega \subseteq (0, 1)$. Any realization of this process can be exhibited as a lattice path in the ($X$, $Y$)-plane, where a particle moves from the origin one step to the right (along $X$-axis) if the incoming observation is 0 and one step above (along $Y$-axis) if it is 1. A *stopping rule* can be viewed as a sequence of functions $\phi_k$, where $\phi_k$ is a function of $(Z_1,\ldots,Z_k)$. Each $\phi_k$ takes the value 0 or 1; given $z_1,\ldots,z_k$, $\phi_k(z_1,\ldots,z_k) = 1$ indicates that we take one more observation and $\phi_k(z_1,\ldots,z_k) = 0$ indicates that we stop at this stage. A point $\alpha = (x, y)$ is a *continuation point* if there exists one sequence of realization $(z_1, z_2,\ldots,z_{x+y})$ leading to $\alpha$ such that $\phi_j(z_1,\ldots,z_j) = 1$ $\forall j \leq x + y$. A point $\alpha = (x, y)$ is a *boundary point* if there exists one sequence of realization $(z_1, z_2,\ldots,z_{x+y})$ leading to $\alpha$ such that $\phi_j(z_1,\ldots,z_j) = 1$ $\forall j < x + y$ and $\phi_j(z_1,\ldots,z_{x+y}) = 0$. A point may be a boundary point or a continuation point depending on the path. A point is an *accessible point* if it is either a boundary point or a continuation point. Points which are not accessible are inaccessible points. For any boundary point $\alpha = (x, y)$, $P(\alpha)$ denotes the probability of stopping at $\alpha$ and is given by

$$P(\alpha) = p^y q^x \sum_{\substack{(z_1,\ldots,z_{x+y}) \\ \text{leading to } (x,y)}} \{1 - \phi_{x+y}(z_1,\ldots,z_{x+y})\}$$

$$= K(\alpha)p^y q^x \text{ (say)} \tag{1}$$

where $K(\alpha)$ is the number of accessible paths from the origin to the point $\alpha$.

A stopping rule yielding the boundary points together with their probabilities $P(\alpha)$ shall be called a sampling plan $P$. We say that $P$ is closed iff $\sum_{\alpha \in B} P(\alpha) = 1$ identically in $p \in \Omega$, $B$ denoting the set of all boundary points of $P$. This refers to eventual termination with probability one. Only closed sampling plans are of interest to the practical experimenter and we shall assume so unless otherwise mentioned.

Given a closed plan $P$, we say that a parametric function $g(p)$ is unbiasedly estimable if there exists a function $f(\alpha)$ such that

$$g(p) = E_p(f(\alpha)) = \sum_{\alpha \in B} f(\alpha)P(\alpha), \ \forall p \in \Omega. \tag{2}$$

When (2) holds, $f(\alpha)$ is said to define an unbiased estimate of $g(p)$ and it is a proper estimate of $g(p)$ if $f(\alpha) \in$ range of $\{g(p)\text{: } p \in \Omega\}$ for every $\alpha \in B$. Otherwise, it is said to be improper. We straightaway insist on non-negative estimability of $g(p)$ (i.e., we demand $f(\alpha) \geq 0$) whenever $g(p) \geq 0$, $\forall p \in \Omega$. The reasons for this shall be clear as we proceed. In the same vein, for unbiased estimation of $1/p$, we insist that the estimate $f(\alpha)$ be proper viz., $f(\alpha) \geq 1$, $\forall \alpha \in B$.

## Remark 1

Given an arbitrary sampling plan, examining its closure is not always an easy task. Consider plans having boundaries determined through two infinite sequences of points $(0, a_0)$, $(1, a_1)$, $(2, a_2)$,... and $(b_0, 0)$, $(b_1, 1)$, $(b_2, 2)$,.... Here $1 \leq a_0 \leq a_1 \leq a_2 \ldots$ and $1 \leq b_0 \leq b_1 \leq b_2 \leq \ldots$ are two infinite sequences of positive integers. Such plans have been termed *doubly simple* (see Wolfowitz, 1946). For such plans, closure holds whenever $\liminf_{n\to\infty} A(n)/\sqrt{n} < \infty$ where $A(n)$ refers to the number of accessible points of index $n$. However, an arbitrary unbounded sampling plan need not be doubly simple and, hence, the condition $\liminf_{n\to\infty} A(n)/\sqrt{n} < \infty$ can be substantially improved for other types of unbounded plans. As a matter of fact, plans with $A(n) = 0(n)$ can also be closed. The point to be noted is that the actual value of $A(n)$ is *not always* an important factor to decide on closure or otherwise of a plan. Once an accessible point is reached by a path, only the nature of the remaining part of the sampling plan ahead of this point is relevant for the path to hit a boundary point, and hence, to lead eventually to closure of the plan. The reader is referred to Sinha and Bhattacharya (1982) for examples of various types of unbounded closed plans and other details. The notion of a transformed plan due to Sinha and Sinha (1975) which is also relevant in this context is explained in the fourth section.

**Efficient Sampling Plans**

DeGroot (1959), under certain regularity conditions, established the validity of the Rao-Cramer lower bound for the variance of an unbiased estimate of any estimable parametric function $g(p)$ based on a sequential sampling design. The concept of efficient sampling plans for unbiased estimation of $g(p)$, as introduced by him, refers to a closed sampling plan $P$ together with an unbiased estimate $f(\cdot)$ such that the sampling variance of $f(\cdot)$ attains its relevant lower bound (which of course depends on $g(p)$ and the particular plan $P$). He observed that the only efficient sampling plans are the family of Inverse Binomials when $g(p)$ is linear in $1/p$. Of course, trivially the family of Binomials is also efficient when $g(p)$ is linear in $p$. All other plans may be termed as inefficient. An efficient plan may be seen to maximize the efficiency per unit observation for all $p \in (0, 1)$.

The sampling plans often studied in the literature implicitly (or explicitly) envisage that the decision to stop at a point (or continue) depends only on the point reached (rather than the path traversed in reaching that point). This leaves out a variety of plans obtained by quite interesting and practically suggested stopping rules. A quick example of such a plan is one in which we stop as soon as we obtain two consecutive successes (let us call this plan Plan P1). In this case there would be some boundary points which are exclusively so, namely, the points on the line $Y = X + 2$. There would be other points which would be continuation or boundary points depending on the path or route followed in reaching them. To differentiate the classical sampling plans from such plans, we shall call the former *boundary point* plans and the plans of the type P1 as *route* plans. These two types together form the class of all conceivable plans.

By easy modification of arguments in De Groot (1959) it can be shown that the Rao-Cramer bound remains valid for route plans. Moreover, in case the parameter space $\Omega$ is an open subset of (0, 1), the regularity conditions may be replaced by their *local* versions. These indicate that the only parametric functions efficiently estimable are of the form $(a + bq)/(p - \beta q)$ ($a$, $b$ being arbitrary real numbers and $\beta$ being an integer $\geq -1$). These include $p$ and $1/p$ in particular. The corresponding efficient plans are given by $P(\beta, c) = \{\alpha = (x, y): y = x\beta + c\}$ with $\beta$, $c$ integer, $c \geq 0$ and $\beta \geq 1$. Such a plan is closed if $q \leq 1/(\beta + 1)$ when $\beta > 0$, and $\forall p \in (0, 1)$ otherwise. These results have been derived recently by Dutta (1980), who designates such plans as *Generalized Inverse Binomial Plans*.

As regards the inefficient plans, we demonstrate in the fifth section that a large number of them are indeed *sufficient* for the efficient plans.

## Sequential Unbiased Estimation of $1/p$

The special problem of sequential unbiased estimation of $1/p$ has been initiated in Gupta (1967) and since then treated extensively in the literature. The central problem has been to characterize all sequential sampling plans which provide unbiased estimation of $1/p$. It may be noted that the analogous problems of unbiased estimation of $1/q$, $1/pq$, etc. can be handled in a similar way.

Gupta (1967) stated a very simple sufficient condition for a sequential sampling plan $P$ to provide an unbiased estimate of $1/p$:

(i) Sufficient condition: if the closed plan $P$ with boundary $B = \{r_i = (x_i, y_i), i = 1, 2, \ldots\}$ be such that by changing its boundary points from $r_i$ to $r_i' = (x_i, y_i + 1)$, we get a closed plan $P'$ with boundary $B' = \{r_i' = (x_i, y_i + 1), i = 1, 2, \ldots\}$, then $1/p$ is estimable for the plan $P$. An unbiased estimate is given by $f(r) = K'(r')/K(r)$, $r \in B$, where $K'(r')$ is the number of paths from the origin to $r' \in B'$.

Sinha and Sinha (1975) studied the problem in a greater detail and, among other things, put forward the notion of a transformed plan which can be described as follows. For a given plan $P$ with the set $B$ of boundary points $\alpha$, let $(x', y')$ be an arbitrary but fixed point in the $XY$-plane. Then the transformed plan $P^T(x', y')$, corresponding to $(x', y')$, with the set $B^T(x', y')$ of boundary points $\alpha^T(x', y')$ is defined by the following three conditions:

I. Every $\alpha^T$ belonging to $B^T$ also belongs to $B$ necessarily.

II. The points $\{(x, y) : x \geq x', y \geq y'\}$ constitute the *totality* of all points (accessible, boundary and inaccessible) of $P^T$.

III. Every boundary point $\alpha \in B$ is either a boundary point $\alpha^T \in B^T$ or an inaccessible point in $P^T$.

Given the plan $P$ with boundary points $B$, the rules for obtaining the boundary points $\alpha^T \in B^T$ are as follows:

(a) if $(x', y') \in B$, i.e., if $\alpha = (x', y')$, then $\alpha^T = \alpha$ is the only boundary point of $B^T$;

(b) if $(x', y') \notin B$, then $inf\{\alpha : \alpha = (x, y'), x > x'\}$, for $(x, y') \in B$, is the only point on '$Y = y'$' that belongs to $B^T$;

(c) if $(x', y') \notin B$, then $inf\{\alpha : \alpha = (x', y), y > y'\}$, for $(x', y) \in B$, is the only point on '$X = x'$' that belongs to $B^T$;

(d) if $(x', y') \notin B$, any boundary point $\alpha \in B$ also belongs to $B^T$ if and only if it can be reached by a path from $(x', y')$. Otherwise, it is an inaccessible point of $P^T$.

It may be noted that whenever the point $t = (x', y')$ is an accessible point of $P$, we have

$$p^{y'} q^{x'} = \sum_{\alpha \in B^T(x', y')} t(\alpha) p^y q^x$$

$$\text{i.e., } 1 = \sum_{\alpha \in B^T(x', y')} t(\alpha) p^{y-y'} q^{x-x'} \tag{3}$$

where $t(\alpha)$ = total number of ways of passing from $t$ to $\alpha$ only through the accessible points of $P^T(x', y')$. Even when $t = (x', y')$ is an inaccessible point of $P$, we may use the above definition of $t(\alpha)$ for all $\alpha \in B^T(x', y')$.

The transformed plan $P^T(x', y')$ is defined to be closed *only when* the identity (3) above holds, no matter whether $(x', y')$ is accessible or not. With reference to the problem of unbiased estimation of $1/p$, Sinha and Sinha (1975) came up with the following separate necessary and sufficient conditions.

(ii) Necessary condition: the sampling plan must be unbounded along the $X$(failure)-direction.

(iii) Sufficient conditions: (a) if no point on the line $Y = 1$ is inaccessible, then $1/p$ is estimable. (b) let $(x_0, 1)$ be the first inaccessible point on the line $Y = 1$. If the transformed plan $P^T(x_0, 1)$ is closed, then $1/p$ is estimable.

It has been demonstrated in Sinha and Sinha (1975) that the sufficient conditions (i) and (iii)(b) are equivalent, and conjectured that the sufficient condition (i) is necessary as well. In Sinha and Bhattacharya (1982), useful notions of finite-step and infinite-step generalizations of the Inverse Binomials have been introduced, and the following results have been deduced. See also Sinha and Bose (1985) in this context.

(iv) All finite-step generalizations of the Inverse Binomials provide unbiased estimation of $1/p$.

(v) Every infinite-step generalized Inverse Binomial, whenever closed, provides unbiased estimation of $1/p$.

Incidentally, an infinite-step generalized Inverse Binomial plan is closed if and only if $\liminf_{n \to \infty} d(n)/n = 0$ where $(n - d(n), d(n))$ is the coordinate position of the boundary point on the line $X + Y = n$ $(n = 1, 2, \ldots)$. For a proof, see Bhattacharya and Sinha (1982), Bose and Sinha (1984).

The conjecture relating to a characterization of all sampling plans providing unbiased estimation of $1/p$ has been settled in the affirmative in Sinha and Bose (1985). The result is stated below.

**Theorem 1**

A plan $P$ provides unbiased estimation of $1/p$ if and only if the plan $P'$ defined in the sufficient condition (i) is closed.

**Connections between Efficient and Inefficient Plans**

In this section, we demonstrate that a large number of inefficient sampling plans are indeed sufficient for the efficient plans. These results have been established in Bose and Sinha (1984).

The concept of sufficiency in comparing statistical experiments is well known. Roughly speaking, an experiment $E$ resulting in a random variable $X$ having law of distribution $F_\theta(\cdot)$ is said to be sufficient for another experiment $E'$ resulting in a r.v. $Y$ having law of distribution $G_\theta(\cdot)$ if, given an observation $x$ on $X$, it is possible to generate an observation $y$ on $Y$ using a known randomization procedure, i.e., a known law of distribution $Z(\cdot|x)$, which is independent of $\theta$. If the above holds, we say that $X$ is *sufficient* for $Y$ and write $X > Y$.

Clearly, when $X > Y$, it is enough to observe $X$ to generate $Y$, if needed. Moreover, it is known (Blackwell and Girshick, 1954) that when $X > Y$, for any estimable parametric function $g(\theta)$, given any unbiased estimate based on $Y$, one can construct an unbiased estimate based on $X$ which is as good (in the sense of having equal or smaller variance). Applied to the present set up, this would mean that any plan, whenever sufficient for a given inefficient plan, would provide smaller variance (but certainly larger ASN) than the latter.

The following general results on comparison of sampling plans for sufficiency consideration are interesting and illuminating. We consider two arbitrary closed plans $P^*$ and $P$, and state conditions under which $P^* > P$. In which follows $B^*(B)$ denotes the set of boundary points of $P^*(P)$. We also assume that each of $\Omega(P)$ and $\Omega(P^*)$, the parameter space for closure of $P$ and $P^*$, is the entire interval (0, 1).

Before we state the results we mention the notion of completeness in this context. Writing $P^*(\alpha^*) = K^*(\alpha^*)p^{y^*}q^{x^*}$ for $\alpha^* = (x^*, y^*) \in B$, a plan $P^*$ is said to be complete if $\sum_{\alpha^* \in B^*} f(\alpha^*)P^*(\alpha^*) = 0, \forall p \in \Omega$ implies $f(\alpha^*) = 0, \forall \alpha^* \in B^*$. The following result (necessity due to Girshick, Mosteller and Savage (1946), sufficiency due to Lehmann and Stein (1950)) gives a characterization of such plans which are useful in the sequel.

**Theorem 2**

A plan $P^*$ is complete if and only if the following hold:

(a) The plan is simple (i.e., the continuation points of $P^*$ on the line $X + Y = n$ form an interval, $\forall\, n \geq 1$).

(b) The removal of any boundary point destroys closure of the plan.

Following Bose and Sinha (1984), a series of results can be stated.

**Theorem 3**

(i) A necessary condition for $P^* > P$ is that for every $\alpha = (x, y) \in B$, $p^y q^x$ is estimable under $P^*$.

(ii) If $P^*$ is complete, then (i) ensures that $P^* > P$.

Bose and Sinha (1984) observed that if $P^*$ is not complete, then the estimability of $p^y q^x$ under $P^*$ for every $\alpha \in B$ may not necessarily yield $P^* > P$. They also noted that the completeness of $P^*$ is not necessary for it to be sufficient for $P$.

It is clear from the above result that the estimability of $p^y q^x$ for an arbitrary point $\alpha = (x, y) \in B$ of $P$ with reference to $P^*$ arises naturally. Wolfowitz (1946) established its estimability in case $\alpha$ is an accessible point of $P^*$, though $p^y q^x$ may be estimable even otherwise. The following theorem provides a necessary condition.

In what follows, a point $\alpha$ is defined to line below $\alpha^*$ if $\alpha$ lies in the rectangle formed by the two axes and the point $\alpha^*$. A point $\alpha^*$ lies above $\alpha$ if it lies in the positive quadrant formed by $\alpha$ as the origin.

**Theorem 4**

A necessary condition for estimability of $p^y q^x$ under a plan $P^*$ is the existence of at least one boundary point of $P^*$ above $(x, y)$.

As a consequence, we have the following corollary on necessary conditions for $P^*$ to be sufficient for $P$.

**Corollary 1**

Two necessary conditions for $P^*$ to be sufficient for $P$ are:

(i) For every $\alpha \in B$, $\exists\ \alpha^* \in B^*$ above $\alpha$.

(ii) For every $\alpha^* \in B^*$, $\exists\ \alpha \in B$ above $\alpha^*$.

However, as noted in Bose and Sinha (1984), (i) and (ii) together with even estimability of $p^y q^x$, $\forall\ \alpha \in B$, are not enough to assert $P^* > P$.

We now state a sufficient condition for the estimability of $p^y q^x$ under a plan $P^*$ based on the notion of transformed plans as explained in the last section. Treating $(x, y)$ as the origin, we can derive a transformed form of $P^*$ to be denoted as $P^{**}(x, y)$. In this new plan, the paths emerge from the new origin, and get merged into accessible points or escape them. Note that if $(x, y)$ is itself a boundary point of $P^*$, the transformed plan does not get started at all. Clearly the set of boundary points of $P^*$ above $(x, y)$ is regarded as the set of boundary points of $P^{**}(x, y)$.

**Theorem 5**

Whenever $P^{**}(x, y)$ is closed, $p^y q^x$ is estimable.

We conclude this section with another simple sufficient condition for $P^* > P$. Let $K^{**}(\alpha)$ be the number of accessible paths of $P^*$ from origin to $\alpha$ without hitting any other $\alpha' \in B$, leading to $\alpha$ as a continuation point of $P^*$.

**Theorem 6**

$K(\alpha) = K^{**}(\alpha)$, $\forall\ \alpha \in B$ implies $P^* > P$.

Specialized to the problem of obtaining plans sufficient for the Binomials, we have the following results.

**Theorem 7**

(a) A closed plan $P^*$ is sufficient for the Inverse Binomial plan $P(0, c)$ if and only if there exists no boundary point of $P^*$ below the line $Y = c$.

(b) A closed plan $P^*$ is sufficient for the fixed Binomial plan of size $n$ if and only if there is no boundary point of $P^*$ below the line $X + Y = n$.

As a consequence of (a), we have the following result for the First Waiting Time plan.

(c) A plan $P^*$ with no boundary points on the $X$-axis is sufficient for the plan $P(0, 1)$. Only such plans are sufficient for $P(0, 1)$.

## Concluding Remarks

(i) In a recent paper, Bhandari and Bose (1989) have derived conditions on the nature of unbiasedly estimable functions $g(p)$. They have demonstrated that $g$ has to be continuous if it is unbiasedly estimable. Further, if $g$ is nondifferentiable, then it is *not* unbiasedly estimable by a bounded estimator with *finite* expected stopping time for all $p$. This shows that $g(p) = min(p, 1 - p)$ is *not* estimable by any finite (or bounded) sampling plan though there are plenty of unbounded sampling plans useful for this purpose. An open problem in this context is the following:

> Does there exist any proper unbiased estimate of $min(p, 1 - p)$?

(ii) The following problem is also of considerable interest. Fix an integer $n$ and consider the class of all Bernoulli sampling plans $P$ such that for boundary points of the type $\alpha = (x, y)$, $\int_\omega E_p(x + y)d\psi(p) \leq n$ for a prior distribution $\psi(\cdot)$ on p. Does there exist a sampling plan in this class which is the *best* for estimation of $p$? Here *bestness* refers to minimum prior expectation of posterior variance. In particular, one would be curious to know if the Binomial plan is the best for all or some priors $\psi(\cdot)$.

By a slight modification of the above problem, we may as well search for the best plan among those for which $E_p(x + y) \leq n$, $\forall\ p$. Bhandari et al. (1989) have obtained some partial results in this direction.

(iii) Rustagi (1975) has studied some aspects of estimation of $p$ in the simple Markovian set up. Following this, Sinha and Bhattacharya (1982) initiated a study in the dependent set up in the context of sequential estimation. Further research is needed in this area.

## References

Bhandari, S. K. and Bose, A.: Existence of unbiased estimates in sequential Binomial experiments, to appear in *Sankhya*, A.

Bhandari, S. K., Howlader, H., Weiss, G. and Sinha, B. K.: Search for optimal sampling plans from Bayesian considerations, under preparation.

Blackwell, D. (1947): Conditional expectation and unbiased sequential estimation, *Ann. Math. Statist.* 18, 105-110.

Blackwell, D. (1951): Comparison of experiments, *Proc. Second Berkeley Sym. Math. Statist. Prob.*, Univ. of California Press.

Blackwell, D. (1953): Equivalent comparisons of experiments, *Ann. Math. Statist.* 24, 265-272.

Blackwell, D. and Girshick, M. A. (1954): *Theory of Games and Statistical Decisions*, Wiley, New York.

Bose, Arup. and Sinha, B. K. (1984): Sequential Bernoulli sampling plans reexamined, *Cal. Stat. Assoc. Bull.* 33, 109-120.

DeGroot, M. H. (1959): Unbiased binomial sequential estimation, *Ann. Math. Statist.* 30, 80-101.

Dutta, K. (1980): *A Comparative Study of Certain Sampling Inspection Plans*, Ph.D. thesis, Utkal University.

Girshick, M. A., Mosteller, F. and Savage, L. J. (1946): Unbiased estimates for certain binomial sampling problems with applications, *Ann. Math. Statist.* 17, 13-23.

Gupta, M. K. (1967): Unbiased estimate for $1/p$, *Ann. Inst. Statist. Math.* 19, 413-416.

Lehmann, E. and Stein, C. (1950): Completeness in the sequential case, *Ann. Math. Statist.* 21, 376-385.

Rustagi, J. S. (1975): Inference in a distribution related to a $2 \times 2$ Markov chain, in *Statistical Inference and Related Topics*, ed. M. L. Puri, 257-271.

Sinha, B. K. and Bhattacharya, B. B. (1982): Some further aspects of sequential estimation of $1/p$, mimeograph series, North Carolina State University.

Sinha, B. K. and Bose, A. (1985): Unbiased sequential estimation of $1/p$; Settlement of a conjecture, *Ann. Inst. Statist. Math.* 37, 455-460.

Sinha, B. K. and Sinha, B. K. (1975): Some problems of unbiased sequential binomial estimation, *Ann. Inst. Statist. Math.* 27, 245-258.

Wasan, M. T. (1964): Sequential optimum procedures for unbiased estimation of a binomial parameter, *Technometrics* 6, 259-272.

Wolfowitz, J. (1946): On sequential binomial estimation, *Ann. Math. Statist.* 17, 489-493.

Wolfowitz, J. (1947): Efficiency of sequential estimates, *Ann. Math. Statist.* 18, 215-230.

# SUFFICIENCY

S. Yamada, Tokyo University of Fisheries

and

H. Morimoto, Osaka City University

## Introduction

We undertake with this title a brief survey of various definitions of *sufficiency*, with some of their properties and relationship between them.

Works on this theme are found in a sequence, if not so much as a stream, of developments from the sixties through eighties. We consider such works as attempts at mathematical conceptualization of the statistical notion of sufficiency, and try to examine how far they have been successful in capturing the intuitive and logical content of the notion. Emphasis has been naturally put on the more recent developments, but some earlier results had to be touched upon as long as they make a part of historical or logical background.

A reason for this choice of a theme is that sufficiency today is not as prolific a subject as in early days, making it difficult to draw a *recent trend* out of the publications in last few years. Only a few titles with the word *sufficiency* appear each year in Current Index to Statistics, mostly with their main interest in neighboring though closely related subjects, e.g., ancillarity, information and comparison of experiments. They will be better treated separately under the respective titles, rather than thrown together into such a short survey as this one.

Out of the remaining papers in sufficiency proper, being still fewer in number, we could pick out some fairly recent results to form an additional section on *Basu Theorems.*

Neither a monograph nor a bibliography on this subject recently came into our attention. So the early bibliography by Basu & Speed (1975) as well as the survey *Partial sufficiency* (Basu, 1978) is still partially sufficient (at least) to a reader.

## Statistical Notion and Mathematical Definitions

Sufficiency as a statistical notion means the property of a statistic retaining all the relevant information contained in the whole sample. As is well known, it first appeared in Fisher (1920) (see Stigler, 1973, for historical background) which pointed out that an estimate of a parameter can be regarded to *sum up the whole of the information respecting the parameter which a sample provides* if, for any of its given value, the conditional distribution of any other estimate is independent of the parameter. This idea of expressing the notion by means of conditional probability developed into Fisher's (1922) first definition of sufficiency. A statistic $T$ is called sufficient if:

(A) The conditional distribution of the sample when given $T$ does not depend on the parameter.

Subsequently various aspects of the same notion concerning with specific class of inference and decision problems found different expressions in the following definitions. A statistic $T$ is called sufficient if:

(B) The distribution of the sample can be reconstructed from that of $T$ through randomization, or, mathematically, a stochastic kernel (*Blackwell sufficiency*, Blackwell, 1951).

(C) For every decision problem, given a decision function based on the sample, there exists a decision function based on $T$ which is at least as good as the former (*Decision sufficiency* attributed to Bohnenblust, Shapley and Sherman. See Blackwell, 1951).

(D) For any prior distribution of the parameter, the posterior is a function of the sample through $T$ (*Bayesian sufficiency*, Kolmogorov, 1942).

Meanwhile the Definition (A) underwent measure-theoretic sophistications through Halmos & Savage (1949) and Bahadur (1954) giving rise to the following:

**Definition 1**

Let $E = (X, \mathcal{A}, P)$ be a statistical experiment and $\mathfrak{B}$ be a subfield (more precisely, a sub-$\sigma$-field of $\mathcal{A}$). $\mathfrak{B}$ is called sufficient if for every $A$ in $\mathcal{A}$ there exists a $\mathfrak{B}$-measurable function $P(A/\mathfrak{B})(x)$ which satisfies, for all $B$ in $\mathfrak{B}$ and $p$ in $P$,

$$p(A \cap B) = \int_B P(A/\mathfrak{B})(x)\,dp.$$

A statistic is called sufficient if the subfield induced by it is sufficient.

Notice that this is more general than (A), as it applies to subfields in general, including in its scope those subfields which are not induced by a statistic. Also, it allows the cases where $P(A/\mathfrak{B})(x)$ is not a measure on $\mathcal{A}$. Though $P(A/\mathfrak{B})(x)$ is called conditional *probability*, it is not guaranteed to be a measure by the Radon-Nikodym Theorem, on which this definition is based.

This is the standard definition of sufficiency, most commonly used at present. We also will adopt it here, but will refer to it as *Sufficiency*, with the initial capital S, so as to avoid confusion. Subfield versions are available also for all other definitions. They are to be understood whenever references are made to the definitions.

The very general and measure-theoretical way in which Sufficiency is defined made it possible to prove many useful results with full rigour and under the widest possible conditions. In particular, it implies the conditions (B) and (D) without any restrictions, while (A) and (C) easily follow in the cases where regular conditional probabilities exist.

**Dominated Case**

The success of Sufficiency was especially remarkable in the dominated case. $E$ is called dominated if there exists a $\sigma$-finite measure $m$ on $\mathcal{A}$ wrt. which each $p$ in $P$ has a density $dp/dm$. In this case, it follows that:

1) $X$ is covered by a countable family of mutually disjoint subsets, called kernels, of the supports $S(p) = \{x;\ dp/dm > 0\}$ of measures $p$ in $P$. Those measures constitute a countable subfamily $P'$ of $P$, which is equivalent to $P$.

2) There exists a *pivotal measure* $n$, a convex combination of the measures in $P'$. Each $p$ in $P$ has a density wrt. $n$.

3) A subfield $\mathfrak{B}$ is Sufficient if and only if the density $dp/dn$ is $\mathfrak{B}$-measurable for each $p$ in $P$ (Neyman Factorization Theorem).

4) If a subfield includes another subfield which is Sufficient, then the former is also Sufficient.

5) There exists the minimal Sufficient subfield, the smallest subfield wrt. which all the densities $dp/dn$, $p \in P$, are measurable.

The existence of the minimal Sufficient statistic is also proved under a slight additional restriction that $P$ is separable wrt. the total variation distance (Lehmann & Scheffé, 1950).

The term *minimal Sufficient* requires slightly technical clarifications. Burkholder (1961) proved that the following two properties of a Sufficient subfield $\mathfrak{B}$ are equivalent to each other:

i) $\mathfrak{B} \subset \mathfrak{C}\ [P]$ for every Sufficient subfield $\mathfrak{C}$, and

ii) $\mathfrak{B} \subset \mathfrak{C}\ [P]$ for every Sufficient subfield $\mathfrak{C}$ such that $\mathfrak{C} \subset \mathfrak{B}\ [P]$.

$\mathfrak{B}$ is called minimal Sufficient when it has these properties. On the other hand, a Sufficient statistic is called minimal if it is a function of every other Sufficient statistic except on a $P$-null set which may depend upon the latter statistic.

This minimality does not coincide with the minimality of the Sufficient subfield which the statistic induces. As a result, the minimal Sufficient statistic and subfield may not coincide with each other even when both exist. In this connection, all logically possible kinds of counter examples are actually available (see Bahadur, 1955; and Landers & Rogge, 1972).

In case Sufficiency is replaced by pairwise Sufficiency in i) and ii), then i) does not follow from ii), so that *smallest pairwise Sufficiency* and *minimal pairwise Sufficiency* have to be differentiated.

**Undominated Cases**

Thus in the dominated case Sufficiency exhibits all the good features to qualify itself for a mathematical embodiment of the statistical notion of sufficiency. However, it came to be known already around 1960 that some of the features are not carried over to the general case. Notably, general validity of 4) and 5) were disproved by the counter examples given by Burkholder (1961, for 4) and Pitcher (1957, for 5), respectively. The phenomena of the failure of 4) and 5) are accordingly called Burkholder and Pitcher pathologies.

Various intermediate conditions more general than domination have been devised in order to avoid these pathologies. Here we present two such conditions, namely, majorization and weak domination. Reader is referred to Luschgy & Mussmann (1985) for details of these and other conditions.

An experiment $E$ is called majorized if there exists a *majorizing measure* m on $\mathcal{A}$ wrt. which each $p$ in $P$ has a density $dp/dm$.

$E$ is called weakly dominated if the majorizing measure $m$ is further assumed to be localizable (for the definition of localizability see Diepenbrock, 1971, or Ghosh et al., 1981).

The majorized case is more or less the most general case in which positive results are being obtained at present. The non-majorized cases are the places mainly for counter examples, but for some early, universal type of theorems by Bahadur (1954, 1955b), Burkholder (1961) and others.

Weak domination is more general than domination, as localizability of a measure follows from $\sigma$-finiteness, and is equivalent to some other conditions which appeared in literature, such as *compactness* (Pitcher, 1965), *coherence* (Hasegawa & Perlman, 1974), etc.

There is a simple but suggestive special case of weak domination, called the discrete case. $E$ is called discrete if $X$ is an uncountable space, $\mathcal{A}$ is the power set, each $p$ in $P$ is a discrete probability and the only $P$-null set is the empty set. It is Professor D. Basu himself who pointed out with J.K. Ghosh (1967) that the problem of sampling from finite populations falls in this category and thus became one of the pioneers of the study of sufficiency in the undominated cases.

These conditions have been only partly successful in removing the pathologies, insofar as the minimal Sufficient subfield was proved to exist in the weakly dominated case, but not in the majorized case in general (Pitcher, 1965, and Hasegawa and Perlman, 1974). Burkholder pathology persists even in the weakly dominated case.

The reason for this difference between dominated and undominated cases becomes apparent if a parallelism to the passage from 1) through 5) is tried out for the majorized case. It follows that:

1′) $X$ is now covered by an uncountable family of almost disjoint *kernels*. This family is called a maximal decomposition (Diepenbrock, 1971). As before, the kernels are subsets of the supports of measures $p$ in $P$. Those measures constitute an uncountable equivalent subfamily $P'$ of $P$.

2′) A *pivotal measure* $n$ can be defined as the sum of the measures in $P'$ restricted to the respective kernels.

3′) A subfield $\mathfrak{B}$ is pairwise Sufficient and contains the supports $S(p)$ for all $p$ in $P$ (*pairwise Sufficiency with supports*, abbreviated as PSS), if and only if the density $dp/dn$ is $\mathfrak{B}$-measurable for each $p$ in $P$ (Analogue of Neyman Factorization Theorem, Ramamoorthi & Yamada, 1982).

4′) If a subfield includes another subfield which is PSS, then the former is also PSS.

5′) There exists the smallest subfield which is PSS (Ghosh et al., 1981).

Thus, instead of Sufficiency in the dominated case, here we arrive at PSS, a property in between Sufficiency and pairwise Sufficiency. Notice further that the likelihood ratios are seen in 3′) to be functions of the sample through PSS rather than Sufficiency, which coincides with the former in the dominated case.

On the other hand, if we insist upon retaining all the nice properties of Sufficiency, i.e. (B) through (D) as well as 4) and 5), we have to take to something even more restrictive than domination, as it would require a type of sample space with regular conditional probabilities. Barndorff-Nielsen (1978) points out it and puts forward one such framework: An Euclidean sample space with the Borel field and a dominated $P$, in which only *B-sufficiency* (defined in terms of the existence of regular conditional probability $P(A/T)$ common to all $p$ in $P$) of statistics rather than subfields is treated. This would restrict us almost within the purview of Definition (A) and would mean little more than a return to Fisher's old setup.

**Relationship Between the Definitions**

Much attention has been directed to the relationship between various definitions of sufficiency, especially on the question as to whether Sufficiency follows from other definitions. It is quite rightly so, as Sufficiency is defined solely in measure-theoretic terms and, unlike other definitions, is not directly concerned with specific statistical problems, though it was also originated in an estimation problem. It is relevant to ask whether the requirement for Sufficiency is just appropriately strong, or actually stronger than the requirements for other definitions.

We take up this question as regards decision, Bayes and, in addition, *test sufficiency*, as it is often called in literature. Blackwell sufficiency would require some preliminaries from the comparison of experiments which is beyond our scope.

A subfield $\mathfrak{B}$ is called test sufficient if for any test function there exists a $\mathfrak{B}$-measurable test function whose expectation is identical with the former for all $p$ in $P$.

It was proved in a series of classical results in Bahadur (1955a, b), Blackwell (1951), Kudo (1967) and others that each of the four concepts including Blackwell sufficiency implies pairwise Sufficiency. In the case of decision sufficiency we need some clarification on the precise definition of a decision problem, but we will not go into the details. In the dominated case as pairwise Sufficiency implies Sufficiency, each of the four concepts implies Sufficiency.

Things are again very different in the undominated case. First, pairwise Sufficiency does not imply Sufficiency in general. Secondly, the implication, e.g. *Bayes sufficiency implies Sufficiency* obviously fails in the face of Burkholder pathology, as Sufficiency implies Bayes sufficiency and a subfield including Bayes sufficient subfield is Bayes sufficient. So the implication needs to be modified to a weaker statement: *A Bayes sufficient subfield includes a Sufficient subfield.* The same modifications are made also in regard to decision, test and Blackwell sufficiency.

This modified statement is proved by Ramamoorthi (1980) for decision sufficiency: A decision sufficient subfield includes at least one Sufficient subfield in it.

The statement *a test sufficient subfield includes a Sufficient subfield* is obviously more difficult to follow, as test sufficiency is weaker than decision sufficiency. Indeed, since the paper of Brown (1975) which says that it holds true for the discrete case, little progress has been seen, but for a recent proof of *PSS does not imply test sufficiency* for the weakly dominated case by Kusama & Fujii (1987). Even this statement, not at all surprising, cannot be readily proved for more general cases.

The questions concerning Bayes sufficiency are even more technical, as Bayes sufficiency involves a measurable structure on $P$, and appears to be weaker than all other definitions. In the extreme case, it is no more than pairwise Sufficiency if $P$ has the discrete $\sigma$-field. It follows from test sufficiency if both $X$ and $P$ have countably generated $\sigma$-fields, and from decision sufficiency in the general case (Ramamoorthi, 1980. Incidentally Blackwell sufficiency also follows from decision sufficiency). On the other hand, a rather natural *counter example*, in which P is a standard Borel space, is available to show that Bayes sufficiency does not imply Sufficiency (Blackwell and Ramamoorthi, 1982).

Then what does Bayes sufficiency at all imply? Suppose $\mathcal{A}$ and a subfield $\mathcal{B}$ are countably generated. Then $\mathcal{B}$ is Bayes sufficient if and only if it is Sufficient for almost all $p$ in $P$ wrt. every prior measure on $P$ (Ramamoorthi, 1980).

**LeCam's Framework of *L*-Space and *M*-Space**

An entirely different approach has been proposed by LeCam (1964) to bypass the difficulties discussed above by means of function spaces and further developed towards various directions (by e.g. Littaye-Petit, Piednoir & Van Cutsem, 1969; Siebert, 1979; Luschgy & Mussmann, 1985; and recently LeCam himself, 1986) through the seventies and eighties.

Let $E = (X, \mathcal{A}, P)$ be an experiment. The band $L(E)$ generated by $P$ in the space of bounded signed measures on $\mathcal{A}$ is called the $L$-space of $E$. If $E$ is majorized and $n$ is a majorizing measure equivalent to $P$ (shown to exist by Diepenbrock, 1971), then $L(E)$ coincides with $\{f.n;\ f \in L_1(X, \mathcal{A}, n)\}$, where $f.n$ denotes the bounded signed measure having $f$ as the density wrt. $n$. Assign the total variation topology to $L(E)$, denote by $M(E)$ its topological dual and call it the $M$-space of $E$. Sufficiency is now defined for a sublattice of $M(E)$ as follows: A sublattice $H$ is sufficient if there exists a positive linear projection $\pi$ of $M(E)$ onto $H$ such that $< p, \pi f > \; = \; < p, f >$ for all $f$ in $M(E)$ and $p$ in $P$.

It then follows that a sublattice including a sufficient sublattice is sufficient, and the smallest sufficient sublattice exists. Thus this *sufficiency* appears to be free from both Burkholder and Pitcher pathologies.

Two concepts of *transition* and *deficiency* play important parts in the theory. A transition is defined as a positive linear mapping from the $L$-space of an experiment $F$ to that of another experiment $E$ which preserves the norms of the positive elements. The deficiency of $F$ to $E$ is devised for measuring how much less informative $F$ is than $E$ when they share a same parameter space. Write $E = (X, \mathcal{A}, P)$, $F = (Y, \mathcal{B}, Q)$ with $P = \{p_\theta;\ \theta \in \Theta\}$ and $Q = \{q_\theta;\ \theta \in \Theta\}$ where $\Theta$ is the common parameter space. The deficiency of $F$ to $E$ is defined by

$$d(F, E) = Inf\left\{ [\underset{\theta \in \Theta}{Sup} \| \tau \cdot q_\theta - p_\theta \| ];\ \tau \text{ is a transition from F to E}\right\}.$$

Now take a sublattice $H$ of $M(E)$. There is an experiment $F$ whose $M$-space is $H$, provided $H$ is closed. It is proved that $H$ is sufficient if and only if the deficiency of $F$ to the original experiment $E$ is 0.

Specialize these concepts to the case of $E = (X, \mathcal{A}, P)$ and $F = (X, \mathcal{B}, P)$ where $\mathcal{B}$ is a subfield of $\mathcal{A}$. A transition from $L(F)$ to $L(E)$ is then a generalization of a stochastic kernel from $(X, \mathcal{B})$ to $(X, \mathcal{A})$ and "$d(F, E) = 0$" is a generalization of Blackwell sufficiency. Hence the foregoing paragraph is interpreted as *sufficiency implies Blackwell sufficiency* in the present context.

Instead of starting from $E = (X, \mathcal{A}, P)$ and going to $M(E)$ via $L(E)$, it is also possible to take an abstract $L$-space as $L$ and construct the whole theory directly based on it. Thereby $P$ appears as a set of positive elements $p$ with the norms 1 in $L$, but not $X$ or $\mathcal{A}$. This is more like what LeCam (1964) actually did. Here we have followed the way Siebert (1979) presented the theory.

Torgersen (1979) undertakes a further generalization by including unbounded functions into $M$, and develops an estimation theory which has a theorem: Every estimable function admits a UMVU if and only if a quadratically complete sufficient statistic exists.

Such abstract developments render highly refined appearance to the theory, though the departure from the basis of the sample space invites critical comments.

It is not very easy to compare this theory to the measure theoretic treatment, as the concepts do not necessarily correspond to each other. When we try to locate a counterpart of a subfield $\mathfrak{B}$, it is found in $M(E)$ in the form of the sublattice $H(\mathfrak{B})$, the totality of the $\mathfrak{B}$-measurable functions. Whether $H(\mathfrak{B})$ is a sufficient sublattice or not can be decided only when it happens to be closed, so as to admit the projection used to define the sufficiency of a sublattice. And in that event, the sublattice $H(\mathfrak{B})$ is sufficient in $M(E)$ if and only if the subfield $\mathfrak{B}$ is pairwise sufficient in $E$ (Littaye-Petit et al., 1969).

In this correspondence between $\mathfrak{B}$ and $H(\mathfrak{B})$, no criterion inherent in $M(E)$ is readily available to distinguish Sufficiency, PSS and pairwise Sufficiency of $\mathfrak{B}$ on the basis of the properties of $H(\mathfrak{B})$ as a sublattice. So the sufficient sublattices correspond to these three kinds of subfields altogether.

This suggests significance of pairwise Sufficiency, and in particular PSS, as something more than a mathematical tool. Remember that the role played by PSS in the majorized case is very similar to, if not as important as, that of Sufficiency in the dominated case.

In the weakly dominated case, PSS possesses some more properties almost parallel to those of Sufficiency (Yamada, 1980). Suppose that $\mathfrak{B}$ is PSS and $f$ is an integrable function. Then there exists a function $g$ which satisfies $g = E_p[f/\mathfrak{B}]$ a.e. for each $p$ in $P$, and falls only a little short of being $\mathfrak{B}$-measurable. In precise terms, $g$ is measurable wrt. the strong completion of $\mathfrak{B}$, i.e. $\mathfrak{B} \vee \mathcal{N}(P)$ on the support of each $p$ in $P$, though on the whole space it is measurable only wrt. the weak completion $\cap\{\mathfrak{B} \vee \mathcal{N}(p);\ p \in P\}$ ($\mathcal{N}(P)$ and $\mathcal{N}(p)$ mean the families of $P$-null and $p$-null sets, respectively).

This property can then be used to prove analogues of test sufficiency and Rao-Blackwell property for PSS, by providing improved test and estimator which are close to being $\mathfrak{B}$-measurable.

Further attempts have been made at extending these properties to the majorized experiments (Yamada, 1988).

**Basu Theorems**

This means the two renowned theorems of Basu on independence of sufficient and ancillary statistics (see Basu, 1982). Because of their nature of connecting such basic concepts as sufficiency, ancillarity, completeness and independence, related works still appear in literature. We first state the theorems. Assume until otherwise noticed that $T$ is a sufficient statistic.

I. Suppose that $T$ is boundedly complete. Then an ancillary statistic $S$ is independent of $T$ (for all $p$ in $P$).

II. Assume that there is no splitting set. Then a statistic $S$ which is independent of $T$ is ancillary.

A splitting set is defined as "a set $A$ such that $p(A) = 1$ for some $p$'s and 0 for all other $p$'s in $P$" by Koehn & Thomas (1975). A slightly different condition to be assumed in II and some remarks on the conditions are found in Basu (1982) and Basu & Cheng (1981). Bayesian versions of these and related theorems are given in Basu & Pereira (1983).

Recently Goossen (1986), while working on *conditional completeness*, made a remark that the assumption of sufficiency of $T$ in I and II can be replaced by sufficiency of $T$ for $(S, T)$.

Lehmann (1981) gives two theorems as *adaptations of Basu's theorems*, aiming at characterizations of (bounded) completeness. Basu theorems as such are not exactly a characterization, as the independence of all the ancillaries from $T$ does not imply bounded completeness of $T$ unconditionally. The reason for this gap, Lehmann considers, lies in the difference between ancillarity and completeness in their nature, one being concerned with the whole distribution while the other only with the expectations. Notice the modifications accordingly made on each concept to bridge the gap in the theorems thus constructed:

III. $T$ is boundedly complete if and only if every bounded function of $T$ is uncorrelated with every bounded *first order ancillary* (a statistic whose expectation is independent of $p$).

IV. $T$ is $F_1$-complete if and only if every ancillary is independent of $T$ ($F_1$ means the class of all functions $f(T)$ such that $f(T) = E[g/T]$ for some two valued function $g$. $T$ is called $F_1$-complete if $f \in F_1$ and $E_p(f(T)) = 0$ for all $p$ together imply $f(T) = 0$).

Basu theorems are closely related to invariance theory, where conditions for sufficiency, ancillarity and mutual independence of an invariant and an equivariant statistic $S$ are studied.

A recent work of this kind is Eberl (1983), which deals with the $n$-dimensional location model. Let $S$ be the maximal invariant and $T$ be an equivariant statistic in this model. Neither sufficiency of $T$ as such nor its bounded completeness is assumed. It follows that:

V. $T$ is independent of $S$ if and only if $T$ is *invariantly sufficient* (i.e. $p(C/T)$ is independent of $p$ for all invariant sets $C$).

Similar questions are asked and considerable amount of results have been obtained with remarkable applications in more general invariant models like compact or locally compact spaces and/or groups. As they cannot be detailed here, readers are referred to, e.g., Dasgupta (1979) and Ramamoorthi (1990) for such results and remarks on their connection with Basu theorems.

**References**

Bahadur, R. R. (1954): Sufficiency and statistical decision functions, *Ann. Math. Stat.* 25, 423–462.

Bahadur, R. R. (1955a): A characterization of sufficiency, *Ann. Math. Stat.* 26, 286–293.

Bahadur, R. R. (1955b): Statistics and subfields, *Ann. Math. Stat.* 26, 490–497.

Barndorff-Nielsen, O. (1978): *Information and Exponential Families in Statistical Theory*, John Wiley & Sons.

Basu, D. (1978): Partial sufficiency: A review, *J. Stat. Pl. Inf.* 2, 1–15.

Basu, D. (1982): Basu theorems, *Encyclopedia of Statistical Sciences*, 1, 193–196, John Wiley & Sons.

Basu, D. and Cheng, S. C. (1981): A note on sufficiency in coherent models, *Int. J. Math. Math. Sci.* 4, 571-582.

Basu, D. and Ghosh, J. K. (1967): Sufficient statistics in sampling from a finite universe, *Proc. 36th Session Int. Stat. Inst.*, 850–859.

Basu, D. and Pereira, C. A. B. (1983): Conditional independence in statistics, *Sankhyā* 45, Ser. A, 324–337.

Basu, D. and Speed, T. P. (1975): Bibliography of sufficiency (mimeographed), Manchester.

Blackwell, D. (1951): Comparison of experiments, *Proc. 2nd Berkeley Symp. Math. Stat. Prob.* 93–102.

Blackwell, D. and Ramamoorthi, R. V. (1982): A Bayes but not classically sufficient statistic, *Ann. Stat.* 10, 1025–1026.

Brown, L. D. (1975): On a theorem of Morimoto concerning sufficiency for discrete distributions, *Ann. Stat.* 3, 1180–1182.

Burkholder, D. (1961): Sufficiency in the undominated case, *Ann. Math. Stat.* 32, 1191–1200.

Dasgupta, S. (1979): A note on ancillarity and independence via measure preserving transformations, *Sankhyā* 41, Ser. A, 117–123.

Diepenbrock, F. R. (1971): *Characterisierung einer allgemeineren Bedingung als Dominiertheit mit Hilfe von lokalisierbaren Massen*, Thesis, University of Münster.

Eberl Jr., W. (1983): Invariantly sufficient equivariant statistics and characterizations of normality in translation classes, *Ann. Stat.* 11, 330–336.

Fisher, R. A. (1920): A mathematical examination of the methods of determining the accuracy of an observation by the mean error, and by the mean square error, *Monthly Notices R. Astr. Soc.* 80, 759–770.

Fisher, R. A. (1922): On the mathematical foundations of theoretical statistics, *Phil. Trans. R. Soc. Lond.* A 222, 309–368.

Ghosh, J. K., Morimoto H. and Yamada, S. (1981): Neyman factorization and minimality of pairwise sufficient subfields, *Ann. Stat.* 9, 514–530.

Goossen, K. (1986): On sufficiency, conditional completeness and the theorems of Basu, *Stat. Dec.* 4, 85–96.

Halmos, P. R. and Savage, L. J. (1949): Application of the Radon-Nikodym theorem to the theory of sufficient statistics, *Ann. Math. Stat.* 20, 225–241.

Hasegawa, M. and Perlman, M. D. (1974): On the existence of a minimal sufficient subfield, *Ann. Stat.* 2, 1049–1055. Correction (1975): *Ann. Stat.* 3, 1371–1372.

Koehn, U. and Thomas, D. L. (1975): On statistics independent of a sufficient statistic: Basu's lemma, *Amer. Stat.* 29, 40–42.

Kolmogorov, A. N. (1942): Sur l'estimation statistique des parametres de la loi de Gauss, *Izv. Akad. Nauk SSSR* Ser. Mat. 6, 3–32.

Kudo, H. (1967): On sufficiency and complete class property of statistics (in Japanese), *Sugaku* 8, 129–138.

Kusama, T. and Fujii, J. (1987): A note on test sufficiency in weakly dominated statistical experiments, *Tokyo J. Math.* 10, 133–137.

Landers, D. and Rogge, L. (1972): Minimal sufficient $\sigma$-fields and minimal sufficient statistics. Two counterexamples, *Ann. Math. Stat.* 43, 2045–2049.

LeCam, L. (1964): Sufficiency and approximate sufficiency, *Ann. Math. Stat.* 35, 1419–1455.

LeCam, L. (1986): *Asymptotic Methods in Statistical Decision Theory*, Springer.

Lehmann, E. L. (1981): An interpretation of completeness and Basu's theorem, *J. Amer. Stat. Assoc.* 76, 335–340.

Lehmann, E. L. and Scheffé, H. (1950): Completeness, similar regions and unbiased estimation, Pt. 1, *Sankhyā* 10, 305–340.

Littaye-Petit, M., Piednoir, J.-L. and van Cutsem, B. (1969): Exhaustivitè, *Ann. Inst. H. Poincaré* 5, 289–322.

Luschgy, H. and Mussmann, D. (1985): Equivalent properties and completion of statistical experiments, *Sankhyā* 47, 174–195.

Pitcher, T. S. (1957): Sets of measures not admitting necessary and sufficient statistics or subfields, *Ann. Math. Stat.* 28, 267–268.

Pitcher, T. S. (1965): A more general condition than domination for sets of probability measures, *Pacific J. Math.* 15, 597–611.

Ramamoorthi, R. V. (1980): *Sufficiency, Pairwise Sufficiency and Bayes Sufficiency in Undominated Experiments*, Thesis, Indian Statistical Institute.

Ramamoorthi, R. V. (1990): Sufficiency, ancillarity and independence in invariant models, *J. Stat. Pl. Inf.* (to appear).

Ramamoorthi, R. V. and Yamada, S. (1982): Neyman factorization theorems for experiments admitting densities, *Sankhyā* 45, Ser. A, 168–180.

Siebert, E. (1979): Pairwise sufficiency, *Z. Wahrsch. verw. Geb.* 46, 237–246.

Stigler, S. M. (1973): Studies in the history of probability and statistics, XXXII. Laplace, Fisher and the discovery of the concept of sufficiency, *Biometrika* 60, 439–445.

Strasser, H. (1985): *Mathematical Theory of Statistics*, De Gruyter, Berlin.

Torgersen, E. N. (1979): On complete sufficient statistics and uniformly minimum variance unbiased estimation, *Symp. Math. Ist. Naz. Alta. Mat.* 22, 299–305.

Yamada, S. (1980): On completions of sigma-fields and pairwise sufficiency, *Sankhyā* 42, Ser. A, 185–200.

Yamada, S. (1988): M-space of majorized experiment and pivotal measure, *Stat. Dec.* 6, 163–174.

# FOUNDATIONS OF STATISTICAL QUALITY CONTROL

Richard E. Barlow, University of California at Berkeley

and

Telba Z. Irony[1], University of California at Berkeley

**Abstract**

The origins of statistical quality control are first reviewed relative to the concept of statistical control. A recent *Bayesian* approach developed at AT&T laboratories for replacing Shewart-type control charts is critiqued. Finally, a compound Kalman filter approach to an inventory problem, closely related to quality control and based on Bayesian decision analysis, is described and compared to other approaches.

## Statistical Control

The control chart for industrial statistical quality control was invented by Dr. Walter A Shewhart in 1924 and was the foundation for his *Economic Control of Quality of Manufactured Product*–his 1931 book. (A highly recommended recent reference is Deming, 1986.) On the basis of Shewhart's industrial experience, he formulated several basic and important ideas. Recognizing that all production processes will show variation in product if measurements of quality are sufficiently precise, Shewhart described two sources of variation; namely

i) variation due to *chance causes* (called *common causes* by Deming, 1986);

ii) variation due to *assignable causes* (called *special causes* by Deming, 1986).

Chance causes are inherent in the system of production while assignable causes, if they exist, can be traced to a particular machine, a particular worker, a particular material, etc. According to both Shewart and Deming, if variation in product is only due to chance causes, then the process is said to be in *statistical control*. Nelson (1982) describes a process in statistical control as follows: "A process is said to have reached a state of *statistical control* when changes in

---

[1]Now at George Washington University, Washington, D.C. 20006.

This research was partially supported by the U.S. Air Force Office of Scientific Research (AFOSR-90-0087) to the University of California at Berkeley.

measures of variability *and* location from one sampling period to the next are no greater than statistical theory would predict. That is, assignable causes of variation have been detected, identified, and eliminated." Duncan (1974) describes chance variations: "If chance variations are ordered in time or possibly on some other basis, they will behave in a random manner. They will show no cycles or runs or any other defined pattern. *No specific variation to come can be predicted from knowledge of past variations.*" Duncan, in the last sentence, is implying statistical independence and not statistical control.

Neither Shewhart nor Duncan have given us a mathematical definition of what it means for a process to be in statistical control. The following example shows that statistical independence depends on the knowledge of the observer and, therefore, we think it should *not* be a part of the definition of statistical control.

### Example

The idea of *chance causes* apparently comes from or can be associated with Monte Carlo experiments. Suppose I go to a computer and generate $n$ *random* quantities normally distributed with mean 0 and variance 1. Since I know the distribution used to generate the observed quantities, I would use a $N(0,1)$ distribution to predict the $(n+1)^{st}$ quantity yet to be generated by the computer. For me, the process is *random* and the generated $n$ random quantities provide no predictive information. However, suppose I show a plot of these $n$ numbers to my friend and I tell her how the numbers were generated except that I neglect to tell her that the variance was 1. Then for her, $x_{n+1}$ is not independent of the first $n$ random quantities because she can use these $n$ quantities to estimate the process variance and, therefore, better predict $x_{n+1}$.

What is interesting from this example is that for one of us the observations are from an independent process while for the other the observations are from a dependent process. But of course (objectively) the plot looks exactly the same to both of us. The probability distribution used depends on the state of knowledge of the analyst. I think we both would agree however that the process is in statistical control.

All authors seem to indicate that the concept of statistical control is somehow connected with probability theory although not with any specific probability model. We think de Finetti (1937, 1979) has given us the concept which provides the correct mathematical definition of statistical control.

### Definition: Statistical control

We say that a product process is in *statistical control* with respect to some measurement variable, $x$, on units $1, 2,\dots,n$ if and only if in our judgement

$$p(x_1, x_2,\dots,x_n) = p(x_{i_1}, x_{i_2},\dots,x_{i_n})$$

for all permutations $\{i_1, i_2,\dots,i_n\}$ of units $\{1, 2,\dots,n\}$. That is, the units are *exchangeable* with respect to $x$ in our opinion. This definition has two implications: namely that the order in which measurements are made is not important and, secondly, as a result, all marginal distributions are the same. It does not, however, imply that measurements are independent.

In addition, the process *remains* in statistical control if, in our judgement, future units are *exchangeable* with past units relative to our measurement variable.

The questions which concern all authors on quality control are:

(1) How can we determine if a production process is in statistical control?

and

(2) Once we have determined that a production process is in statistical control, how can we detect a departure from statistical control if it occurs?

The solution offered by most authors to both questions is to first plot the data. A plot of the measurements in time order is called a *run chart*. Run charts are also made of sample averages and sample ranges of equal sample sizes at successive time points. The grand mean is plotted and *control limits* are set on charts of sample averages and sample ranges. The process is judged to be in statistical control if

i) there are no obvious trends, cycles or runs below or above the grand mean;

ii) no sample average or sample range falls outside of *control limits*.

Samples at any particular time are considered to constitute a *rational sample* (i.e., in our terminology, to be exchangeable with units not sampled at that time). The only question is that of exchangeability of *rational samples* over time. In practice, *control limits* are based on a probability model for the rational samples and *all* observed sample averages and ranges over time.

The marginal probability model can, in certain cases, also be inferred from the judgement of exchangeability. If measurements are in terms of attributes; i.e., $x_i = 1$ (0) if the $i^{th}$ unit is good (bad) and if the number of such measurements is conceptually unbounded, then it follows from de Finetti's representation theorem that

$$p(x_i = 1) = \int_0^1 p(x_i = 1|\theta)p(\theta)\,d\theta = \int_0^1 \theta p(\theta)\,d\theta$$

for some measure $p(\theta)d\theta$ and further, that $x_1$, $x_2,\ldots,x_n$ are conditionally independent given $\theta$. In this case $\theta$ can be interpreted as the long run "chance" that a unit is good; i.e., $(\sum x_1)/n$ tends to $\theta$ with subjective probability one as $n$ increases without limit. *Chance* in this case is considered a parameter – not a probability.

In general, however, exchangeability alone is too weak to determine a probability model and additional judgements are required to determine marginal probability distributions. Let $x_1, x_2, \ldots, x_n$ be exchangeable measurement errors. If, in addition, we judge measurement errors to be *spherically symmetric*; i.e., $p(x_1, x_2, \ldots, x_n)$ is invariant under rotations of the vector $(x_1, x_2, \ldots, x_n)$ and this for all $n$, then it follows that the joint probability function is a *mixture* of normal distributions and $x_i$ given $\sigma^2$ is $N(0, \sigma^2)$ while $x_1, x_2, \ldots, x_n$ given $\sigma^2$ are conditionally independent. Also $(\sum x_i^2)/n$ tends to a limit, $\sigma^2$, with subjective probability one. For more details see Dawid (1986).

The problem of determining and justifying *control limits* remains. It was this problem which led Hoadley (1981) to develop his quality measurement plan critiqued in the next section. The usual method for computing *control limits* (e.g. Nelson, 1982) violates the likelihood principle. Basu (1988) has argued convincingly against such methods.

## A Critique of the Quality Measurement Plan

A quality auditing method called the quality measurement plan or QMP was implemented throughout AT&T technologies in 1980 (see Hoadley, 1981). The QMP is a statistical method for analyzing discrete time series of quality audit data relative to the expected number of defects given standard quality. It contains three of the audit ingredients: defects assessment, quality rating and quality reporting.

A quality audit is a system of inspections done continually on a sampling basis. Sampled product is inspected and defects are assessed whenever the product fails to meet engineering requirements. The results are combined into a rating period and compared to a quality standard which is a target value of defects per unit. It reflects a trade-off between manufacturing cost, operating costs and customer need.

Suppose there are $T$ rating periods: $t = 1, \ldots, T$ (current period). For period $t$, we have the following data from the audit:

$n_t$ = audit sample size;

$x_t$ = number of defects in the audit sample;

$s$ = standard number of defects per unit;

$e_t$ = expected number of defects in the sample when the quality standard is met; $e_t = sn_t$;

$$I_t = \frac{x_t}{e_t} = \text{quality index (measure of the defect rate).}$$

$I_t$ is the defect rate in units of standard defect rate. For instance, if $I_t = 2$, it means that twice as many defects as expected have been observed.

The statistical model used in QMP is a version of the Empirical Bayes model. The assumptions are the following:

1. $x_t$ has a Poisson distribution with mean $n_t\lambda_t$, i.e. $(x_t|n_t\lambda_t) \sim Poi(n_t\lambda_t)$ where $\lambda_t$ is the *true* defect rate per unit in time period $t$. If $\lambda_t$ is reparametrized on a quality index scale, the result is:

   $$\theta_t = \lambda_t/s = \text{true quality index.}$$

   In other words, $\theta_t = 1$ is the standard value. Therefore, we can write:

   $$(x_t|\theta_t) \sim Poi(e_t\theta_t).$$

2. For each rating period $t$, there is a true quality index $\theta_t$. $\theta_t$, $t = 1,\ldots,T$ is a random sample from a Gamma distribution with mean $\theta$ and variance $\gamma^2$. $\theta$ is called the *process average* and $\gamma^2$ is called the *process variance*. We can write $(\theta_t|\theta,\gamma^2) \sim Gamma(\theta^2/\gamma^2,\ \theta/\gamma^2)$. In this model, both $\theta$ and $\gamma^2$ are unknown.

3. $\theta$ and $\gamma^2$ have a joint prior distribution $\rho(\theta,\ \gamma^2)$.

The parameter of interest is $\theta_T$ given the past data, $d_{T-1}$ and current data, $x_T$. Here $d_{t-1} = (x_1,\ x_2,\ldots,x_{T-1})$ and $d_0$ is a constant.

The model assumes that the process average, $\theta$, although unknown, is fixed; i.e., the model assumes exchangeability. In reality $\theta$ may be changing. In order to handle this, the QMP procedures uses a moving window of six periods of data.

A suitable way to describe and to analyze the QMP model is via *probabilistic influence diagrams.* Probabilistic influence diagrams have been described by Shachter (1986) and Barlow and Pereira (1990).

A probabilistic influence diagram is a special kind of graph used to model uncertain quantities and the probabilistic dependence among them. It is a network with directed arcs and no directed cycles. Circular nodes (probabilistic nodes) represent random quantities and arcs into random quantities indicate probabilistic dependence. An influence diagram emphasizes the relationships among the random quantities involved in the problem and represents a complete probabilistic description of the model. The solution for the QMP model, i.e., the posterior distribution of $\theta_T$ given the past data, $d_{T-1}$ and current data, $x_T$ can be achieved through the use of influence diagrams operations, namely, node merging, node splitting, node elimination and arc reversal. These operations are described in Barlow and Pereira (1990). Figure 1 is an influence diagram representation corresponding to the QMP model.

The joint distribution for random quantities in the QMP model is completely defined by the influence diagram above. The absence of arrows into node $(\theta, \gamma^2)$ means that we start with the unconditional joint distribution of $\theta$ and $\gamma^2$. The arrows originating at note $(\theta, \gamma^2)$ and ending at nodes $\theta_t$ $(t = 1,\ldots,T)$ indicate that the distributions of $\theta_t$ are conditional on $\theta$ and $\gamma^2$. This means that the process is considered exchangeable, that is, the process average, $\theta$, is constant over time. Finally, each node $x_t$ is the sink of an arrow starting at node $\theta_t$ meaning that the distribution of the random quantity $x_t$ is conditional on $\theta_t$ for each $t = 1,\ldots,T$.

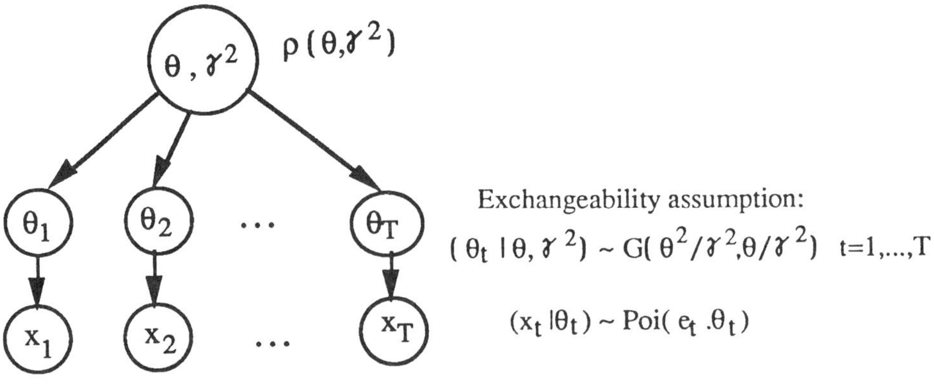


**Figure 1**

The QMP chart is a control chart for analyzing defect rates. Quality rating in QMP is based on posterior probabilities given the audit data. It provides statistical inference for the true quality process. Under QMP, a box and whisker plot (Figure 2) is plotted each period. The box plot is a graphical representation of the posterior distribution of $\theta_T$ given $d_{T-1} = (x_1,\ldots,x_{T-1})$ and $x_T$. The standard quality on the quality index scale is one. Two means twice as many defects as expected under the standard. Hence, the larger the quality index, the worse the process.

The posterior probability that the true quality index is less than the top whisker ($I_{99\%}$) is 99%. The top of the box ($I_{95\%}$), the bottom of the box ($I_{5\%}$) and the bottom whisker ($I_{1\%}$) correspond to probabilities of 95%, 5% and 1%, respectively.

The $x$ is the observed value in the current sample, the heavy dot is the Bayes estimate of $\theta$ and the dash is the Bayes estimate of the current quality index ($\theta_T$), a weighted average between $x_T$ and $\theta$.

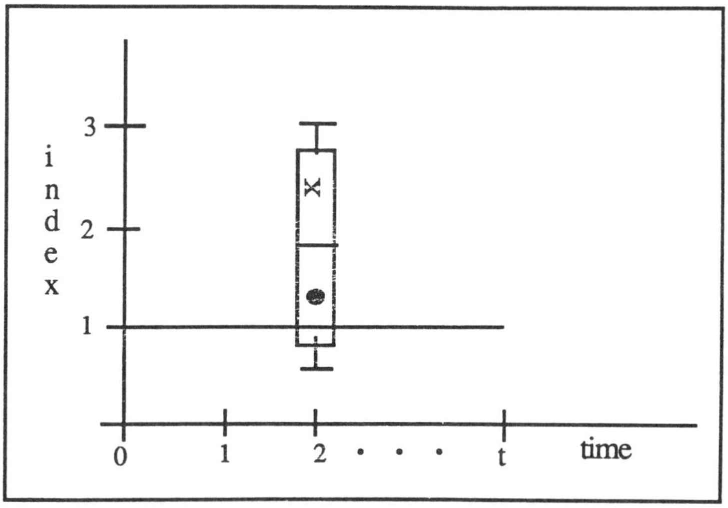


Figure 2

In a complete QMP chart (with all boxes), the dots are joined to show trends, i.e., it is assumed implicitly that the quality index $\theta_t$ may be changing from period to period.

**1. Exception reporting**

The objective of quality rating is to give a specific rule that defines quality exceptions and a measure (e.g., probability) associated with an exception. For QMP there are two kinds of exceptions:

a. A rating class is Below Normal (BN) if $I_{99\%} \geq 1$, i.e. if $P(\theta_T > 1) \geq 99\%$.

b. A rating class is on Alert if $I_{99\%} \leq 1 \leq I_{95\%}$, i.e., if $95\% \leq P(\theta_T > 1) \leq 99\%$. (See Figure 3.)

Products that meet these conditions are highlighted in an exception report.

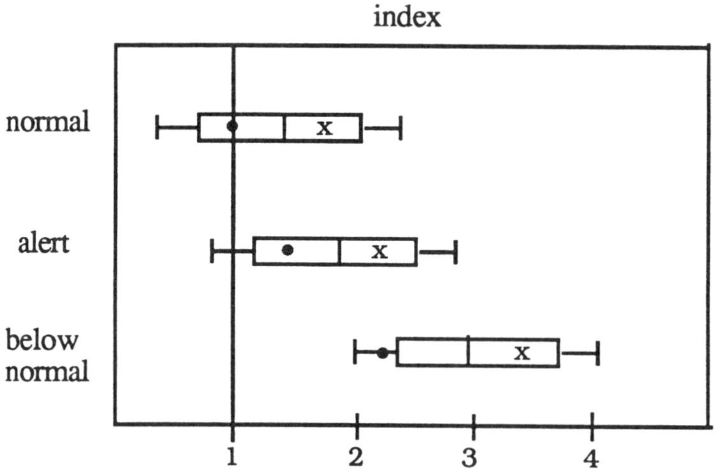


Figure 3

**2. Posterior distribution of current quality**

In order to get the exact solution for QMP, we have to compute the posterior distribution of $\theta_T$ given $d_{T-1} = (x_1,\ldots,x_{T-1})$ and $x_T$. Hoadley (1981) describes a complicated mathematical "solution" to this model. It can be best understood through the following sequence of influence diagrams:

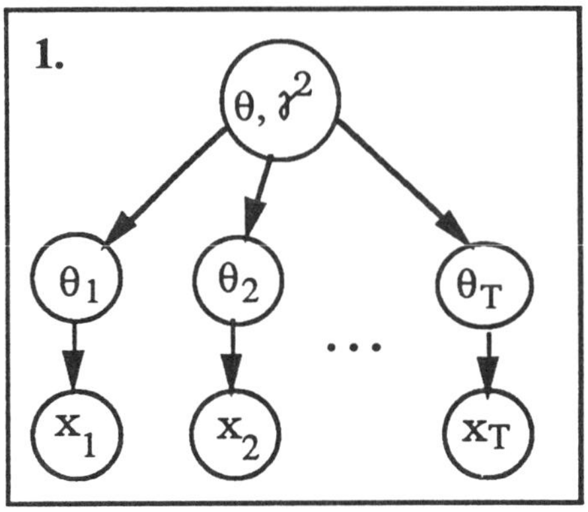


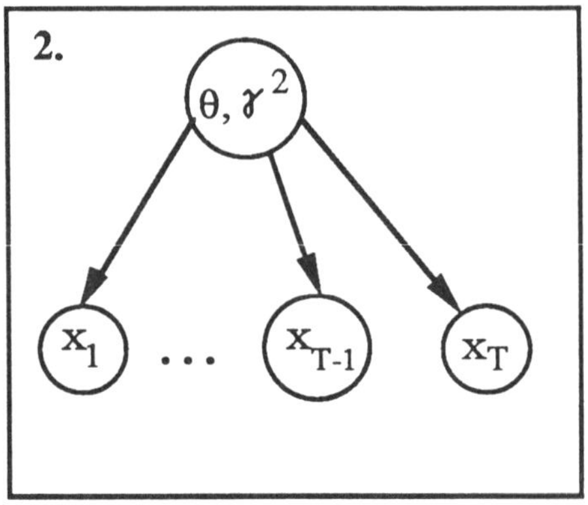

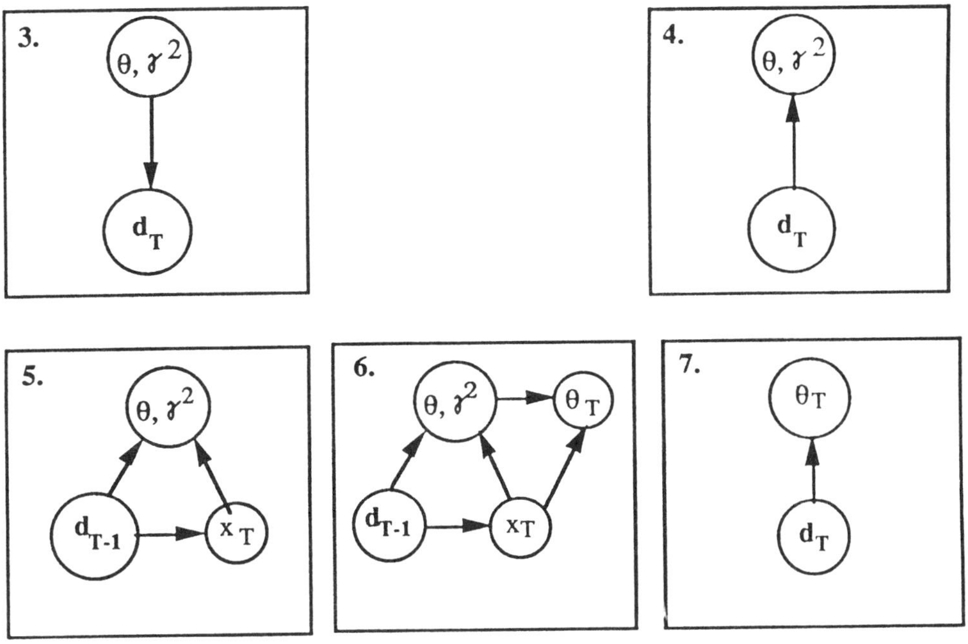


Figure 4

Diagram 1: Starting model: $(\theta,\ \gamma^2)\ \sim\ \rho(\theta,\ \gamma^2)$.

$(\theta_t|\theta,\gamma^2)\ \sim\ Gamma\left(\frac{\theta^2}{\gamma^2}, \frac{\theta}{\gamma^2}\right)$ for $t = 1,\ldots,T$. $(x_t|\theta_t)\ \sim\ Poi(e_t\theta_t)$.

Diagram 2: Nodes $\theta_1,\ldots,\theta_T$ are eliminated through integration.

$(x_t|\theta,\gamma^2)\ \sim$ Negative Binomial (Aitchison and Dunsmore, 1975).

Diagram 3: Nodes $x_1,\ x_2,\ldots,x_T$ are merged, i.e., the joint distribution of $d_T = (x_1,\ x_2,\ldots,x_T)$ is computed.

$(d_T|\theta,\gamma^2)\ \sim\ \prod_{t=1}^{T}$ Negative Binomials.

Diagram 4: The arc that goes from node $(\theta,\gamma^2)$ to node $d_T$ is reversed, i.e., Bayes theorem is used to compute the posterior distribution of $\theta$ and $\gamma^2$ given $d_T$. $\rho(\theta,\gamma^2|d_T)$: posterior for $\theta$ and $\gamma^2$ given $d_T$.

Diagram 5: Node $d_T$ is split. The joint distribution of $(x_1,\ x_2,\ldots,x_T)$ is written as the distribution of $d_{T-1} = (x_1,\ x_2,\ldots,x_{T-1})$ and the conditional distribution of $x_T$ given $d_{T-1}$.

Diagram 6: Node $\theta_T$ is added again into the model. The distribution of $\theta_T$ given $(\theta,\gamma^2)$ and $x_T$ is determined.

Diagram 7: Node $(\theta,\gamma^2)$ is eliminated through integration.

As we can see from the diagrams, the quality indexes $\theta_1,\ldots,\theta_T$ are eliminated in order to compute the distribution of the data, $d_T$, given $\theta$ and $\gamma^2$ and then, to compute the posterior distribution of $\theta$ and $\gamma^2$ given the data, $d_T$. Nevertheless, the parameter of interest is the current quality index, $\theta_T$, which has to be re-introduced into the influence diagrams. This procedure is not correct. According to this, $x_T$ is influencing $\theta_T$ twice in diagram 6. On one hand, directly (there is an arrow from $x_T$ to $\theta_T$), and on the other hand, through the posterior distribution of $\theta$ and $\gamma^2$ given $d_T$. In other words, node $\theta_T$ is eliminated (in influence diagram 2) and is added again (in influence diagram 6) and this is not the way one should solve an inference problem.

Even if this procedure were correct, the posterior distribution of $\theta_T$ would be a complex triple integral depending on the prior distribution assessed for $\theta$ and $\gamma^2$. This integral would have to be inverted in order to compute the QMP box chart. In other words, the exact solution is mathematically intractable, especially when many rating classes have to be analyzed each period. The result is a complicated algorithm (Hoadley, 1981) that computes all the parameters that are needed in order to construct the Gamma distribution for $\theta_T|d_T$. Hoadley's model assumes exchangeability, i.e., statistical control. Hence it does not provide an alternative to statistical control which can be used to decide whether or not the process is still in statistical control at the current time period. In the absence of an alternative model to exchangeability a better solution would have been to simply plot the standardized likelihoods (gamma densities) for $\theta_t$ at each time period based on the Poisson model. This would implicitly assume the $\theta_t$'s independent a priori.

**A Kalman Filter Model for Inventory Control**

As we have seen, the problem of quality control is to determine if and when a process has gone out of *statistical control*. The main difficulty with classical quality control procedures and also with the QMP model is that the models used assume the process is in statistical control and consider no alternative models to this situation. For coherent decision making, it is necessary to determine logical alternative models corresponding to a process out of statistical control.

In a paper dealing with inventory control (Barlow, Durst and Smiriga, 1984), a Kalman filter model was discussed from a decision theory point of view which could also be used for quality control problems. The paper describes an integrated decision procedure for deciding whether a diversion of Special Nuclear Material (SNM) has occurred. The problem is especially relevant for statistical analysis because it concerns (a priori) low probability events which would have high consequence if any occur. Two possible types of diversion are considered: a block loss during a single time period and a cumulative trickle loss over several time periods. The methodology used is based on a compound Kalman filter model.

Perhaps the simplest Kalman filter model is

$$\begin{aligned} y(t) &= \theta(t) + v(t) \\ \theta(t) &= \theta(t\text{-}1) + w_\theta(t), \end{aligned} \tag{1}$$

where $y(t)$ is the measured inventory at time period $t$ and $\theta(t)$ is the actual inventory level. Our uncertainty with respect to measuring error is modeled by $v(t)$ while $w_\theta(t)$ models our uncertainty about the difference in the actual amounts processed between time period $t$-1 and $t$.

The $y(t)$ process will be in statistical control in the sense of the first section if and only if $w_\theta(t) \equiv 0$ for all $t$. For the inventory problem it seems reasonable to use (1) to model the process in the absence of any diversions. Later we will extend this model to account for possible diversions.

The compound Kalman filter model allows a decision maker to decide at each time period whether the data indicate a diversion. A block loss, by definition, will be a substantial amount which, it is hoped, will be detected at the end of the period in which it occurs. A trickle loss, on the other hand, is a smaller amount which is not expected to be detected in a single occurrence. A trickle loss may consist of a diversion or process holdup (or both), while a block loss is always a diversion. Two models are given for the process during each time period; in one, a block loss is assumed to have occurred, while in the other, only the usual trickle loss takes place. Since there are two models at each time period, a fully Bayesian analysis would required $2^n$ models at the end of $n$ time periods, which is computationally untenable. A simple approximation is made which rests on the assumption that a block loss is a low-probability event. With this approximation only two models need be considered at each period, with all inference conditional on the assumption of no block loss in past periods (which has probability virtually equal to 1 as long as we have never come close to deciding that a block loss has occurred). By comparing these two models, we decide whether a block loss has occurred, and if we decide that it has an investigation is initiated. Since trickle loss, at least in the form of process holdup, is always assumed to occur, we will never decide that no trickle loss has occurred. We will either decide that a trickle diversion has occurred over several past periods, or we will decide that we as yet are unconvinced that a trickle loss beyond the normal holdup has occurred.

In Figure 5, $\beta(1)$, $\beta(2)$,..., etc., denote the amount of possible but unknown block losses during their respective time periods. The amount of possible but unknown trickle losses are denoted by $\tau(1)$, $\tau(2)$,..., etc. In our approach, we shall have two models: one model for block loss, say $M_B$, and one model for trickle loss, say $M_T$. We believe that model $M_B$ holds with probability $p(M_B)$ and model $M_T$ with probability $1 - p(M_B)$. Given data $D$, $p(M_B|D)$ is our updated probability for the block loss model $M_B$. If our updated probability for the block loss model is too high, then we will decide to investigate the possibility of a block loss. A decision regarding possible trickle loss, on the other hand, is based on the probability that loss beyond the normally expected holdup has occurred over several time periods; i.e.

$$P\{\tau(1) + \ldots + \tau(t) > c \mid D\}$$

where $c$ is the normally expected holdup over t time periods. Thus, as indicated in Figure 5, our decision sequence is the customary one; at each time period we either decide that a

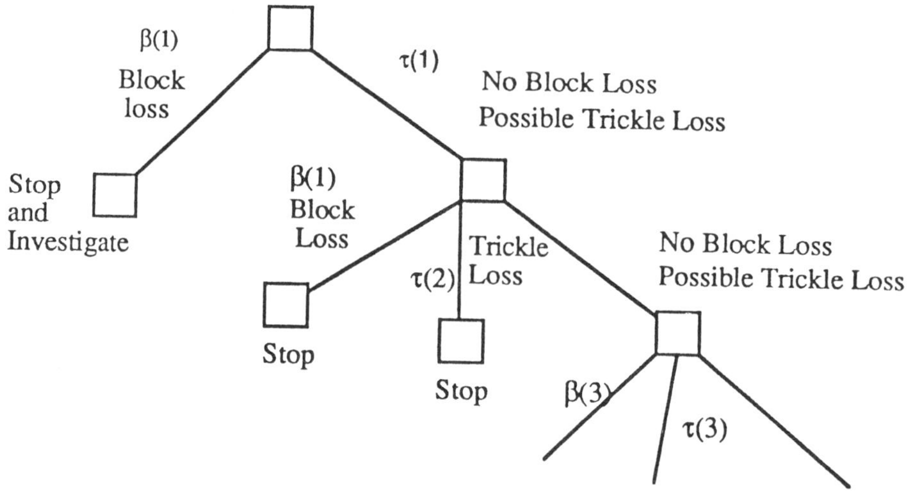


Figure 5 Diagram of possible decision sequences relative to diversion of special nuclear material

substantial block loss has occurred in the most recent period, that an unusually large trickle loss has been occurring in the past few periods, or that no block loss is likely to have occurred and that trickle loss is within acceptable limits. Our decision procedure does not formally permit the conclusion that a block loss has occurred other than within the most recent period, but it is shown that certain trickle alarms indicate the presence of an undetected block loss in some past period.

In order to clearly illustrate the salient features of these models, consider the simplified model (1) with only one measurement each period. At time $t$, $\theta(t)$

is the quantity of interest, but we can only observe $y(t)$. We assume that all variables in (1) are normally distributed.

The simplified trickle model $M_T$ is:

$$y(t) = \theta(t) + v(t),$$

$$\theta(t) = \theta(t-1) - \tau(t) + w_\theta(t), \tag{2}$$

$$\tau(t) = \tau(t-1) + w_\theta(t).$$

The simplified Kalman filter block model $M_B$ is:

$$y(t) = \theta(t) - \beta(t) + v(t),$$

$$\theta(t) = \theta(t-1) - \tau(t) + w_\theta(t), \tag{3}$$

$$\tau(t) = \tau(t-1) + w_\theta(t).$$

For the $MB$ model, assume that $\beta(0)$ is also normally distributed.

The values of distribution parameters, even in our simplest model, must be carefully set. Too little initial uncertainty about possible trickle loss may make the model surprisingly unresponsive to large unexpected losses. A set of distribution parameters can be entirely self-consistent, seem on casual inspection quite sensible, and still produce undesirable behavior of the detection procedure. Thus distribution parameters should not be set arbitrarily or casually, but only after a careful assessment of process and loss uncertainties which takes into account the effect of the parameters on the resulting decision procedure.

The compound Kalman filter model provides a detection process which can compete with currently popular methods. Large block losses are detected handily, while somewhat smaller block losses are often detected later by the trickle model. Trickle losses consistently in excess of the expected holdup are detected rapidly, and smaller trickle losses are detected as the total amount of trickle loss becomes large.

With standard quality control methods, decisions must be made with a test of fixed significance level; otherwise, the frequentist interpretation of the test does not hold. Since we are dealing with probability distributions, we are not limited to setting a critical threshold and a critical probability. In fact, simulations indicate that it is best to take into account all the information given by the posterior probabilities. The results of a single hypothesis test, although a convenient summary, may be misleading. The user of these methods is encouraged to examine the probabilities of multiple critical regions, something which is not possible with standard quality control methods.

**References**

Aitchison, J. and Dunsmore, I. R. (1975): *Statistical Prediction Analysis*, Cambridge University Press.

Barlow, R. E., M. J. Durst, and N. G. Smiriga (1984): A Kalman filter model for determining block and trickle SNM losses, in *Low-Probability High-Consequence Risk Analysis*, eds. R. A. Waller and V. T. Covello, Plenum Press, New York, pp. 161-179.

Barlow, R.E. and Pereira, C. A. (1990): Conditional independence and probabilistic influence diagrams, in *Topics in Statistical Dependence, IMS Lecture Notes Series*, eds. H. W. Block, A. R. Sampson, and T. H. Savits, Institute of Mathematical Statistics, Hayward, Calif., pp. 1-14.

Basu, D. (1988): Statistical Information and Likelihood–A Collection of Critical Essays, *Lecture Notes in Statistics*, Number 45, Springer Verlag.

Dawid, A. P. (1986): A Bayesian view of statistical modelling, in *Bayesian Inference and Decision Techniques*, eds. P. Goel and A. Zellner, Elsevier Science Publishers, B. V., pp. 391-404.

Deming, W. J. (1986): *Out of the Crisis*, MIT Press, Cambridge, MA.

de Finetti, B. (1937): La Prevision: ses lois logiques, ses sources subjectives, *Annales de l'Institut Henri Poincare* 7, 1-68. Translated by Kyburg Jr. and H. E. Smokler (1980): *Studies in Subjective Probability*, 2nd ed., Robert E. Krieger Pub. Co., Huntington, New York.

de Finetti, B. (1979): Probability and exchangeability from a subjective point of view, *International Statistical Review* 47, 129-135.

Duncan, A. J. (1974): *Quality Control and Industrial Statistics*, 4th ed., Richard D. Irwin, Homewood, IL.

Hoadley, B. (1981): The quality measurement plan (QMP), *The Bell System Technical Journal* 60(2), 215-273.

Nelson, L. S. (1982): Control charts, *Encyclopedia of Statistical Sciences*, eds. Kotz and Johnson, J. Wiley Interscience, pp. 176-182.

Shachter, R. (1986): Evaluating influence diagrams, *Operations Research* 34, 871-882.

Shewart, W. A. (1931): *Economic Control of Quality Manufactured Product*, D. Van Nostrand, New York.

## PREQUENTIAL DATA ANALYSIS

A.P. Dawid, University College London, Department
of Statistical Science, England

### Abstract

The basic theory of the prequential approach to data analysis is described, and illustrated by means of both simulation experiments and applications to real data-sets.

### Introduction

The prequential approach to the problems of theoretical statistics was introduced by Dawid (1984). It is based on the idea that statistical methods should be assessed by means of the validity of the predictions that flow from them, and that such assessments can usefully be extracted from a sequence of realized data-values, by forming, at each intermediate time-point, a forecast for the next value, based on an analysis of earlier values. The main emphasis is on probability forecasting, requiring that one describe current uncertainty about the predictand by means of a fully specified probability distribution. However, point forecasts, or other forms of prediction, can also be accommodated.

The purpose of the above paper was to indicate the fertility of the prequential point of view for furthering understanding of traditional concerns of theoretical statistics, such as consistency and efficiency. However, the prequential approach is essentially data-analytic. As such, it is particularly well suited to empirical investigation of the structure and properties of real-world observations, and their sources. In this paper, we shall discuss some of the ways in which prequential assessment may be applied in practical problems, including goodness-of-fit, model choice and density estimation. These methods are illustrated, by means of simulation experiments and applications to real data.

### Prequential Assessment

Let $\underset{\sim}{Y} = (Y_1, Y_2, \ldots)$ be a potentially infinite sequence of observables, and $\underset{\sim}{Y}^{(k)} = (Y_1, Y_2, \ldots, Y_k)$. We consider methods of forming, for each $k = 1, 2, \ldots$, a *prediction*, $\hat{y}_k$, for $Y_k$, based on past data $\underset{\sim}{Y}^{(k-1)} = \underset{\sim}{y}^{(k-1)}$; or, more generally, of deciding on an *action* $a_k$ on the basis of $\underset{\sim}{y}^{(k-1)}$, when subject to a loss $L_k(y, a)$ if $Y_k = y$ and $a_k = a$. Such a method $M$ having been applied for $k =$ 1 to $n$, and resulting in actions $(a_1, a_2, \ldots, a_n)$, its performance might be assessed by means of its total *prequential loss*

$$L_n^*(M) = \sum_{k=1}^{n} L_k(y_k, a_k),$$

which measures the success of its earlier forecasts; and comparison amongst methods on this basis provides a guide (albeit imperfect) as to their likely relative future performance.

Starting from a parametric family of such methods, $\mathcal{M} = \{M_\theta: \theta \in \mathcal{T}\}$, with $M_\theta$ specifying $a_k = a_k(\underset{\sim}{y}^{(k)}; \theta)$, each $\theta$-value is thus assessed by

$$L_n^*(\theta) = \sum_{k=1}^{n} L_k\left(y_k, a_k(\underset{\sim}{y}^{(k-1)}; \theta)\right).$$

The *optimizing strategy* $\hat{\mathcal{M}}$ based on $\mathcal{M}$ then uses, for selection of $a_{n+1}$, $M_{\hat{\theta}_n}$, where $\hat{\theta}_n$ minimizes $L_n^*(\theta)$ ($n = 0, 1,\ldots$; modification for small $n$ may be required). This itself needs to be assessed by its prequential loss

$$L_n^*(\hat{\mathcal{M}}) = \sum_{k=1}^{n} L_k\left(y_k, a_k(\underset{\sim}{y}^{(k-1)}; \hat{\theta}_{k-1})\right),$$

which will typically exceed $L_n^*(\hat{\theta}_n)$.

Prequential assessment of past predictive performance is very close in spirit to the method of *cross-validation* (Stone, 1974) but bases its prediction for $Y_k$ on all *previous* outcomes, rather than on all outcomes *distinct* from $Y_k$. In both methods, the intention is to avoid the bias involved in letting $Y_k$ contribute to its own prediction, and so to produce an honest assessment of uncertainty.

**Probability Forecasting**

One way to choosing the action $a_k$, after observing $\underset{\sim}{Y}^{(k-1)} = \underset{\sim}{y}^{(k-1)}$, is to specify a *predictive distribution* $P_k$ for $Y_k$, and to choose $a_k$ to minimize the *predictive expected loss*

$$\int L_k(y_k, a)\, dP_k(y_k).$$

Specification of such a sequence of predictive distributions $(P_k)$, for any data, constitutes a *probability forecasting system* (PFS), and is equivalent to choosing a joint distribution $P$ for the sequence $\underset{\sim}{Y}$. Under broad regularity conditions, it then follows that, with $P$-probability 1, $\lim_{n\to\infty} sup(L_n^*(M) - L_n^*(M')) < \infty$, where $M$ is given by the above method, and $M'$ is an arbitrary method. Thus if *Nature* is regarded as generating $\underset{\sim}{Y}$ from $P$, then using $P$ as a PFS to construct an action sequence will be optimal, for any loss function.

A PFS $P$ for $\underset{\sim}{Y}$, or its associated sequence $(P_k)$ of predictive distributions of $Y_k$ given $\underset{\sim}{Y}^{(k-1)} = \underset{\sim}{y}^{(k-1)}$, can be assessed directly if we take the *action* $a_k$ to be the choice of a distribution $Q_k$ for $Y_k$, and use a *proper scoring rule* $S_k(y, Q_k)$, i.e. such that, for any distribution $P_k$ for $Y_k$, $E_{P_k}[S_k(Y_k, Q_k)]$ is minimized in $Q_k$ when $Q_k = P_k$ (Dawid, 1986). Then the optimal sequence of *actions* is just the sequence $(P_k)$. The assessment becomes particularly simple if we use the *logarithmic scoring rule* $S_k(y, P_k) = -log\, f_k(y)$, $f_k$ being the density of $P_k$. We then obtain $L_n^*(P) = -log\ f(\underset{\sim}{y}^{(n)})$, $f$ being the implied joint density for $\underset{\sim}{Y}^{(n)}$ under $P$. That is, we can, and henceforth shall, assess and compare PFS's by means of their *prequential log-likelihoods.*

It is interesting to note that, if the distributions $P$ and $Q$ for $\underset{\sim}{Y}$ are mutually absolutely continuous, then $L_n^*(P) - L_n^*(Q)$ will (with probability 1 under either $P$ or $Q$) remain bounded, and may oscillate between positive and negative values. In this case we shall never achieve an ultimate preference for either PFS, and it seems that we remain forever in a quandary as to which to use for further forecasts. However, a result of Blackwell and Dubins (1962) shows that, in this case, the forecasts produced by $P$ and $Q$ will be asymptotically indistinguishable, so that the choice is unimportant. This is an instance of *Jeffreys's Law* (Dawid, 1984): observationally indistinguishable statistical approaches must be in essential agreement on their assertions about observables.

If $\mathcal{P} = \{P_\theta\colon \theta \in \mathcal{T}\}$ is a parametric family of PFS's, with predictive densities $f_i(y_i; \theta)$, the optimizing strategy $\hat{\mathcal{P}}$ based on $\mathcal{P}$ describes $Y_{n+1}$ as having density $f_{n+1}(y_{n+1}; \hat{\theta}_n)$; $\hat{\theta}_n$ being the maximum likelihood estimator based on data $\underset{\sim}{y}^{(n)}$. The success of this *plug-in MLE* strategy must itself, however, be judged by means of its own prequential log-likelihood, viz.

$$log \prod_{i=1}^{n} f_i(y_i; \hat{\theta}_{i-1}),$$

rather than

$$log \prod_{i=1}^{n} f_i(y_i; \hat{\theta}_n).$$

Similarly we can judge any other such *statistical forecasting system* (SFS), based on the same model or on another. A SFS might involve plugging-in some estimate of $\theta$ from past data, as above; Bayesian or fiducial elimination of $\theta$; or any other suitable (standard or *ad hoc*) procedure. However, any such strategy will itself always be describable as a PFS, and hence as a joint distribution for $\underset{\sim}{Y}$. This allows standard probability theory to be applied in theoretical studies of the performance of a SFS for data generated from $P_\theta \in \mathcal{P}$, and opens up a fresh approach to the traditional problems of statistical theory (Dawid, 1984). In general, (efficient-estimate) plug-in and Bayesian SFS's are asymptotically optimal. The latter yield prequential likelihoods expressible in the form $\int f(\underset{\sim}{y}^{(n)}; \theta)\pi(\theta)d\theta$, which has computational advantages, as well as being insensitive to reordering of the data.

## Empirical Assessment

Sometimes an absolute assessment is required as to whether a PFS $P$ adequately describes data $\underset{\sim}{y}$. If the $Y_i$ are continuous real variables, and $F_i$ denotes the distribution function of $Y_i$ under $P_i$, then $\underset{\sim}{U} = (U_1, U_2, \ldots)$, where $U_i = F_i(Y_i)$, should be independently uniform on [0,1] if $\underset{\sim}{Y}$ arises from $P$, and so a variety of tests can be based on the observed values $\underset{\sim}{u}$. To assess uniformity, we might examine the $u$-plot, i.e. the empirical c.d.f. of the $u$'s, which should be close to the line of unit slope. This could be tested formally using, say, the Kolmogorov-Smirnov statistic. One should also inspect the $(u_i)$ for any sign of non-independence, trend, or dependence on omitted variables. A simple indicator of trend is provided by the *uniform conditional test* (Cox and Lewis, 1966) or $y$-plot, which forms the empirical c.d.f. of $(y_j)$, where $y_j = \sum_{i=1}^{j} x_i \big/ \sum_{i=1}^{n} x_i$, with $x_i = -log(1 - u_i)$. These $y$'s are uniform order-statistics under $P$, and this can again be tested formally.

If the $Y_i$ are 0–1 variables, we can form *calibration plots* in which, for various $\pi \in [0, 1]$, the observed relative frequency of $Y_i = 1$ over the set of occasions having $\Pi_i = \pi$ (where $\Pi_i = P_i(Y_i = 1)$) is plotted against $\pi$. This should give an approximate diagonal line. More formally, we can construct test-statistics such as $Z = \Sigma(Y_i - \Pi_i)/[\Sigma\Pi_i(1 - \Pi_i)]^{1/2}$, the sum possibly being restricted to a suitable subset of the data. Under very weak conditions, *not* requiring independence, $Z$ and similar standardized statistics will be asymptotically standard normal under $P$ (Seillier and Dawid, 1987) and independent of statistics based on disjoint subsets. An observed value $z$ can thus be referred to standard normal tables, or a sum of squares of $z$'s based on $k$ disjoint subsets to chi-square tables with $k$ degrees of freedom.

It is noteworthy that all the methods described above are applicable given only the two sequences, of outcomes and of their probability forecasts, and make no reference to the structure of $P$ over outcomes not observed. This is in accord with the *Prequential Principle* (Dawid, 1984).

If P is itself constructed as a SFS based on a parametric model $\mathcal{P} = \{P_\theta\}$, it turns out, again under mild conditions, that the asymptotic distributions of the test-statistics considered above continue to hold under any $P_\theta \in \mathcal{P}$ (Seillier et al., 1988). Consequently, these methods can be used to test the overall goodness-of-fit of a parametric model.

If the distribution or model being used fails to describe the data, it may be possible to *massage* it to provide a better fit. Thus suppose that the $(u_i)$ above look like a random sample, but from a non-uniform distribution. This distribution could itself be estimated, either parametrically or nonparametrically (as in *Density estimation* below). If the estimate based on $\underset{\sim}{u}^{(n)}$ is $G_n$, then $Y_{n+1}$ could be forecast by requiring that $F_{n+1}(Y_{n+1})$ has distribution $G_n$, rather than uniform. Alternatively, serial correlation, or other suspected structure, in the $(u_i)$ could be estimated and allowed for. In the $(0 - 1)$ case, if previous occasions on which the same probability forecast as $p_{n+1}$ was issued had resulted in a proportion $q$ of 1's, then $p_{n+1}$ might be replaced by $q$. Such adaptive recalibration methods can improve the performance of a badly chosen initial model, although there can be no guarantee that they will, since the recalibration is based on the past but applied to the future.

**Model Choice**

Given a choice between two competing models, say $\mathcal{P} = \{P_\theta\}$ and $\mathcal{Q} = \{Q_\theta\}$, we can first replace each of these by an appropriate SFS, say $\hat{P}$ and $\hat{Q}$, respectively. We might then optimize the choice between these at each time-point. Thus if it were $\hat{P}$, say, rather than $\hat{Q}$, that gave the larger prequential likelihood (or smaller total prequential loss) to the data $\underset{\sim}{y}^{(k)}$ at time $k$, the probability forecast for $Y_{k+1}$ would be that based on $\hat{P}$. Of course, such a two-stage optimization strategy needs assessing afresh in its own right. The method extends to more stages, and to an arbitrary collection of models at each stage, but clearly less trust can be placed in prequential analyses iterated to more stages: even though the prequential approach avoids obvious bias at each stage, no finite set of data can support more than a certain amount of investigation without throwing up misleading messages.

In place of repeated optimization, one can take a Bayesian approach, assigning prior weights $\alpha$ and $1 - \alpha$ to $\mathcal{P}$ and $\mathcal{Q}$. After observing $\underset{\sim}{y}^{(k)}$, with prequential joint density $f(\underset{\sim}{y}^{(k)})$ under $\hat{P}$ and $g(\underset{\sim}{y}^{(k)})$ under $\hat{Q}$, $\alpha$ is replaced by $\alpha_k = \alpha f(\underset{\sim}{y}^{(k)})/[\alpha f(\underset{\sim}{y}^{(k)}) + (1 - \alpha)g(\underset{\sim}{y}^{(k)})]$, and the forecast density for $Y_{k+1}$ is then the mixture $\alpha_k f_{k+1} + (1 - \alpha_k)g_{k+1}$. The overall prequential likelihood for this strategy is simply $\alpha f(\underset{\sim}{y}^{(n)}) + (1 - \alpha)g(\underset{\sim}{y}^{(n)})$. Again the method extends simply to more models and more stages.

If one has a finite or countable collection of alternative models, and the data arise from some distribution in one of these, either of the above methods will be consistent and asymptotically optimal, in the sense that their forecasts will tend to those given by the true distribution, and at the fastest possible rate. However, for finite data-sets, the forecasts under the two methods may look rather different. In either case, if the true distribution is contained in a model of high-dimensionality, early analysis will generally tend to favor incorrect models of low dimensionality. This is intuitively sensible, since, early on, the mis-modelling bias may well be less of a problem than the imprecision involved in trying to estimate many parameters.

As an alternative to allowing such transient behavior to be entirely data-driven, as above, one might build it in directly, by setting out with a strategy for choosing, at each stage, the complexity of the model to be fitted and how it is to be used for prediction. Different strategies, all yielding consistent estimates of the true model (and which use each fixed model efficiently) will all be *asymptotically* equally good. However, their *transient* behaviors, which may be long-lasting, can be very different, with some yielding much larger prequential log-likelihoods (or, more generally, much smaller prequential losses) than others even though these discrepancies will be bounded as the sample size goes to infinity. More empirical and theoretical work is needed to indicate good forms for such strategies. A sensible super-strategy could be built up from a low-dimensional parametrized family of such strategies, using optimizing or Bayesian methods. This could combine good transient behavior with sensitivity to the data and avoidance of data-mining.

**Non-parametric Approximation**

Many non-parametric problems, such as density estimation or fitting a stationary time-series, can be approached through a sequence of finitely parametrized methods, such as fitting histogram or kernel density estimates with adjustable bin width, or autoregressive models of various finite orders. One can then apply the techniques of the previous section, even though none of the models used is now expected to contain the distribution generating the data. The component models will generally each be characterized by some quantity, such as kernel width ($w$) or autoregressive order ($p$), which controls the balance between over-fitting (tracking noise in the data) and over-smoothing (not picking up the signal). Prequential choice of such a quantity will start out with a preference for smoothing (large $w$, small $p$), and then, as the data-sequence grows longer and can support more detailed modelling, gradually move towards fitting the past data more and more closely ($w \to 0$, $p \to \infty$). Such a method will often be prequentially consistent for a wide range of generating distributions, and can provide sensible answers based on finite data-sets, by making the predictively optimal compromise between fitting and smoothing.

Investigation of the structure of good *strategies*, for choosing the model to fit at each stage, is still more vital in this context, since the behavior described as *transient* in the previous section now extends to infinity! Again, much further empirical and theoretical work is required to illuminate this problem area.

**Simulations**

**1. Time-series modelling.**

Autoregressive models of varying order $k$ ($0 \leq k \leq 8$) were fitted to several simulated time-series of 500 observations, and their prequential likelihoods calculated using both optimization (plugging-in current least-squares estimates) and Bayesian methods (using a *non-informative* prior), always excluding the first 15 observations. Results were as follows.

(i) Independent standard normal variates: $Y_t = \epsilon_t$; Prequential Log-Likelihoods

| $k$ | : | 0 | 1 | 2 | 3 | 4 | 5 | 6 | 7 | 8 |
|---|---|---|---|---|---|---|---|---|---|---|
| Optimization | : | -715.5 | -717.1 | -721.0 | -724.0 | -728.0 | -728.7 | -735.1 | -739.1 | -740.5 |
| Bayes | : | -712.8 | -713.9 | -717.6 | -719.4 | -722.0 | -722.4 | -726.0 | -728.4 | -730.1 |

The strategy of optimizing over $k$ chose $k = 0$ at all points, except one, beyond the 57th observation, and chose $k = 1$ at all the exceptional points. This strategy itself had a prequential log-likelihood of -714, better than that for any fixed $k$.

The Bayes strategy (using equal prior probabilities) finished by assigning probability 0.75 to $k = 0$ and 0.25 to $k = 1$. Its prequential log-likelihood too was -714.

(ii) Autoregression: $Y_t = 0.1Y_{t\text{-}1} - 0.3Y_{t\text{-}2} + 0.2Y_{t\text{-}3} + \epsilon_t$; Prequential Log-Likelihoods

| $k$ | 0 | 1 | 2 | 3 | 4 | 5 | 6 | 7 | 8 |
|---|---|---|---|---|---|---|---|---|---|
| Optimization | -723.9 | -725.3 | -705.5 | -700.1 | -701.2 | -701.3 | -704.4 | -708.6 | -709.3 |
| Bayes | -722.8 | -724.1 | -703.5 | -698.3 | -699.1 | -700.6 | -702.7 | -705.9 | -707.6 |

There is a clear preference for the true order, with under-fitting being more heaving penalized than overfitting. Optimizing over $k$ chose $k = 2$ up to observation 40, $k = 3$ thereafter. This strategy had a prequential log-likelihood of -700, indistinguishable from that of $k = 3$. The Bayes strategy ended by assigning probability 0.63 to $k = 3$, 0.29 to $k = 4$ and 0.07 to $k = 5$, and itself had a prequential log-likelihood of -700.

(iii) Moving average: $Y_t = 0.5\epsilon_t - 0.2\epsilon_{t\text{-}1}$; Prequential Log-Likelihoods

| $k$ | 0 | 1 | 2 | 3 | 4 | 5 | 6 | 7 | 8 |
|---|---|---|---|---|---|---|---|---|---|
| Optimization | -370.1 | -354.8 | -355.3 | -357.7 | -355.5 | -356.9 | -358.7 | -360.3 | -364.5 |
| Bayes | -368.9 | -353.4 | -353.2 | -355.7 | -353.2 | -355.3 | -357.1 | -359.1 | -362.2 |

The true process can be expressed as an infinite-order autoregression: $Y_t = -0.4Y_{t-1} - 0.16Y_{t-2} - 0.064Y_{t-3} - \ldots + 0.5\epsilon_t$. The optimal autoregressive fit to 500 observations, however, gave $k = 1$ (optimization) or $k = 2$ (Bayes), closely followed by $k = 4$ (for which the estimated coefficient of lag 4 was -0.139, compared with the true value of -0.026). Optimizing over $k$ gave $k = 1$ at all points, except for observations 16 to 33 (for which $k$ was 0) and most points between observations 460 and 486 (with $k = 4$). This strategy had prequential log-likelihood of -355.5. The Bayes strategy assigned probabilities 0.27 to $k = 1$, 0.33 to $k = 2$, 0.03 to $k = 3$, 0.32 to $k = 4$, and 0.04 to $k = 5$, and itself had a prequential log-likelihood of −354.3.

**2. Density estimation.**

Simple histogram-type density estimators were constructed from data-values in [0,1], based on a division of the unit interval into k equal sub-intervals. For each initial sub-sequence of data, the current density estimate was used to forecast the next observation. This was repeated for $1 \leq k \leq K$.

(i) A random sample of size 1000 from the uniform distribution on [0,1] yielded the following overall prequential log-likelihoods (up to $K = 10$);

| $k$ | : | 1 | 2 | 3 | 4 | 5 | 6 | 7 | 8 | 9 | 10 |
|---|---|---|---|---|---|---|---|---|---|---|---|
| log-likelihood | : | 0 | -3.7 | -6.4 | -8.8 | -12.5 | -16.0 | -18.8 | -22.3 | -25.1 | -32.4 |

The deterioration in performance when fitting more intervals than needed (viz. 1) is clear.

The optimizing strategy, formed by selecting, at each point, that value for $k$ yielding the highest prequential likelihood to date, always chose $k = 1$, except at a number of points up to the 52nd observation, for which $k = 2$ was chosen.

(ii) A random sample of size 3000 was generated from the symmetric unimodel density

$$f_1(x) = \frac{1}{2}\pi sin(\pi x) \qquad (0 \leq x \leq 1).$$

With $K = 20$, the optimal $k$ based on all the data was 14, the prequential log-likelihoods for $k = 10$ to 15 being, respectively, 399.3, 389.2, 401.3, 396.9, 402.2 and 399.0. When optimizing over $k$ at all points, the first and last appearances of various values, and their frequencies, were:

| $k$ | : | 1 | 2 | 3 | 4 | 5 | 6 | 7 |
|---|---|---|---|---|---|---|---|---|
| First used | : | 1 | 6 | 21 | 77 | 229 | 148 | 319 |
| Last used | : | 20 | 43 | 154 | 138 | 388 | 395 | 842 |
| Frequency | : | 18 | 6 | 109 | 16 | 46 | 182 | 39 |

| $k$ | : | 8 | 9 | 10 | 11 | 12 | 13 | 14 | >14 |
|---|---|---|---|---|---|---|---|---|---|
| First used | : | 1423 | 400 | 1457 | - | 1435 | - | 2856 | - |
| Last used | : | 1423 | 2133 | 2335 | - | 2915 | - | 3000 | - |
| Frequency | : | 1 | 1118 | 434 | 0 | 892 | 0 | 139 | 0 |

The general message of the above simulations would seem to be that, even for large data sets, it is generally far more effective to fit a very simple model that is approximately true, rather than one which contains the true distribution (or comes close to doing so), but is of highish dimension.

## Applications

### 1. Weather forecasting.

Jain (1983) analyzed a 53-year sequence of daily precipitation records from Morogoro, Tanzania, as discussed in Stern and Coe (1984). The model $\mathcal{P}$

for the conditional probability $p_t$ of rain on day $t$ (coded as $Y_t = 1$), given past outcomes, was a non-stationary two-state second-order generalized linear Markov Chain:

$$logit\ p_t(\theta) = a_{ij0} + \sum_{k=1}^{4}\left[a_{ijk}\ sin(kt') + b_{ijk}\ cos(kt')\right]$$

where $t' = (2\pi t/366)$, $i$ and $j$ are the outcomes of days $t - 2$ and $t - 1$, and $\theta$ consists of the $a$'s and $b$'s. The parameters were estimated recursively, with initial estimates fitted, using maximum likelihood, to the first 700 data-points, and the probability forecasts $\hat{p}_t$ of the resulting *plug-in* strategy compared with the actual outcomes ($t > 700$). Calibration plots and test-statistics were constructed for various subsets of the data, corresponding to the months of the year, and to specified outcomes of the three previous days. Table I gives, for each month, the overall proportion $\bar{y}$ of rainy days, and the average forecast probability $\bar{\hat{p}}$. The final line gives values of the test statistic

$$z_B = \frac{\Sigma(y_i - \hat{p}_i)^2 - \Sigma\hat{p}_i(1 - \hat{p}_i)}{\left[\Sigma\hat{p}_i(1 - \hat{p}_i)(1 - 2\hat{p}_i)^2\right]^{\frac{1}{2}}}$$

for assessing departure from expectation of the within-month *Brier Score* $\Sigma(y_i - \hat{p}_i)^2$. These should be approximately independent standard normal variables under the model $\mathcal{P}$. The combined chi-square of 78 on 12 degrees of freedom clearly indicates poor model fit, and closer scrutiny reveals that the model is noticeably under-forecasting rain in April, and when the third previous day was wet.

TABLE I

| Month | : | J | F | M | A | M | J | J | A | S | O | N | D |
|---|---|---|---|---|---|---|---|---|---|---|---|---|---|
| $\bar{y}$ | : | .21 | .22 | .33 | .54 | .31 | .10 | .06 | .04 | .09 | .10 | .17 | .22 |
| $\bar{\hat{p}}$ | : | .20 | .20 | .36 | .47 | .31 | .09 | .05 | .04 | .09 | .09 | .16 | .20 |
| $z_B$ | : | 1.57 | 4.68 | 2.15 | 5.00 | 1.80 | 2.19 | 2.26 | 0.40 | 0.42 | 0.66 | 1.26 | 2.93 |

## 2. Medical diagnosis.

Seillier (1982) analyzed 58 cases of jaundice, caused either by hepatitis ($Y = 1$) or by cirrhosis ($Y = 0$). Various logistic models to discriminate between the two diagnoses were considered, using regressor variables chosen from a set of ten symptoms ($A$, $B$, $C$, $D$, $E$, $F$, $X1$, $X2$, $X3$, $X4$) and a location indicator $Q$.

Each model was fitted by maximum likelihood to the first $k$ cases ($k = 30, 31, \ldots, 57$), and used to provide a probability forecast $\hat{p}_{k+1}$ for $Y_{k+1}$, based on its associated regressor variables. The assessment of each model was then based on its overall Brier score $\sum_k (y_k - \hat{p}_k)^2$. The results are shown in Table II, which also gives $\bar{\hat{p}}$, for comparison with $\bar{y} = 0.29$.

TABLE II

| Variables | Brier Score | $\bar{\hat{p}}$ |
|---|---|---|
| A + B + C + D + E + F + X1 + X2 + X3 + X4 + Q | 4.7 | 0.25 |
| A + B + C + D + E + F + X1 + X2 + X3 + X4 | 3.8 | 0.36 |
| A + C + D + E + X1 + X2 + X3 + X4 + Q | 4.6 | 0.24 |
| A + C + D + E + X1 + X2 + X3 + X4 | 3.8 | 0.36 |
| A + D + E + X1 + X2 + X3 + X4 + Q | 4.0 | 0.22 |
| A + D + E + X1 + X2 + X3 + X4 | 3.4 | 0.39 |
| A + D + X1 + X2 + X3 + X4 + Q | 4.6 | 0.32 |
| A + D + X1 + X2 + X3 + X4 | 3.6 | 0.38 |
| A + D + X1 + X3 + X4 + Q | 3.0 | 0.23 |
| A + D + X1 + X3 + X4 | 2.3 | 0.31 |
| A + D + X1 + X3 + Q | 3.4 | 0.24 |
| A + D + X1 + X3 | 3.0 | 0.33 |
| D + X1 + X3 + Q | 3.0 | 0.25 |
| D + X1 + X3 | 2.9 | 0.39 |
| D + X1 + Q | 4.8 | 0.24 |
| D + X1 | 4.3 | 0.37 |
| D + Q | 5.3 | 0.22 |
| D | 5.1 | 0.36 |

Fitting all variables leads to poor predictions on this size data-set, as does fitting only two or three. The most successful model, as measured by its Brier score, is $A + D + X1 + X3 + X4$, which also has $\bar{\hat{p}}$ closest to $\bar{y}$. It is of interest that, for any collection of symptom variables, adding in the location indicator $Q$ leads to *worse* predictions. This offers some empirical support for the arguments of Dawid (1976) that suitable diagnostic models should be robust over a range of locations.

## 3. Educational scaling.

Opie (1983) conducted an analysis to see whether items in an educational testing item-bank fitted the Rasch model, under which $P$(student $i$ gets item $j$ correct) $= e^{\alpha_i+\beta_j}/(1 + e^{\alpha_i+\beta_j})$. The data-set contained responses to 60 test items from 150 students. At an intermediate stage, a number of items, 1 to $k-1$ say, have been accepted, and item $k$ is under test. For $m = 75$ to 150, the parameters are estimated (by maximum likelihood) from the responses of students 1 to $m$ on items 1 to $k$, omitting that of student $m$ on item $k$. The fitted probability for this omitted response can then be calculated, and the process repeated with m increased by 1. Comparison of these forecast probabilities with the actual responses (where these were not missing) then allows assessment of the fit of item $k$ to the model.

For testing item 60, with all other items included, the probabilities were grouped into 8 intervals, with counts, average probability and relative frequency of a right answer as given in Table III.

TABLE III

| Group ($g$) | Count ($n_g$) | Average probability ($\pi_g$) | Relative frequency ($\bar{y}_g$) |
|---|---|---|---|
| 0.0 - 0.1 | 14 | 0.07 | 0.07 |
| 0.1 - 0.15 | 16 | 0.12 | 0 |
| 0.15 - 0.2 | 11 | 0.17 | 0.09 |
| 0.2 - 0.3 | 12 | 0.25 | 0.25 |
| 0.3 - 0.4 | 8 | 0.33 | 0 |
| 0.4 - 0.5 | 6 | 0.44 | 0 |
| 0.5 - 0.6 | 4 | 0.55 | 0.5 |
| 0.6 - 1.0 | 4 | 0.86 | 0.5 |

If the item fits the model, then $\sum_g n_g(\bar{y}_g-\pi_g)^2 \div \pi_g(1-\pi_g)$ should be approximately distributed as chi-square with 8 degrees of freedom. The observed value of 15.8 is significant at 5%, suggesting a failure of calibration on this item, and thus its non-conformity with the Rasch model.

**4. Software reliability.**

Littlewood et al. (1986) have made a thorough comparison of a number of model-based prediction systems for prequential probability forecasting of the successive inter-failure times of complex software systems. The data comprised 136 inter-failure times ranging between 0 and 6150 seconds, and the models used all incorporated reliability growth (improved performance after each bug-fix). Some forecasting systems used optimization, some were Bayesian, others combined the two methods. The results are summarized in Table IV.

TABLE IV

| System | $u$-plot $K$-$S$ distance (sig. level) | $y$-plot $K$-$S$ distance (sig. level) |
|---|---|---|
| 1. JM | .190 (1%) | .120 (NS) |
| 2. BJM | .170 (1%) | .116 (NS) |
| 3. GO | .153 (2%) | .125 (10%) |
| 4. L | .109 (NS) | .069 (NS) |
| 5. BL | .119 (NS) | .075 (NS) |
| 6. LNHPP | .081 (NS) | .064 (NS) |
| 7. LV | .144 (5%) | .110 (NS) |
| 8. KL | .138 (5%) | .109 (NS) |
| 9. W | .075 (NS) | .075 (NS) |
| 10. D | .159 (2%) | .093 (NS) |

Systems 1, 2 and 3 are all based on essentially the same model, as are 4, 5 and 6. It appears that the method of data analysis is less important here than choosing a good model. Measured by prequential likelihood, the optimal system was 6. The authors also considered adaptive recalibration of the above systems, as well as Bayesian and optimizing strategies for combining them, leading in all cases to improvements in performance.

## Conclusion

The prequential method is broad in range, simple in concept, and based on a firm theoretical foundation. However its implementation leaves plenty of scope for variations, and is currently more art than science. Further work should lead to an improved understanding, and give guidance on good strategies of applying the method. Efficient computational methods or approximations will also be essential for routine application.

## References

Blackwell, D. and Dubins, L. E. (1962): Merging of opinions with increasing information, *Ann. Math. Statist.* 33, 882-886.

Cox, D. R. and Lewis, P. A. W. (1966): *Statistical Analysis of Series of Events*, Methuen, London.

Dawid, A. P. (1976): Properties of diagnostic data distributions, *Biometrics* 32, 647-658.

Dawid, A. P. (1984): Statistical theory: the prequential approach (with discussion), *J. Roy. Statist. Soc.* A 147, 278-292.

Dawid, A. P. (1986): Probability forecasting, *Encyclopedia of Statistical Sciences*, eds. S. Kotz, N. L. Johnson and C. B. Reid, Wiley-Interscience, Vol. 7, 210-218.

Jain, R. (1983): *Probabilistic Weather Forecasting*, M.Sc. dissertation, Department of Statistical Science, University College London.

Littlewood, B., Abdel Ghaly, A. A., and Chan, P. Y. (1986): Evaluation of competing software reliability predictions, *IEEE Trans. Software Eng.* SE-12, 950-967.

Opie, G. A. (1983): *Educational Scaling*, B.Sc. dissertation, Department of Statistical Science, University College London.

Seillier, F. (1982): *Selection Aspects in Medical Diagnosis*, M.Sc. dissertation, Department of Statistical Science, University College London.

Seillier, F. and Dawid, A. P. (1987): On testing the validity of probability forecasts, *Research Report 57*, Department of Statistical Science, University College London.

Seillier, F., Sweeting, T. J., and Dawid, A. P. (1988): Prequential tests of model fit, *Research Report 61*, Department of Statistical Science, University College London.

Stern, R. D. and Coe, R. (1984): A model-fitting analysis of daily rainfall data (with discussion), *J. Roy. Statist. Soc.* A 147, 1-34.

Stone, M. (1974): Cross-validatory choice and assessment of statistical predictions (with discussion). *J. Roy. Statist. Soc.* B 36, 111-147.

## BAYESIAN NONPARAMETRIC INFERENCE

Thomas S. Ferguson, University of California at Los Angeles,

Eswar G. Phadia, Willian Paterson College,

and

Ram C. Tiwari, University of North Carolina

### Introduction

Problems of statistical inference with an infinite dimensional parameter space, usually a space of probability distributions over a set, are of great importance both theoretically and practically. The Bayesian approach to such *nonparametric* problems requires that a probability distribution be placed over this space. Much progress has been made in the past 15 years and the results have been scattered throughout the statistical and probability literature. It is the purpose of this paper to review the progress in this area to date with special emphasis on random probability measures and on results that have appeared since the review article of Ferguson (1974).

The central class of distributions for use in these problems is the class of Dirichlet processes. Developments in the basic theory of such processes are reviewed in the next section. The settling of Doksum's conjecture by James and Mosimann is observed in the third section on tailfree and neutral processes. Progress in the application of mixtures of Dirichlet processes to the Bayesian analysis of empirical Bayes problems, bio-assay and density estimation is presented in the fourth section. The far-reaching extension of the basic techniques to problems with partially censored data is reviewed in the fifth section, with application to reliability and the Cox proportional hazard model. The use of random distributions in empirical Bayes estimation, initiated by Hollander and Korwar, has been extensively developed and is reviewed in the sixth section. In the seventh section, the problems of inconsistency of the Bayes estimates in Dalal's symmetric Dirichlet model, discovered by Diaconis and Freedman, are presented. In the final section, various other Bayesian nonparametric techniques and applications are briefly touched upon.

### The Dirichlet Process

Let $\mathfrak{X}$ be a set, let $\mathcal{A}$ be a $\sigma$-field of subsets of $\mathfrak{X}$, and let $\alpha$ be a finite nonnull measure on $(\mathfrak{X}, \mathcal{A})$. Among the various methods for putting prior distributions on the set of all probability distributions over $(\mathfrak{X}, \mathcal{A})$, the Dirichlet process is still central. As defined in Ferguson (1973), a *Dirichlet process with parameter* $\alpha$, denoted $\mathfrak{D}(\alpha)$, is a random process, $P$, indexed by elements of $\mathcal{A}$ with the property that for all positive integers $k$, and every measurable partition

$A_1,\ldots, A_k$ of $\mathfrak{X}$, the random vector $(P(A_1),\ldots,P(A_k))$ has a $k$-dimensional Dirichlet distribution with parameter $(\alpha(A_1),\ldots,\alpha(A_k))$. The basic result for this process is:

**Theorem 1 (Ferguson, 1973)**

If $P$ is a Dirichlet process with parameter $\alpha$, and if, given $P$, $X_1,\ldots,X_n$ is a sample from $P$, then the posterior distribution of $P$ given $X_1,\ldots,X_n$ is a Dirichlet process with parameter $\alpha + \Sigma\delta(X_j)$, where $\delta(x)$ represents the distribution giving mass one to the point $x$.

Two proofs of the existence of such a process were given, one non-constructive using the Kolmogorov consistency conditions, and the other constructive, in which $P$ is a sum of a countable number of point masses whatever be $\alpha$. That a Dirichlet process has a representation that is discrete a.s. even if $\alpha$ is continuous is a striking fact that has been the subject of several papers, e.g., Blackwell (1973), Berk and Savage (1979), Basu and Tiwari (1982). A new construction simpler than that of Ferguson has been given by Sethuraman and Tiwari (1982).

**Theorem 2 (Sethuraman and Tiwari, 1982)**

Let $Y_1$, $Y_2,\ldots$ be i.i.d. with a beta distribution, $\mathbb{B}e(M,1)$ $M > 0$, let $Z_1$, $Z_2,\ldots$ be i.i.d. $F_0$, and let $\{Y_j\}$ and $\{Z_j\}$ be independent. Define $P_1 = (1 - Y_1)$, and $P_n = Y_1 \ldots Y_{n-1}(1 - Y_n)$ for $n > 1$. Then, $P = \Sigma P_j\delta(Z_j)$ is a Dirichlet process with parameter $\alpha = MF_0$.

Throughout, we shall use $M = \alpha(\mathfrak{X})$ to represent the total mass of $\alpha$, and $F_0 = \alpha/M$ to be the *prior guess at P*. The latter phrase stems from the fact that from the definition, $P(A)$ has a beta distribution, $\mathbb{B}e(\alpha(A), M-\alpha(A))$, so that $EP(A) = \alpha(A)/M = F_0(A)$. In particular, the posterior guess at $P$ given a sample from $P$ is, according to Theorem 1, $F_n = p_nF_0 + (1 - p_n)\hat{F}_n$, where $\hat{F}_n$ is the empirical process and $p_n = M/(M + n)$. As a consequence, suppose that it is required to estimate with squared error loss the mean $\mu = \int x dP(x)$ of an unknown distribution $P$ on the real line based on a sample $X_1,\ldots,X_n$, with prior $P \in \mathfrak{D}(MF_0)$, where $F_0$ has finite first moment. Then, $\mu$ is finite a.s. and

$$E(\mu \mid X_1,\ldots,X_n) = p_n\mu_0 + (1 - p_n)\bar{X}_n,$$

where $\mu_0$ is the mean of $F_0$, and $\bar{X}_n$ is the sample mean. (In subsequent discussions, Bayes procedures are assumed to be taken with respect to squared error loss, unless stated otherwise.)

In regard to this simple problem, there was an error in Ferguson (1974) in stating that $\mu$ is finite a.s. if and only if $F_0$ has a finite first moment. That the *only if* part is false was pointed out in Doss and Sellke (1982), who obtain the following results on the tail behavior of $P$. Let $F(t) = P((-\infty, t])$.

**Theorem 3 (Doss and Sellke, 1982)**

If $F \in \mathcal{D}(MF_0)$, then

$$exp(-h_1(t)) \leq 1 - F(t) \leq exp(-h_2(t))$$

for sufficiently large $t$ a.s.

where $h_1(t) = 2\ log\ |\ log(1 - F_0(t))|/(1 - F(t))$ and $h_2(t) = \{(1 - F_0(t)) \times [log(1 - F_0(t))]^2\}^{-1}$.

As an example of this behavior, Yamato (1984) obtains the distribution of $\mu$ when $F_0$ is a Cauchy distribution.

**Theorem 4 (Yamato, 1984)**

If $F \in \mathcal{D}(MF_0)$ where $F_0$ is a Cauchy distribution, then the random variable $\mu = \int xdF(x)$ has the same Cauchy distribution.

In Cifarelli and Regazzini (1979) and in Hannum, Hollander and Langberg (1981), methods of finding the distribution of the mean of a Dirichlet process are reported.

A number of simple applications were presented in Ferguson (1973) such as estimating a distribution function or a median, mean or variance. In the two-sample problem of estimating $P(X > Y)$, the Mann-Whitney-Wilcoxon rank-sum statistic was seen to appear naturally. A number of other similar applications have appeared since that time. We mention a few.

Yamato (1975) obtains a Bayes estimate for $d(F, G) = \int (F(x) - G(x))^2 d(F(x) + G(x))/2$, based on independent samples from $F$ and $G$ which are given independent Dirichlet priors. Campbell and Hollander (1978) provide estimates of the rank of $X_1$ among $X_1,\ldots,X_n$ based on $X_1,\ldots,X_s$, $s < n$, when sampling from a Dirichlet process $F$. Hollander and Korwar (1980) find a Bayes estimate of $\Delta(x) = G^{-1}(F(x)) - x$, a measure of the difference between $F$ and $G$ at $x$, based on independent samples from each, with $G$ known and $F$ having a Dirichlet prior. Dalal and Phadia (1983) consider the problem of estimating $\tau = E\{sign((X - X')(Y - Y'))\}$, a measure of dependence for a bivariate distribution, where $(X, Y)$ and $(X', Y')$ are independent samples from the distribution. The Bayes estimate is computed using a Dirichlet prior in 2-dimensions, and Kendall's tau is seen to appear naturally. Zalkikar, Tiwari and Jammalamadaka (1986) obtain a Bayes estimate for $\Delta(F) = P(Z > X + Y)$, where $X$, $Y$, $Z$ are i.i.d. chosen from $F$, based on a sample from $F$, which is given a Dirichlet prior.

These are all examples of estimation problems. The difficulty of using Dirichlet priors in hypothesis testing problems was mentioned in Ferguson (1973), but Susarla and Phadia (1976) show how to test $H_0$: $F \leq F_0$ for a given distribution function $F_0$ using a Bayes approach. The idea is to replace the usual zero/one loss function with the smoother loss $L(F, a_0) = \int (F - F_0)^+ dW$ and

$L(F, a_1) = \int (F - F_1)^- dW$, where $a_0$ (resp. $a_1$) is the action accept (resp. reject) $H_0$, and $W$ is an arbitrary weighting measure. This idea also extends to multiple decision problems.

**Relation to Tailfree and Neutral Processes**

Let $\mathcal{P}_1, \mathcal{P}_2, \ldots$ be a sequence of finite measurable partitions of $\mathfrak{X}$ such that for all $n \geq 1$, $\mathcal{P}_{n+1}$ is a refinement of $\mathcal{P}_n$. We say that a random probability measure $P$ on $(\mathfrak{X}, \mathcal{A})$ is *tail-free w.r.t. the sequence* $\{\mathcal{P}_n\}$ if the sets of random variables $\{P(B \mid A)\colon A \in \mathcal{P}_{n-1}, B \in \mathcal{P}_n\}$ for $n = 1, 2, \ldots$ are independent. (Here $\mathcal{P}_0 = \{\mathfrak{X}\}$.) The notion of tailfree processes goes back to Freedman (1963), Fabius (1964) and Kraft (1964). In the dyadic tailfree process, each set of the partition $\mathcal{P}_n$ is cut into two pieces in the partition $\mathcal{P}_{n+1}$.

One drawback of using a tailfree process as a prior is that the behavior of the estimates depends on the choice of the partitions used to describe the process. This is true with one notable exception. The Dirichlet process is tailfree with respect to every sequence of partitions. Moreover, if a process is tailfree with respect to every sequence of partitions then it is either a Dirichlet process or a limit of Dirichlet processes or concentrated on two nonrandom points (Fabius, 1973).

There is another class of prior distributions that shares this property to a lesser degree, the processes neutral to the right, introduced by Doksum (1974). A random distribution function $F(t)$ on the real line is said to be *neutral to the right* if for every $m$ and $t_1 < t_2 < \ldots < t_m$, the random variables $1 - F(t_1)$, $(1 - F(t_2))/(1 - F(t_1)), \ldots, (1 - F(t_m))/(1 - F(t_{m-1}))$ are independent. This is equivalent to saying $Y(t) = -log(1 - F(t))$ has nonnegative independent increments. The basic theorem is:

**Theorem 5 (Doksum, 1974)**

If $F$ is neutral to the right, and if $X_1, \ldots, X_n$ is a sample from $F$, then the posterior distribution of $F$ given $X_1, \ldots, X_n$ is neutral to the right.

Basically, a process neutral to the right is tailfree with respect to every sequence of partitions $\{\mathcal{P}_n\}$ such that $\mathcal{P}_{n+1}$ is obtained from $\mathcal{P}_n$ by splitting the rightmost element, $(t_n, \infty)$ into two pieces, $(t_n, t_{n+1}]$, $(t_{n+1}, \infty)$. Thus, a Dirichlet process on the real line is neutral to the right, and neutral to the left, etc. Doksum (1974) conjectured that this property characterizes the Dirichlet process. This has been settled affirmatively.

**Theorem 6 (James and Mosimann, 1980)**

If $F$ is neutral to the right and neutral to the left, then $F$ is a Dirichlet process or a limit of Dirichlet processes or concentrated on two nonrandom points.

For another characterization of the Dirichlet process in terms of Johnson's sufficiency postulate or *learn-merge invariance*, see Böge and Möcks (1986).

## Applications of Mixtures of Dirichlet Processes

In the paper of Antoniak (1974), a number of Bayesian statistical problems with Dirichlet process priors were discussed whose solution involved posterior mixtures of Dirichlet processes, in particular empirical Bayes, bio-assay, regression, discrimination, and classification problems. The computational difficulties involved were such that Antoniak treated only very small size problems. Since then, Monte Carlo methods due to Kuo (1986) have been developed making Bayes solutions to these problems feasible. See Dalal (1978) and Dalal and Hall (1980) for a discussion of approximation of arbitrary random probability measures by mixtures of Dirichlets.

### 1. Bayes empirical Bayes

Consider first the Bayes empirical Bayes problem. In the usual empirical Bayes setting, it is assumed that unobservable parameters $\theta_j$, $j = 1,\ldots,n$ are taken independently from an unknown distribution G, and that associated with each $\theta_j$, a random variable $X_j$ is chosen independently from a distribution with density $f_j(x|\theta_j)$ for $j = 1,\ldots,n$. The problem is to estimate one or more of the $\theta_j$. Most procedures use $X_1,\ldots,X_n$ to obtain an estimate $G_n$ of $G$ first and then estimate $\theta_j$ as the Bayes estimate with respect to the prior $G_n$. In the Bayes approach to the empirical Bayes problem, a prior distribution is placed on $G$. Berry and Christensen (1979) take $G$ to be a Dirichlet process, $\mathfrak{D}(\alpha)$. Following Antoniak, the posterior distribution of $G$ is a mixture of Dirichlet processes with parameter $\alpha + \Sigma\delta(\theta_j)$ and mixing distribution $H(\underline{\theta} \mid \underline{X})$, the posterior distribution of the $\theta_j$ given the $X_j$, in symbols,

$$G \mid \underline{X} \in \int \mathfrak{D}\left(\alpha + \sum_{j=1}^{n} \delta(\theta_j)\right) dH(\underline{\theta} \mid \underline{X}). \tag{1}$$

In view of the computation difficulties, even in the simple case where $f_j(x|\theta)$ is a binomial distribution with probability of success $\theta$ and sample size depending on $j$, Berry and Christensen suggest a couple of rough approximations to the Bayes rule that are easy to evaluate.

Monte Carlo approximation of the exact Bayes estimate was considered by Kuo (1986a, 1986b). Let $H(\underline{\theta})$ denote the unconditional marginal distribution of $\underline{\theta}$,

$$dH(\underline{\theta}) = \prod_{j=1}^{n} (M + j - 1)^{-1}\left(\alpha + \sum_{i=1}^{j-1} \delta(\theta_i)\right)(d\theta_j), \tag{2}$$

as given in Blackwell and MacQueen (1973). Then, from a formula of Lo (1984) for the posterior distribution of $\underline{\theta}$ given $\underline{X}$,

$$dH(\underline{\theta}|\underline{X}) = \prod_{j=1}^{n} f_j(X_j|\theta_j)\,dH(\underline{\theta}) \Bigg/ \left[\int \prod_{j=1}^{n} f(X_j|\theta_j)\,dH(\underline{\theta})\right], \tag{3}$$

the exact Bayes estimate of $\theta_n$, say, may be written,

$$\hat{\theta}_n(\underline{X}) = \int \theta_n dH(\underline{\theta}|\underline{X}) = \frac{\int \theta_n \Pi f_j(X_j|\theta_j)\,dH(\theta_j)}{\int \Pi f_j(X_j|\theta_j)\,dH(\theta_j)}. \tag{4}$$

The obvious Monte Carlo method in which vectors $\underline{\theta}^1,\ldots,\underline{\theta}^N$ are generated i.i.d. from the distribution (2) and then used to approximate the two integrals in the right side of (4) does not work well. In the method of Kuo, Monte Carlo is used only to decide which of the $\theta_j$ are equal to which others according to (2). Then the $n$-dimensional integrals in the right side of (4) reduce to a product of 1-dimensional integrals $dF_0(\theta)$, which can often be integrated exactly, for example, if $F_0(\theta)$ is taken as a conjugate prior of $f(x|\theta)$.

## 2. Bayesian bio-assay

As another application, consider the bio-assay problem. Let $F(t)$ denote the probability of a positive response for a subject treated at dose level $t$. It is assumed that $F(t)$ increases with $t$. Suppose that $n_j$ subjects are treated at dose level $t_j$ and that $Y_j$ is the number of positive responses, $j = 1,\ldots,L$. It is assumed that the $Y_j$ are independent binomial variables with probability $F(t_j)$ of success. The problem is to estimate $F$. The Bayes approach to this problem goes back to Kraft and Van Eeden (1964) who use a dyadic tailfree process as a prior. Ramsey (1972) uses a Dirichlet process prior and obtains the modal estimates of F by maximizing the finite dimensional joint density of the posterior distribution. (This seems to be the first description of the Dirichlet process; unfortunately, it is in a problem where the posteriors are not Dirichlet.)

Bhattacharya (1981) develops a large sample procedure for approximating the finite-dimensional distributions of the posteriors as a normal mixture of Dirichlet distributions. Disch (1981) considers the problem of estimating quantiles of a potency curve with Dirichlet process priors, and avoids the difficult computational tasks by suggesting approximations similar to those made by Berry and Christensen in the empirical Bayes problem. However, the methods of Kuo may be applied to this problem as well. For related work, see Kuo (1983, 1988) and Ammann (1984).

### 3. Bayesian density estimation

Another application of mixtures of Dirichlet processes is to estimate a density, $f(x)$, based on a sample of size $n$ from $f$. Lo (1984) puts a prior on $f$, by writing $f(x) = \int K(x, u)dG(u)$ and letting $G$ have a Dirichlet process prior, $\mathcal{D}(\alpha)$. He obtains the posterior distribution of $G$ as a mixture of Dirichlet processes and uses this to obtain formulas for the Bayes estimate of $f$. One of his applications is to the two-parameter normal kernel $K = \phi(x|\mu,\sigma)$. This example was expanded in Ferguson (1983) who, using the representation of Sethuraman and Tiwari (1982), described $f(x)$ as a mixture of normal densities, $\Sigma P_j\phi(X|\mu_j,\sigma_j)$, where the $P_j$ are as in Theorem 2 and the $(\mu_j,\sigma_j)$ are a sample from the four-parameter conjugate prior for the normal. Kuo's method was seen to provide a simple and effective means of performing the computations for large data sets. The estimates are seen to provide evidence for two suggestions: (1) for using a variable kernel estimate with wider windows at the tails, and (2) for using shrinkage estimates on the observations, namely bringing observations in toward the center, proportional to their distance from the center. In the paper of Kumar and Tiwari (1989), Kuo's method is applied to estimating a mixture of exponential densities.

Gaussian processes may also be used to generate densities. In the approach of Leonard (1978), a density on the interval $(a, b)$ is written as $exp\{g(t)\}/\int exp\{g(x)\}dx$ where $g$ is a given Gaussian process. An alternate approach is provided by Thorburn (1986), in which the density is written as $exp\{g(t)\}$ where $g(t)$ is a Gaussian process conditional on $\int exp\{g(x)\}dx = 1$.

### Application to Censored Data and Reliability

An important extension of nonparametric Bayes theory is to the treatment of censored data. The problem of estimating an unknown cdf $F$ based on censored data is usually formulated as follows. Let $X_1,\ldots,X_n$ be a sample from $F$, and let the censoring points, $Y_1,\ldots,Y_n$, be random variables independent of the $X$'s. The observations are $Z_j = min(X_j, Y_j)$, and $d_j = I(X_j \leq Y_j)$, $j = 1,\ldots,n$, where $I(A)$ represents the indicator function of the set $A$. The problem is to estimate $F$ based on the observations. The usual nonparametric estimate is the product limit estimate, due to Kaplan and Meier (1958).

The first completely Bayes approach to this problem was made by Susarla and Van Ryzin (1976) who use a Dirichlet process as a prior for $F$. Let $u_1 < u_2 < \ldots < u_k$ be the distinct observations among $Z_1,\ldots,Z_n$; let $\lambda_j$ denote the number of censored observations at $u_j$; let $k(t)$ denote the number of $u_j \leq t$; and let $h_k$ be the number of $Z_j > u_k$.

### Theorem 7 (Susarla and Van Ryzin, 1976)

If $F \in \mathcal{D}(\alpha)$, then the posterior expectation of the survival function, $1 - F(t)$, given the observations is

$$E(1 - F(t)|data) = \frac{\alpha(t) + h_{k(t)}}{M + n} \prod_{j=1}^{k(t)} \frac{\alpha(u_j) + h_j + \lambda_j}{\alpha(u_j) + h_j} \tag{5}$$

where $\alpha(t) = \alpha(t, \infty)$, and $M = \alpha(\mathbb{R})$.

This estimate reduces to the Kaplan-Meier estimate as the prior information, $M$, goes to zero. If there are no censored observations, the product term vanishes and we get the Bayes estimator of Ferguson (1973). Blum and Susarla (1977) complemented this result by showing that the posterior distribution of $F$ given the data is a mixture of Dirichlet processes with specified transition and mixing measures.

This research was generalized to prior distributions neutral to the right by Ferguson and Phadia (1979). With Dirichlet process priors, the updating mechanism of going from prior to posterior is easy for uncensored observations and difficult for censored observations. For prior processes neutral to the right, it is the other way around. Thus, the generality provided by priors neutral to the right make them the natural priors to use for censoring problems. Also, it should be noted that the estimate in Theorem 1 does not depend on the distributions of the $Y_j$. Indeed, this should be the case when a Bayesian analysis is performed; in fact, as Ferguson and Phadia point out, the $Y_j$ may be considered as constants, allowing treatment of problems in which future $Y_j$ may depend upon past observations.

However, if $X_j$ and $Y_j$ are allowed to be dependent, the marginal distribution of $X$ may not be identifiable. Nevertheless, a Bayesian treatment of the problem is possible and has been carried out by Phadia and Susarla (1983), by assuming a Dirichlet process prior for the joint distribution of $(X, Y)$. They derive the Bayes estimate of the joint distribution, which of course need not be consistent. See also Arnold et al. (1984). Tsai (1986) adopts a different approach by taking the joint distribution of $(Z, d)$ to be a Dirichlet process on $\mathbb{R} \times \{0,1\}$, and making an independence-like assumption that makes the marginal distribution of $X$ identifiable, and the Bayes estimate of $F$ consistent. Since the marginal distribution of $F$ is not Dirichlet under this assumption, his resulting Bayes estimate is quite distinct from that of Susarla and Van Ryzin in the independent case.

For a review of the area up to 1980, see Phadia (1980b). For consistency of (5) and the product limit estimate, see Susarla and Van Ryzin (1978b) and Phadia and Van Ryzin (1980). For related results, see Gardiner and Susarla (1982, 1983), Colombo, Costantini and Jaarsma (1985), Rao and Tiwari (1985), Johnson and Christensen (1986) and Berliner and Hill (1988).

## 1. Application to reliability theory

A useful generalization of the gamma process for statistical problems has been introduced independently by Dykstra and Laud (1981) and Lo (1982). Given a nondecreasing left-continuous function $\alpha$ on $[0, \infty)$ with $\alpha(0) = 0$, $V(t)$ is said to be a gamma process with parameter $\alpha$ if $V(t)$ is a process with independent increments such that for all $t > 0$ the distribution of $V(t)$ is $\mathcal{G}(\alpha(t), 1)$, the gamma distribution with shape parameter $\alpha(t)$ and scale parameter 1. Given a nonnegative function $\beta$ on $[0, \infty)$, the weighted gamma process with parameters $\alpha$ and $\beta$, $\mathcal{G}(\alpha, \beta)$, is then defined as the process $r(t) = \int_{[0,t]} \beta(s)\,dV(s)$. Its elementary properties include

**Theorem 8 (Dykstra, Laud and Lo)**

If $r \in \mathcal{G}(\alpha, \beta)$, then $r$ is a process with independent increments, $E(r(t)) = \int_{[0,t]} \beta(s)\,d\alpha(s)$, and $Var(r(t)) = \int_{[0,t]} \beta^2(s)\,d\alpha(s)$.

Dykstra and Laud use this process (which they call an extended gamma process) as a prior distribution on the hazard rate function in nonparametric reliability problems; that is, they assume that the survival function, $S(t) = 1 - F(t)$, has the form $S(t) = exp\{-\int_{[0,t]} r(s)\,ds\}$, where $r \in \mathcal{G}(\alpha, \beta)$.

**Theorem 9 (Dykstra and Laud)**

If $r \in \mathcal{G}(\alpha, \beta)$, then $ES(t) = exp\{-\int_{[0,t]} log(1 + \beta(s)(t - s))\,d\alpha(s)\}$. If $X_1,\ldots,X_n$ is a sample from $S$, then the posterior distribution of $r$ given the censored data $X_1 \geq x_1,\ldots,X_n \geq x_n$ is $\mathcal{G}(\alpha, \beta^*)$, where $\beta^*(t) = \beta(t)/(1 + \beta(t)\Sigma_{i=1}^n (x_j - t)^+)$.

They also show that the distribution of $r$ given an uncensored sample is a mixture of weighted gamma processes, and examples are given showing the computational problems involved can be solved. This approach gives probability one to the absolutely continuous distributions, and Bayes estimates of the hazard rate and the cdf are derived.

Since in the above construction the gamma process has nondecreasing sample paths, the resulting survival distribution has increasing failure rate (IFR). Ammann (1984, 1985) puts this approach in a more general setting by representing the hazard rate as a function of the sample paths of nonnegative processes with independent increments which consist of an increasing component as well as a decreasing component. This results in a broad class of priors over a space of absolutely continuous distributions which contain IFR, DFR and U-shaped failure rate survival distributions. Ammann finds the posterior Laplace transforms of these processes based on data that may contain censored observations, and applies his approach to the competing risk model as well.

The Bayesian analysis discussed above may be extended to incorporate a covariate using the Cox proportional hazard model as was done by Kalbfleisch (1978). Independent observations $X_1,\ldots,X_n$ are made with respective covariate vectors $w_1,\ldots,w_n$ according to the survival distribution,

$$S(x \mid w) = S_0(x)^{\beta' w}$$

where $\beta$ is the vector of regression parameters, and $S_0(x)$ is the baseline survival distribution. While the main interest in covariate analysis centers around the estimation and hypothesis testing of $\beta$, considering $S_0(x)$ as a nuisance parameter, it is still of interest to estimate $S_0(x)$ by itself. Writing $S_0(x) = exp\{-\Lambda(x)\}$, Kalbfleisch takes $\Lambda(x)$ to have a gamma process prior, and carries out the estimation of $\beta$ by determining the marginal distribution of the observations as a function of $\beta$ with $S_0(x)$ integrated out. Thus, the treatment is semi-parametric and semi-Bayesian. This approach was generalized to allow $1 - S_0(x)$ to be an arbitrary process neutral to the right by Wild and Kalbfleisch (1981). For related results, see Padgett and Wei (1981) and Mazzuchi and Singpurwalla (1985).

**Empirical Bayes Estimation**

Bayesian methods have been found to be useful in the non-Bayesian treatment of empirical Bayes problems. Suppose we are at the $n + 1^{st}$ stage of an experiment, and information is available not only from the current stage but also from the $n$ previous stages. Let $F_1, F_2, \ldots, F_{n+1}$ be $n+1$ distributions on the real line, and for $j = 1, \ldots, n+1$, let $\underline{X}_j = (X_{j1}, \ldots, X_{jm_j})$ be a sample of size $m_j$ from $F_j$. As a prior, we assume that $F_1, \ldots, F_{n+1}$ are a sample from the Dirichlet $\mathcal{D}(\alpha)$ where $\alpha = MG_0$. We wish to estimate $F_{n+1}(t)$ with squared error loss.

$$L(F_{n+1}, \tilde{F}) = \int (F_{n+1}(x) - \tilde{F}(x))^2 dW(x) \tag{6}$$

for some finite measure $W$. If we know $M$ and $G_0$, this becomes a straightforward Bayes problem whose solution is

$$\tilde{F}_{n+1}(t) = q_{n+1} G_0(t) + (1 - q_{n+1}) \hat{F}_{n+1}(t) \tag{7}$$

where $q_j = M/(M + m_j)$ and $\hat{F}_j(t)$ is the sample distribution function based on $\underline{X}_j$. If $\alpha$ is unknown, we cannot use this estimate, but we may use $\underline{X}_1, \ldots, \underline{X}_n$ to help estimate $M$ and $G_0$.

Korwar and Hollander (1976) and Hollander and Korwar (1977) consider the case where $M$ is known and $G_0$ is unknown. They estimate $G_0(t)$ by the average of the sample distribution functions of $\underline{X}_1, \ldots, \underline{X}_n$, and propose the following empirical Bayes estimator of $F_{n+1}$:

$$H_{n+1}(t) = q_{n+1} \sum_{j=1}^{n} \hat{F}_j(t)/n + (1 - q_{n+1}) \hat{F}_{n+1}(t). \tag{8}$$

We say that this sequence of estimates is asymptotically optimal relative to a class of Dirichlet process priors if the Bayes risk of $H_{n+1}$ given $\alpha$, call it $r(\alpha, H_{n+1})$, converges to the Bayes risk of the Bayes estimate (7), call it $r(\alpha)$, whatever be $\alpha$ in the class. Since asymptotic optimality is a weak property, one wants rates of convergences. Korwar and Hollander prove:

**Theorem 10 (Hollander and Korwar, 1977)**

$$r(\alpha, H_{n+1}) = r(\alpha)\Big[1 + q_{n+1} \textstyle\sum_{j=1}^{n}(1 - q_j)^{-1}/n^2\Big]$$

When all the $m_j$ are equal, say to $m$, this reduces to $r(\alpha)(1 + M/(mn))$. Thus, $\{H_{n+1}\}$ is asymptotically optimal relative to the class of Dirichlet priors with fixed $M$, and the rate of convergence is $0(1/n)$. Hollander and Korwar also treat the empirical Bayes estimation of a mean, with similar results.

In their paper on testing hypotheses, Susarla and Phadia (1976) also consider the empirical Bayes extension of their problem using the method of Hollander and Korwar. In addition, they allow $M$ as well as $G_0$ to be unknown, and, using an estimate of $M$ based on the estimate of Korwar and Hollander (1973), exhibit an empirical Bayes estimate that is asymptotically optimal relative to the class of all Dirichlet priors. The extension of the Hollander and Korwar result to unknown $M$ was made in the equal sample size case by Zehnwirth (1981), using a new estimate of $M$. The estimate is as follows. Let $F_n$ denote the $F$-statistic in the one-way analysis of variance based on $\underline{X}_1,\ldots,\underline{X}_n$, ($F_n$ = ratio of the mean sum of squares between populations to the mean sum of squares within populations).

**Theorem 11 (Zehnwirth, 1981)**

$$m/(1 - F_n) \rightarrow M \text{ in probability as } n \rightarrow \infty.$$

The extension to empirical Bayes estimation of a distribution function based on censored data was made by Susarla and Van Ryzin (1978a) when all sample sizes, $m_j$, are 1, obtaining asymptotically optimal estimates at rate $0(1/n)$. Since the proposed estimate was not necessarily nondecreasing, Phadia (1980a) suggested using a simpler somewhat better estimate of $G_0$, which has the desirable property that the resulting empirical Bayes estimate is nondecreasing. This problem has also been treated by Ghorai (1981), taking a gamma process for $-log(1 - F(t))$ and obtaining asymptotically optimal estimates at rate $0(1/n)$.

In the uncensored case, Ghosh, Lahiri and Tiwari (1989) propose an empirical Bayes estimator of $F_{n+1}$ that uses both the past as well as the current data for estimating $G_0$. Their proposed estimator is given by (7) with $G_0$ replaced by

$$\hat{G}_0(t) = \sum_{j=1}^{n+1}(1 - q_j)\hat{F}_j(t) \Big/ \sum_{j=1}^{n+1}(1 - q_j). \tag{9}$$

Letting $\tilde{H}_{n+1}$ denote the resulting estimator, they derive the following result.

**Theorem 12 (Ghosh et al., 1989)**

$$r(\tilde{H}_{n+1}, \alpha) = r(\alpha)\left[1 + q_{n+1}\left(\sum_{j=1}^{n+1}(1 - q_j)\right)^{-1}\right]. \tag{10}$$

That this is a uniform improvement on the estimator in Theorem 10 is easily seen using Schwartz' inequality. Moreover Ghosh et al. have established the optimality of the weights used in (9), namely that the Bayes risk of $\tilde{H}_{n+1}$ is smaller than the Bayes risk of any other estimator that is a linear combination of the $\hat{F}_j$. In addition, they make a similar improvement to Zehnwirth's estimator of $M$ by allowing it to depend upon $\underline{X}_{n+1}$ as well as by allowing the sample sizes to differ.

We comment briefly on other papers in the area. Hollander and Korwar (1976) treats the empirical Bayes estimation of $P(X > Y)$ in a two-sample problem. Phadia and Susarla (1979) treat the same problem allowing right censored data, Ghorai and Susarla (1982) consider the empirical Bayes estimation of a density using Lo's estimate. Ghosh (1985) and Tiwari and Zalkikar (1985a, b) consider empirical Bayes estimation problems for general estimable parameters of degree one and two. Tiwari, Jammalamadaka and Zalkikar (1988) treat the empirical Bayes version of the paper of Gardiner and Susarla (1983).

**Random Symmetric Distributions; Problems of Consistency**

An extension of the family of Dirichlet processes to the family of Dirichlet invariant processes was introduced by Dalal (1979a). Let $\mathcal{G} = \{g_1,\ldots,g_k\}$ be a fixed finite group of measurable transformations from $\mathfrak{X}$ into itself. Let $\alpha$ be a $\mathcal{G}$-invariant finite non-null measure on $\mathfrak{X}$. A random probability measure $P$ on $(\mathfrak{X}, \mathcal{A})$ is said to be a *Dirichlet invariant process with parameter* $\alpha$, in symbols $P \in \mathcal{D}\mathcal{G}(\alpha)$, if $P$ is $\mathcal{G}$-invariant (surely) and if for every partition $(A_1,\ldots,A_m)$ of $\mathfrak{X}$ made up of measurable invariant sets, $(P(A_1),\ldots,P(A_m)) \in \mathcal{D}(\alpha(A_1),\ldots,\alpha(A_m))$. Dalal and others (Tiwari, 1988; Hannum and Hollander, 1983) give constructive definitions along the following lines. Let $P \in \mathcal{D}(\alpha)$ and define $P^*$ as $P^*(A) = (1/k)\Sigma_{g \in \mathcal{G}}P(gA)$. Then the distribution of $P^*$ depends only upon $\alpha^*$, where $\alpha^*(A) = (1/k)\Sigma_{g \in \mathcal{G}}\alpha(gA)$, and $P^* \in \mathcal{D}\mathcal{G}(\alpha^*)$.

When $\mathcal{G}$ consists of only the identity transformation, $\mathcal{D}\mathcal{G}(\alpha)$ corresponds to the usual Dirichlet process, $\mathcal{D}(\alpha)$. When $\mathcal{G}$ is generated by $g(x) = -x$, $\mathcal{D}\mathcal{G}(\alpha)$ gives probability one to distributions that are symmetric about zero. Dalal (1979a) derives several properties of the Dirichlet invariant process and applies the theory to the estimation of a distribution function known to be symmetric about a known point, $\theta$. The analysis is extended in Dalal (1979b) to the case where $\theta$ is unknown but given a prior distribution $\nu$ independent of $P$. See Dalal (1980) for an expository article on these problems.

An important analysis of these results, both theoretically and practically, has been given by Diaconis and Freedman (1986a, b). Such estimates may not be consistent throughout the support of the prior, as detailed in Theorem 13 below. The first example of an inconsistent Bayes estimate was given by Freedman (1963). A simple example of this phenomenon, Ferguson (1973), may be described as follows.

Let the prior distribution of $F$ be the mixture, $F = p_0 H + (1 - p_0)\mathfrak{D}(\alpha)$, where $p_0$, the prior probability of $H$, is 1/2, where $H$ is the uniform distribution on the interval (0, 1), and where $\alpha = MH$ with $M = 1$. The support of $F$ is the set of all distributions on [0, 1]. The distribution of the distinct observations among a sample $X_1,\ldots,X_n$ from $F$ is the same when $F = \mathfrak{D}(\alpha)$ as when $F = H$. Thus, as long as the observations are distinct, the posterior distribution of $F$ given $X_1,\ldots,X_n$ is $p_n H + (1 - p_n)\mathfrak{D}(\alpha + \Sigma\delta(X_j))$, where $p_n$, the posterior probability of $H$, is easily computed to be $p_n = n!/(n! + 1)$. If ever two observations are exactly equal, then the possibility of $H$ disappears and $F$ has the posterior distribution $\mathfrak{D}(\alpha + \Sigma\delta(X_j))$. Now, suppose that the true distribution is continuous on (0, 1). No matter how non-uniform this distribution may be, the Bayes estimate of $F$ converges to $\mathfrak{U}(0, 1)$.

Freedman and Diaconis (1983) have a positive result along the lines of this example: If $F$ is a mixture of $\mathfrak{D}(\alpha_j)$ with $\alpha_j = M_j F_j$, and if the $M_j$ are bounded, then the Bayes estimate of $F$ is consistent. In the example above, one can think of $H$ as a Dirichlet process with $M = \infty$, so although $F$ is a mixture of Dirichlets, the $M_j$ are not bounded. In Dalal's model, even if the true distribution is symmetric, the Bayes estimate may oscillate indefinitely between two wrong values.

**Theorem 13 (Diaconis and Freedman, 1986a, b)**

Let $\theta$ and $F$ be independent, with $\theta$ having a standard normal distribution, and $F \in \mathfrak{DG}(\alpha)$ symmetric about zero, where $\alpha = MF_0$ with $F_0$ the standard Cauchy distribution. Then there exists a symmetric density, $h(x)$, with a maximum at zero and bounded support, such that if the true distribution of the $X_j$ has density $h$, then the Bayes estimate of $\theta$ does not converge.

Doss (1984) provides a deep extension of the analysis of these problems from symmetric Dirichlet priors to symmetric priors neutral to the right. Doss (1985a, b) considers the problem of estimating a median in a different nonparametric Bayes framework. Let $F(x)$ be a distribution function with median zero, let $\theta$ be a real number, and let $X_1,\ldots,X_n$ be a sample from $F(x - \theta)$. To place a prior distribution on $F$ that chooses median zero distributions with probability one, let $\alpha$ be a finite non-null measure, written as $\alpha = MF_0$, where $F_0$ is a distribution function with median zero, and suppose for simplicity that $F_0$ has no mass at zero. Let $\alpha_-$ and $\alpha_+$ denote the restrictions of $\alpha$ to $(-\infty, 0)$ and $(0, \infty)$ respectively. Choose $F_-$ and $F_+$ independently from $\mathfrak{D}(\alpha_-)$ and $\mathfrak{D}(\alpha_+)$

respectively, and let $F(t) = (F_-(t) + F_+(t))/2$. Thus, $F$ is a random distribution function such that $F(0) = 1/2$; denote the distribution of $F$ by $\mathcal{D}^*(\alpha)$.

**Theorem 14 (Doss, 1985a)**

Let $\theta$ and $F$ be independent, with $\theta \in \nu$ and $F \in \mathcal{D}^*(\alpha)$, and assume that $F_0$ has continuous density $f_0$. Given $\theta$ and $F$, let $\underline{X} = (X_1,\ldots,X_n)$ be a sample from $F(x - \theta)$. Then the posterior distribution of $\theta$ given $\underline{X}$ is

$$d\nu(\theta|\underline{X}) = c(\underline{X})[\Pi^* f_0(X_i - \theta)]\ M(\underline{X},\ \theta)\, d\nu(\theta),$$

where $M(\underline{X},\ \theta)^{-1} = \Gamma(M/2 + n\hat{F}_n(\theta))\Gamma(M/2 + n(1 - \hat{F}_n(\theta)))$, $\hat{F}_n$ is the empirical distribution function of $\underline{X}$, $\Pi^*$ represents the product over the distinct $X_i$, and $c(\underline{X})$ is a normalizing constant.

Doss shows that if the true distribution of the $X_j$ is discrete, the Bayes estimate of $\theta$ is consistent. However, if it is continuous, then the Bayes estimate can converge to a wrong value, it can oscillate indefinitely between two wrong values, or the set of its limit points can be dense in $\mathbb{R}$.

Hannum and Hollander (1983) have derived the Bayes risk of Dalal's (1979a) estimate of the distribution function under $\mathcal{DG}(\alpha)$, and have compared it to the risk of Ferguson's (1973) estimator under $\mathcal{D}(\alpha)$. This enables them to (i) assess the savings in risk obtained by incorporating known symmetry structure in the model, and (ii) provide information about the robustness of Ferguson's estimator against a prior for which it is not Bayes. Yamato (1986, 1987) and Tiwari (1988) used the Dirichlet invariant process prior to derive the Bayes estimator of estimable parameters of an arbitrary degree.

**Other Applications**

Our survey is by no means complete. We mention a few other selected results and applications in this final section. Binder (1982) considers finite population models in which a population $\{Y_1,\ldots,Y_N\}$ consists of a sample from $F \in \mathcal{D}(\alpha)$. A sample, $y_1,\ldots,y_n$, is then taken from $\{Y_1,\ldots,Y_N\}$, and the Bayes estimate of $\Sigma Y_j$ is derived. The asymptotic distributions are found in Lo (1986). Problems of finding confidence bounds for a distribution function have been considered by Breth (1978), who finds recursive methods for computing $\mathcal{P}(u_j \leq F(t_j) \leq v_j$ for $j = 1,\ldots, m)$ for fixed numbers $\{u_j\}$, $\{v_j\}$ and $\{t_j\}$ when $F$ is a Dirichlet process. In a continuation paper, Breth (1979) applies the method to finding confidence intervals for quantiles and the mean, and also treats Bayesian tolerance intervals. Tamura (1988) applies Dirichlet process methods to auditing problems.

**1. Linear Bayes estimation**

The useful idea of restricting attention to a linear space of estimates in Bayesian nonparametric problems is due to Goldstein (1975a, b). Such estimates may require less knowledge of the prior and be much easier to compute than Bayes estimates without much loss of efficiency. As an example, consider the problem of estimating a mean $\mu = \int x dP(x)$ within the class of linear functions, $\hat{\mu} = a + \Sigma b_j X_j$. The Bayes solution is

$$\hat{\mu} = \frac{M}{M+n}\mu_0 + \frac{n}{M+n}\bar{X}_n, \text{ where}$$

$$\mu_0 = E(\mu) \quad \text{and} \quad M = \frac{E(\sigma^2)}{E(\mu^2) - (\mu_0)^2}.$$

Here, $\sigma^2 = \int x^2 dP(x) - \mu^2$ is the variance of the random distribution. This estimate is formally identical to the Bayes estimate with the Dirichlet prior, Theorem 1, with however a new interpretation for the parameter $M$. In addition, the only information needed to be elicited from the prior are the three quantities, $E(\mu)$, $E(\mu^2)$ and $E(\sigma^2)$. These ideas were further developed by Zehnwirth (1985) in treating estimation with censored data, by Poli (1985), who finds the best linear predictor in a multivariate regression model and specializes to a Dirichlet prior and to a normal/Wishart mixture of Dirichlets, and by Kuo (1988) in estimating the potency curve in Bayesian bio-assay.

## 2. Sequential problems

A number of papers treat sequential nonparametric problems from a Bayesian viewpoint. Hall (1976, 1977) in treating sequential search problems with random overlook probabilities allows the distributions of the overlook probabilities to be Dirichlet or a mixture of Dirichlet. Ferguson (1982) discusses $k$-stage lookahead rules and modified rules in some nonparametric sequential estimation problems with Dirichlet priors. Clayton and Berry (1985) treat the finite horizon one-armed bandit with the unknown arm producing observations from a Dirichlet process. In a sequential testing problem, Clayton (1985) assumes that in sampling from $F \in \mathcal{D}(\alpha)$, the payoff if you stop at $n$ is $max(E(X|X_1,\ldots,X_n), \nu) - nc$, where $\nu$ and $c > 0$ are constants. He shows that the optimal stopping rule is bounded if the support of $\alpha$ is bounded, and he conjectures that this is true even if the support of $\alpha$ is unbounded. Christensen (1986) obtains a similar result for the problem of sampling without recall from a distribution $F \in \mathcal{D}(\alpha)$ and constant cost of observation. Betró and Schoen (1987) consider the problem of sampling with recall and constant cost from a distribution $F$ assumed to be a simple homogeneous process neutral to the right.

## 3. Point processes

Lo (1982) considers the problem of estimation of the intensity measure $\gamma$ of a nonhomogeneous Poisson point process based on a random sample from this process. He shows that if the prior distribution for $\gamma$ is a weighted gamma distribution $\mathcal{G}(\alpha, \beta)$, then given a sample $N_1,\ldots,N_n$ of $n$ functions from this process, the posterior distribution of $\gamma$ is again gamma, $\mathcal{G}(\alpha + \Sigma N_j, \beta/(n\beta + 1))$. Lo also shows that the posterior process converges weakly to the Brownian bridge.

Another paper of Lo (1981) describes an application to shock models and wear processes. A device is subject to shocks occurring randomly at times according to a homogeneous Poisson point process $N(t)$ with intensity $\gamma$. The $i^{th}$ shock causes a random amount $X_i$ of damage, assumed to be i.i.d. $F$ on $[0, \infty)$. For the prior distribution, $\gamma$ and $F$ are chosen to be independent, with $\gamma \in$ a gamma distribution $\mathcal{G}(\lambda, \theta)$, and $F \in \mathcal{D}(\alpha)$. In the posterior distribution based on a single observation of $N$ up to time $T$, $\gamma$ and $F$ are still independent, with $\gamma \in \mathcal{G}(\lambda + N(T), \theta + T)$ and $F \in \mathcal{D}(\alpha + N)$. This readily yields Bayes estimates of $\gamma$ and $F$.

Johnson, Susarla and Van Ryzin (1979) present an application to the Bellman-Harris age-dependent branching process. Each individual $x$ born has a random length of life $\lambda_x$ and reproduces at death a random number $\xi_x$ of offspring, where the $(\lambda_x, \xi_x)$ are i.i.d. from $G \times P$. The prior distribution of $G$ and $P$ are taken to be independent Dirichlet processes with parameters $\alpha_1$ and $\alpha_2$, and Bayes estimates of $G$ and $P$ are developed based on an observation of the process through time $T$ starting with one individual.

**References**

Ammann, L. P. (1984): Bayesian nonparametric inference for quantal response data, *Ann. Statist.* 12, 636-645.

Ammann, L. P. (1985): Conditional Laplace transforms for Bayesian nonparametric inference in reliability theory, *Stoch. Proc. and Their Appl.* 20, 197-212.

Antoniak, C. (1974): Mixtures of Dirichlet processes with application to Bayesian nonparametric problems, *Ann. Statist.* 2, 1152-1174.

Arnold, B. C., Brockett, P. L., Torrez, W. and Wright, A. L. (1984): On the inconsistency of Bayesian non-parametric estimators in competing risk/multiple decrement models, *Insurance: Mathematics and Economics* 3, 49-55.

Basu, D. and Tiwari, R. C. (1982): A note on the Dirichlet process, *Statistics and Probability: Essays in Honor of C. R. Rao*, Kallianpur, Krishnaiah and Ghosh, Eds., 89-103.

Berk, R. H. and Savage, I. R. (1979): Dirichlet processes produce discrete measures: an elementary proof, *Contributions to Statistics. Jaroslav Hajek Memorial Volume*, Academia, North Holland, Prague, 25-31.

Berliner, L. M. and Hill, B. M. (1988): Bayesian nonparametric survival analysis, *J. Amer. Statist. Assoc.* 83, 772-779.

Berry, D. A. and Christensen, R. (1979): Empirical Bayes estimation of a binomial parameter via mixtures of Dirichlet process, *Ann. Statist.* 7, 558-568.

Betró, B. and Schoen, F. (1987): Sequential stopping rules for the multistart algorithm in global optimization, *Math. Programming* 38, 271-286.

Bhattacharya, P.K. (1981): Posterior distribution of a Dirichlet process from quantal response data, *Ann. Statist.* 9, 803-811.

Binder, D. A. (1982): Non-parametric Bayesian models for samples from finite populations, *J. Roy. Statist. Soc. B* 44, 388-393.

Blackwell, D. (1973): Discreteness of Ferguson selections, *Ann. Statist.* 1, 356-358.

Blackwell, D. and MacQueen, J. B. (1973): Ferguson distributions via Polya urn schemes, *Ann. Statist.* 1, 353-355.

Blum, J. and Susarla, V. (1977): On the posterior distribution of a Dirichlet process given randomly right censored observations, *Stoch. Processes Appl.* 5, 207-211.

Böge, W. and Möcks, J. (1986): Learn-merge invariance of priors: a characterization of Dirichlet distributions and processes, *J. Mult. Anal.* 18, 83-92.

Breth, M. (1978): Bayesian confidence bands for a distribution function, *Ann. Statist.* 6, 649-657.

Breth, M. (1979): Nonparametric Bayesian interval estimation, *Biometrika* 66, 641-644.

Campbell, G. and Hollander, M. (1978): Rank order estimation with the Dirichlet prior, *Ann. Statist.* 6, 142-153.

Christensen, R. (1986): Finite stopping in sequential sampling without recall from a Dirichlet process, *Ann. Statist.* 14, 275-282.

Cifarelli, D. M. and Regazzini, E. (1979): Considerazioni generali sull'impostazione bayesiana di problemi non parametrici, *Rivista di Matematica per le Scienze Economiche e Socialli* 2 Part I 39- 52, Part II 95-111.

Clayton, M. K. (1985): A Bayesian nonparametric sequential test for the mean of a population, *Ann. Statist.* 13, 1129-1139.

Clayton, M. K. and Berry, D. A. (1985): Bayesian nonparametric bandits, *Ann. Statist.* 13, 1523-1534.

Colombo, A. G., Costantini, D. and Jaarsma, R. J. (1985): Bayes nonparametric estimation of time-dependent failure rate, *IEEE Trans. Reliability* R-34, 109-112.

Dalal, S. R. (1978): A note on the adequacy of mixtures of Dirichlet processes, *Sankhya Ser. A* 40, 185-191.

Dalal, S. R. (1979a): Dirichlet invariant processes and applications to nonparametric estimation of symmetric distribution functions, *Stoch. Proc. Appl.* 9, 99-107.

Dalal, S. R. (1979b): Nonparametric and robust Bayes estimation of location, in *Optimizing Methods in Statistics*, ed. J. Rustagi, Academic Press Inc., 141-166.

Dalal, S. R. (1980): Bayesian nonparametric theory, in *Bayesian Statistics*, eds. J. M. Bernardo et al., University Press, Valencia, Spain, 521-534.

Dalal, S. R. and Hall, G. J., Jr. (1980): On approximating parametric Bayes models by nonparametric Bayes models, *Ann. Statist.* 8, 664-672.

Dalal, S. R. and Phadia, E. G. (1983): Nonparametric Bayes inference for concordance in bivariate distributions, *Comm. Statist. Theor. Meth.* 12, 947-963.

Diaconis, P. and Freedman, D. (1986a): On the consistency of Bayes estimates, *Ann. Statist.* 14, 1-26.

Diaconis, P. and Freedman, D. (1986b): On inconsistent Bayes estimates of location, *Ann. Statist.* 14, 68-87.

Disch, D. (1981): Bayesian nonparametric inference for effective doses in a quantal response experiment, *Biometrics* 37, 713-722.

Doksum, K. (1974): Tailfree and neutral random probabilities and their posterior distributions, *Ann. Prob.* 2, 183-201.

Doss, H. (1984): Bayesian estimation in the symmetric location problem, *Z. W.* 68, 127-147.

Doss, H. (1985a): Bayesian nonparametric estimation of the median; Part I: Computation of the estimates, *Ann. Statist.* 13, 1432-1444.

Doss, H. (1985b): Bayesian nonparametric estimation of the median; Part II: Asymptotic properties of the estimates, *Ann. Statist.* 13, 1445-1464.

Doss, H. and Sellke, T. (1982): The tails of probabilities chosen from a Dirichlet prior, *Ann. Statist.* 10, 1302-1305.

Dykstra, R. L. and Laud, P. (1981): A Bayesian nonparametric approach to reliability, *Ann. Statist.* 9, 356-367.

Fabius, J. (1964): Asymptotic behavior of Bayes estimates, *Ann. Math. Statist.* 35, 846-856.

Fabius, J. (1973): Two characterizations of the Dirichlet distributions, *Ann. Statist.* 1, 583-587.

Ferguson, T. S. (1973): A Bayesian analysis of some nonparametric problems, *Ann. Statist.* 1, 209-230.

Ferguson, T. S. (1974): Prior distributions on spaces of probability measures, *Ann. Statist.* 2, 615-629.

Ferguson, T. S. and Phadia, E. G. (1979): Bayesian nonparametric estimation based on censored data, *Ann. Statist.* 7, 163-186.

Ferguson, T. S. (1982): Sequential estimation with Dirichlet process priors, in *Statistical Decision Theory and Related Topics III*, Vol. 1, eds. S. Gupta and J. Berger, 385-401.

Ferguson, T. S. (1983): Bayesian density estimation by mixtures of normal distributions, in *Recent Advances in Statistics*, eds. H. Rizvi and Rustagi, Academic Press, New York, 287-302.

Freedman, D. (1973): On the asymptotic behavior of Bayes estimates in the discrete case, *Ann. Math. Statist.* 34, 1386-1403.

Freedman, D. and Diaconis, P. (1983): On inconsistent Bayes estimates in the discrete case, *Ann. Statist.* 11, 1109-1118.

Gardiner, J. C. and Susarla, V. (1982): A nonparametric estimator of the survival function under progressive censoring, *Survival Analysis, IMS Lecture Notes-Monograph Series* Vol. 2, eds. J. Crowley and R. A. Johnson, 26-40.

Gardiner, J. C. and Susarla, V. (1983): Weak convergence of a Bayesian nonparametric estimator of the survival function under progressive censoring, *Statistics and Decisions* 1, 257-263.

Ghorai, J. (1981): Empirical Bayes estimation of a distribution function with a gamma process prior, *Comm. Statist. Theory Meth.* A10(12), 1239-1248.

Ghorai, J. K. and Susarla, V. (1982): Empirical Bayes estimation of probability density functions with Dirichlet process prior, Probability and Statistical Inference, *Proc. of the 2nd Pannonian Symp. on Math. Stat.*, eds. Grossmann et al., 101-114.

Ghosh, M. (1985): Nonparametric empirical Bayes estimation of certain functionals, *Comm. Statist. Theory Meth.* 14(9), 2081-2094.

Ghosh, M., Lahari, P. and Tiwari, R. C. (1989): Nonparametric empirical Bayes estimation of the distribution function and the mean, *Comm. Statist. Theory Meth.* 18(1), 121-146.

Goldstein, M. (1975a): Approximate Bayes solutions to some nonparametric problems, *Ann. Statist.* 3, 512-517.

Goldstein, M. (1975b): A note on some Bayesian nonparametric estimates, *Ann. Statist.* 3, 736-740.

Hall, G. J., Jr. (1976): Sequential search with random overlook probabilities, *Ann. Statist.* 4, 807-816.

Hall, G. J., Jr. (1977): Strongly optimal policies in sequential search with random overlook probabilities, *Ann. Statist.* 5, 124-135.

Hannum, R. C., Hollander, M. and Langberg, N. A. (1981): Distributional results for random functionals of a Dirichlet process, *Ann. Prob.* 9, 665-670.

Hannum, R. C. and Hollander, M. (1983): Robustness of Ferguson's Bayes estimator of a distribution function, *Ann. Statist.* 11, 632-639, 1267.

Hollander, M. and Korwar, R. M. (1976): Nonparametric empirical Bayes estimation of the probability that X $\leq$ Y, *Commun. Statist. Theor. Meth.* A5(14), 1369-1383.

Hollander, M. and Korwar, R. M. (1977): Nonparametric estimation of distribution functions, in *Thy. and Appl. of Reliability 1*, Academic Press, New York, 85-107.

Hollander, M. and Korwar, R. M. (1980): Nonparametric Bayesian estimation of the horizontal distance between two populations, *Colloq. Math. Soc. Janos Bolyai 32. Nonparametric Statistical Inference*, Budapest, 409-415.

James, I. R. and Mosimann, J. E. (1980): A new characterization of the Dirichlet distribution through neutrality, *Ann. Statist.* 8, 183-189.

Johnson, R. A., Susarla, V. and Van Ryzin, J. (1979): Bayesian non-parametric estimation for age-dependent branching processes, *Stoch. Proc. & Appl.* 9, 307-318.

Johnson, W. and Christensen, R. (1986): Bayesian nonparametric survival analysis for grouped data, *Can. J. Statist.* 14, 307-314.

Kalbfleisch, J. D. (1978): Nonparametric Bayesian analysis of survival time data, *J. Royal Statist. Soc.* B40, 214-222.

Kaplan, E. L. and Meier, P. (1958): Nonparametric estimation from incomplete observations, *J. Amer. Statist. Assoc.* 53, 457-481.

Korwar R. M. and Hollander, M. (1973): Contributions to the theory of Dirichlet processes, *Ann. Prob.* 1, 705-711.

Korwar, R. M. and Hollander, M. (1976): Empirical Bayes estimation of a distribution function, *Ann. Statist.* 4, 581-588.

Kraft, C. H. (1964): A class of distribution function processes which have derivatives, *J. Appl. Prob.* 1, 385-388.

Kraft, C. H. and van Eeden, C. (1964): Bayesian bio-assay, *Ann. Math. Statist.* 35, 886-890.

Kumar, S. and Tiwari, R. C. (1989): Bayes reliability estimation under a random environment governed by a Dirichlet process, *IEEE Transactions on Reliability*, Aug.

Kuo, L. (1983): Bayesian bio-assay design, *Ann. Statist.* 11, 886-895.

Kuo, L. (1986a): A note on Bayes empirical Bayes estimation by means of Dirichlet processes, *Statist. Prob. Letters* 4, 145-150.

Kuo, L. (1986b): Computations of mixtures of Dirichlet processes, *SIAM J. Sci. Statist. Comput.* 7, 60-71.

Kuo, L. (1988): Linear Bayes estimators of the potency curve in bioassay, *Biometrika* 75, 91-96.

Leonard, T. (1978): Density estimation, stochastic processes and prior information, *J. R. Statist. Soc.* B40, 113-146.

Lo, A. (1981): Bayesian nonparametric statistical inference for shock models and wear processes, *Scand. J. Statist.* 8, 237-242.

Lo, A. (1982): Bayesian nonparametric statistical inference for Poisson point processes, *Z. Wahrsch. verw. Gebiete* 59, 55-66.

Lo, A. (1983): Weak convergence for Dirichlet processes, *Sankhya* 45, 105-111.

Lo, A. (1984): On a class of Bayesian nonparametric estimates: I. Density estimates, *Ann. Statist.* 12, 351-357.

Lo, A. (1986): Bayesian statistical inference for sampling a finite population, *Ann. Statist.* 14, 1226-1233.

Mazzuchi, T. A. and Singpurwalla, N. D. (1985): A Bayesian approach to inference for monotone failure rates, *Statist. & Probab. Letters* 3, 135-141.

Padgett, W. J. and Wei, L. J. (1981): A Bayesian nonparametric estimator of survival probability assuming increasing failure rate, *Comm. Statist. Theor. Meth.* A10(1), 49-63.

Phadia, E. G. (1980a): A note on empirical Bayes estimation of a distribution function based on censored data, *Ann. Statist.* 8, 226-229.

Phadia, E. G. (1980b): Nonparametric Bayesian inference based on censored data–an overview, *Coll. Math. Societ. Janos Bolyai 32. Nonparametric Statistical Inference*, Budapest, 667-686.

Phadia, E. G. and Susarla, V. (1979): An empirical Bayes approach to two-sample problems with censored data, *Comm. Statist. Theor. Meth.* A8(13), 1327-1351.

Phadia, E. G. and Susarla, V. (1983): Nonparametric Bayesian estimation of a survival curve with dependent censoring mechanism, *Ann. Inst. Statist. Math.* 35A, 389-400.

Phadia, E. G. and Van Ryzin, J. (1980): A note on convergence rates for the product limit estimator, *Ann. Statist.* 8, 673-678.

Poli, I. (1985): A Bayesian non-parametric estimate for multivariate regression, *J. Econometrics* 28, 171-182.

Ramsey, F. (1972): A Bayesian approach to bio-assay, *Biometrics* 28, 841-848.

Rao, J. S. and Tiwari, R. C. (1985): Estimation of survival function and failure rate, *Statistics* 16, 535-540.

Sethuraman, J. and Tiwari, R. C. (1982): Convergence of Dirichlet measures and the interpretation of their parameter, in *Statistical Decision Theory and Related Topics III* 2, eds. Gupta and Berger, Academic Press, New York, 305-315.

Susarla, V. and Phadia, E. G. (1976): Empirical Bayes testing of a distribution function with Dirichlet process priors, *Commun. Statist. Theor. Meth.* A5(5), 455-469.

Susarla, V. and Van Ryzin, J. (1976): Nonparametric Bayesian estimation of survival curves from incomplete observations, *J. Amer. Statist. Soc.* 71, 897-902.

Susarla, V. and Van Ryzin, J. (1978a): Empirical Bayes estimation of a distribution (survival) function from right censored observations, *Ann. Statist.* 6, 740-754.

Susarla, V. and Van Ryzin, J. (1978b): Large sample theory for a Bayesian nonparametric survival curve estimator based on censored samples, *Ann. Statist.* 6, 755-768. Addendum (1980) *Ann. Statist.* 8, 693.

Tamura, H. (1988): Estimation of rare errors using expert judgement, *Biometrika* 75, 1-9.

Thorburn, D. (1986): A Bayesian approach to density estimation, *Biometrika* 73, 65-75.

Tiwari, R. C. (1988): Convergence of Dirichlet invariant measures and the limits of Bayes estimates, *Commun. Statist. Theory Meth.* A17(12), 375-393.

Tiwari, R. C. and Zalkikar, J. N. (1985a): Empirical Bayes estimation of functionals of unknown probability measures, *Comm. Statist. Theor. Meth.* 14, 2963-2996.

Tiwari, R. C. and Zalkikar, J. N. (1985b): Empirical Bayes estimate of certain estimable parameters of degree two, *Calcutta Statist. Assoc. Bull.* 34, 179-188.

Tiwari, R. C., Jammalamadaka, S. R. and Zalkikar, J. N. (1988): Bayes and empirical Bayes estimation of survival function under progressive censoring, *Comm. Statist. Theory Meth.* A17(10): 3591-3606.

Tsai, W.-Y. (1986): Estimation of survival curves from dependent censorship models via a generalized self-consistent property with nonparametric Bayesian estimation application, *Ann. Statist.* 14, 238-249.

Wild, C. J. and Kalbfleisch, J. D. (1981): A note on a paper by Ferguson and Phadia, *Ann. Statist.* 9, 1061-1065.

Yamato, H. (1975): A Bayesian estimation of a measure of the difference between two continuous distributions, *Rep. Fac. Sci. Kagoshima University* 8, 29-38.

Yamato, H. (1984): Characteristic functions of means of distributions chosen from a Dirichlet process, *Ann. Probab.* 12, 262-267.

Yamato, H. (1986): Bayes estimates of estimable parameters with a Dirichlet invariant process, *Comm. Statist. Theor. Meth.* 15(8), 2383-2390.

Yamato, H. (1987): Nonparametric Bayes estimates of estimable parameters with a Dirichlet invariant process and invariant U-statistics, *Comm. Statist. Theor. Meth.* 16, 525-543.

Zalkikar, J. N., Tiwari, R. C. and Jammalamadaka, S. R. (1986): Bayes and empirical Bayes estimation of the probability that $Z > X + Y$, *Comm. Statist. Theor. Meth.* 15(10), 3079-3101.

Zehnwirth, B. (1981): A note on the asymptotic optimality of the empirical Bayes distribution function, *Ann. Statist.* 9, 221-224.

Zehnwirth, B. (1985): Nonparametric linear Bayes estimation of survival curves from incomplete observations, *Comm. Statist. Theor. Meth.* 14(8), 1769-1778.

# HIERARCHICAL AND EMPIRICAL BAYES MULTIVARIATE ESTIMATION

Malay Ghosh*, Department of Statistics, University of Florida

**Abstract**

This article reviews and unifies the hierarchical and empirical Bayes approach for estimating the multivariate normal mean. Both the ANOVA and the regression models are considered.

## Introduction

Empirical and hierarchical Bayes methods are becoming increasingly popular in statistics, especially in the context of simultaneous estimation of several parameters. For example, agencies of the Federal Government have been involved in obtaining estimates of per capita income, unemployment rates, crop yields and so forth simultaneously for several state and local government areas. In such situations, quite often estimates of certain area means, or simultaneous estimates of several area means can be improved by incorporating information from similar neighboring areas. Examples of this type are especially suitable for empirical Bayes (EB) analysis. As described in Berger (1985), an EB scenario is one in which known relationships among the coordinates of a parameter vector, say $\underset{\sim}{\theta} = (\theta_1,\ldots,\theta_p)^T$ allow use of the data to estimate some features of the prior distribution. For example, one may have reason to believe that the $\theta_i$'s are iid from a prior $\pi_0(\lambda)$, where $\pi_0$ is structurally known except possibly for some unknown parameter $\lambda$. A *parametric empirical Bayes* (EB) procedure is one where $\lambda$ is estimated from the marginal distribution of the observations.

Closely related to the EB procedure is the hierarchical Bayes (HB) procedure which models the prior distribution in stages. In the first stage, conditional on $\Lambda = \lambda$, $\theta_i$'s are iid with a prior $\pi_0(\lambda)$. In the second stage, a prior distribution (often improper) is assigned to $\Lambda$. This is an example of a two stage prior. The idea can be generalized to multistage priors, but that will not be pursued in this article.

It is apparent that both the EB and the HB procedures recognize the uncertainty in the prior information, but whereas the HB procedure models the uncertainty in the prior information by assigning a distribution (often *noninformative* or *improper*) to the prior parameters (usually called *hyperparameters*), the EB procedure attempts to estimate the unknown hyperparameters, typically by some classical method like the method of moments, method of maximum likelihood etc., and use the resulting estimated priors for inferential purposes. It turns out that the two methods can quite often

*This paper is dedicated to Professor D. Basu on the occasion of his 65th birthday. The research is partially supported by NSF Grant Numbers DMS 8701814 and DMS 8901334.

lead to comparable results, especially in the context of point estimation. This will be revealed in some of the examples appearing in the later sections. However, when it comes to the question of measuring the standard errors associated with these estimators, the HB method has a clear edge over a naive EB method. Whereas, there are no clear cut measures of standard errors associated with EB point estimators, the same is not true with HB estimators. To be precise, if one estimates the parameter of interest by its posterior mean, then a very natural estimate of the risk associated with this estimator is its posterior variance. Estimates of the standard errors associated with EB point estimators usually need an *ingenious approximation* (see, e.g., Morris, 1981, 1983), whereas the posterior variances, though often complicated, can be found exactly.

The above ideas will be made more concrete in the subsequent sections with the aid of examples. Ours is an expository article which compares and contrasts the EB and the HB methods for multivariate normal linear models. The outline of the remaining sections is as follows. In the next section, we address the problem of estimating the multivariate normal mean. EB procedures for such problems are discussed quite adequately in Efron and Morris (1973), Morris (1981, 1983) and Casella (1985). However, the interrelationship between the EB and the HB procedures for such problems is not discussed in these papers. Lindley and Smith (1972) introduced and provided a detailed discussion of the HB approach for estimating the multivariate normal mean. However, there is no mention of the EB approach in their paper.

Deely and Lindley (1981) compared and contrasted the EB and the HB procedures much in the spirit of the discussion in the preceding paragraphs. However, unlike the present article, they did not emphasize simultaneous estimation problems, nor did they incorporate discussion of multivariate normal models.

In the third section, we consider the regression problem. The EB and the HB methods are contrasted both for the balanced and unbalanced linear models. This section is largely a review of the work of Lindley and Smith (1972) as well as Morris (1981, 1983). However, for the unbalanced case, our calculations go beyond those of Lindley and Smith (1972). It is our belief that the present calculations will shed more light on some of the EB approximations of Morris (1983). For the balanced case, the reader is also referred to Berger (1985).

Extensive development of the EB methodology began with Robbins (1951, 1955), who called problems of the above type *compound decision problems.* In Robbins's terminology, an EB procedure is one where $X_1,\ldots,X_p$ are the *past* data about $\theta_1,\ldots,\theta_p$. The past data should be used together with the current data to infer on a current $\theta_i$. The terminological distinction between the EB and compound decision problems will be ignored in this article, and the term *empirical Bayes* will be used to cover problems of both types. Also, Robbins's procedure is a nonparametric EB procedure in contrast to the parametric EB approach taken in this paper.

The term *hierarchical Bayes* was first used by Good (1965). Lindley and Smith (1972) called such priors *multistage priors.* As noted earlier, the latter used the idea very effectively for estimating the vector of normal means, as well as the vector of regression coefficients.

**Estimation of the Multivariate Normal Mean**

This section is devoted to a comparison of the EB and the HB procedures for estimating the multivariate normal mean. We begin with a simple example.

I. Conditional on $\theta_1,\ldots,\theta_p$, let $X_1,\ldots,X_p$ be independent with $X_i \sim N(\theta_i, \sigma^2)$, $i = 1,\ldots,p$, $\sigma^2$ ( > 0) being known. Without loss of generality, assume $\sigma^2 = 1$.

II. The $\theta_i$'s have independent $N(\mu_i, A)$, $i = 1,\ldots,p$ priors. Write $\underset{\sim}{\theta} = (\theta_1,\ldots,\theta_p)^T$, $\underset{\sim}{X} = (X_1,\ldots,X_p)^T$ and $\underset{\sim}{x} = (x_1,\ldots,x_p)^T$.

The posterior distribution of $\underset{\sim}{\theta}$ given $\underset{\sim}{X} = \underset{\sim}{x}$ is then $N((1-B)\underset{\sim}{x} + B\underset{\sim}{\mu}, (1-B)\underset{\sim}{I}_p)$, where $B = (A+1)^{-1}$. Accordingly, the posterior mean (the usual Bayes estimate) of $\underset{\sim}{\theta}$ is given by

$$E(\underset{\sim}{\theta}|\underset{\sim}{X} = \underset{\sim}{x}) = (1-B)\underset{\sim}{x} + B\underset{\sim}{\mu}. \tag{1}$$

In an EB or a HB scenario, some or all of the prior parameters are unknown. In an EB set up, these parameters are estimated from the marginal distribution of $\underset{\sim}{X}$ which in this case is $N(\underset{\sim}{\mu}, B^{-1}\underset{\sim}{I}_p)$. A HB procedure, on the other hand, models the uncertainty of the unknown prior parameters by assigning distributions to them. Such distributions are often called *hyperpriors*. We shall consider the following three cases.

**Case I.** Let $\mu_1 = \ldots = \mu_p = \mu$ (say), where $\mu$ (real) is unknown, but $A$ ( > 0) is known. Based on the marginal distribution of $\underset{\sim}{X}$, $\bar{X}$ is the UMVUE, MLE and the best equivariant estimator of $\mu$. Accordingly, from (1), an EB estimator of $\underset{\sim}{\theta}$ is given by

$$\underset{\sim}{\theta}^{(1)}_{EB} = (1-B)\underset{\sim}{X} + B\bar{X}\underset{\sim}{1}_p. \tag{2}$$

The estimator given in (2) was proposed by Lindley and Smith (1972). They used a HB approach to arrive at the estimator given in (2). The procedure is described below.

Consider the HB model under which (i) conditional on $\underset{\sim}{\theta}$ and $\mu$, $\underset{\sim}{X} \sim N(\underset{\sim}{\theta}, \underset{\sim}{I}_p)$; (ii) conditional on $\mu$, $\underset{\sim}{\theta} \sim N(\mu\underset{\sim}{1}_p, A\underset{\sim}{I}_p)$; (iii) $\mu$ is uniform on $(-\infty, \infty)$. Then the joint (improper) pdf of $\underset{\sim}{X}$, $\underset{\sim}{\theta}$ and $\mu$ is given by

$$f(\underset{\sim}{x}, \underset{\sim}{\theta}, \mu) \propto exp\left[-\frac{1}{2}\|\underset{\sim}{x} - \underset{\sim}{\theta}\|^2\right] A^{-\frac{1}{2}p}\, exp\left[-\frac{1}{2A}\|\underset{\sim}{\theta} - \mu\underset{\sim}{1}_p\|^2\right]. \tag{3}$$

The factor $A^{-\frac{1}{2}p}$ could have been left out in (3), but will be needed for later calculations.

Integrating with respect to $\mu$ in (3), it follows that the joint (improper) pdf of $\underset{\sim}{X}$ and $\underset{\sim}{\theta}$ is

$$f(\underset{\sim}{x}, \underset{\sim}{\theta}) \propto exp\left[-\frac{1}{2}(\underset{\sim}{\theta}^T \underset{\sim}{D} \underset{\sim}{\theta} - 2\underset{\sim}{\theta}^T \underset{\sim}{x} + \underset{\sim}{x}^T \underset{\sim}{x})\right], \tag{4}$$

where $\underset{\sim}{D} = A^{-1}[(A+1)\underset{\sim}{I}_p - p^{-1}\underset{\sim}{J}_p]$ with $\underset{\sim}{J}_p = \underset{\sim}{1}_p \underset{\sim}{1}_p^T$. Recall that $B = (A+1)^{-1}$. It follows from (4) that the posterior distribution of $\underset{\sim}{\theta}$ given $\underset{\sim}{X} = \underset{\sim}{x}$ is $N(\underset{\sim}{D}^{-1}\underset{\sim}{x}, \underset{\sim}{D}^{-1})$. Since $\underset{\sim}{D}^{-1} = (1-B)\underset{\sim}{I}_p + Bp^{-1}\underset{\sim}{J}_p$, one gets

$$E(\underset{\sim}{\theta}|\underset{\sim}{X} = \underset{\sim}{x}) = (1-B)\underset{\sim}{x} + B\bar{x}\underset{\sim}{1}_p; \tag{5}$$

$$V(\underset{\sim}{\theta}|\underset{\sim}{X} = \underset{\sim}{x}) = (1-B)\underset{\sim}{I}_p + Bp^{-1}\underset{\sim}{J}_p. \tag{6}$$

A naive EB approach as noted earlier uses the estimated posterior distribution $N((1-B)\underset{\sim}{x} + B\bar{x}\underset{\sim}{1}_p, (1-B)\underset{\sim}{I}_p)$ to infer about $\underset{\sim}{\theta}$. A comparison of (2) and (5) reveals that the EB and the HB approaches yield the same point estimate of $\underset{\sim}{\theta}$, but the naive EB approach estimates the posterior variance by $(1-B)\underset{\sim}{I}_p$, which is an underestimate when compared to (6). This point is discussed more fully below.

Based on (3), the posterior distribution of $\underset{\sim}{\theta}$ given $\underset{\sim}{x}$ and $\mu$ is $N((1-B)\underset{\sim}{x} + B\mu\underset{\sim}{1}_p, (1-B)\underset{\sim}{I}_p)$. Also, integrating with respect to $\underset{\sim}{\theta}$ in (3), it follows that the joint (improper) distribution of $\underset{\sim}{x}$ and $\mu$ is given by

$$f(\underset{\sim}{x}, \mu) \propto B^{\frac{1}{2}p} \, exp\left[-\frac{1}{2} B\left\|\underset{\sim}{x} - \mu\underset{\sim}{1}_p\right\|^2\right]. \tag{7}$$

It follows from (7) that the posterior distribution of $\mu$ given $\underset{\sim}{X} = \underset{\sim}{x}$ is $N(\bar{x}, (Bp)^{-1})$. Hence, one may note that

$$(1 - B)\underset{\sim}{I}_p = E[V(\underset{\sim}{\theta}|\underset{\sim}{X}, \mu)|\underset{\sim}{X}]; \tag{8}$$

$$\begin{aligned} Bp^{-1}\underset{\sim}{J}_p &= V[B\mu\underset{\sim}{1}_p|\underset{\sim}{X}] = V[(1 - B)\underset{\sim}{X} + B\mu\underset{\sim}{1}_p|\underset{\sim}{X}] \\ &= V[E(\underset{\sim}{\theta}|\underset{\sim}{X}, \mu)|\underset{\sim}{X}]. \end{aligned} \tag{9}$$

Thus a naive EB procedure ignores estimating $V[E(\underset{\sim}{\theta}|\underset{\sim}{X}, \mu)|\underset{\sim}{X}]$ which amounts to ignoring the uncertainty involved in estimating the prior parameters when estimating the posterior variance.

It is shown in Lindley and Smith (1972) that the risk of $\hat{\underset{\sim}{\theta}}_{EB}^{(1)}$ is *not* uniformly smaller than that of $\underset{\sim}{X}$ under the squared error loss $L(\underset{\sim}{\theta}, \underset{\sim}{a}) = \left\|\underset{\sim}{\theta} - \underset{\sim}{a}\right\|^2$.

However, there is a Bayes risk superiority of $\hat{\theta}_{EB}^{(1)}$ over $X$ which is described below.

**Theorem 1**

Consider the model $X|\theta \sim N(\theta, I_p)$ and the prior $\theta \sim N(\mu 1_p, AI_p)$. Let $E$ denote expectation over the joint distribution of $X$ and $\theta$. Then, assuming the matrix loss $L_1(\theta, a) = (a-\theta)(a-\theta)^T$, and writing $\hat{\theta}_B$ as the Bayes estimator of $\theta$ under $L_1$,

$$EL_1(\theta, X) = I_p;$$

$$EL_1(\theta, \hat{\theta}_B) = (1-B)I_p;$$

$$EL_1(\theta, \hat{\theta}_{EB}^{(1)}) = (1-B)I_p + Bp^{-1}J_p. \tag{10}$$

Next assuming the quadratic loss $L_2(\theta, a) = (a-\theta)^T Q(a-\theta)$, where $Q$ is a known non-negative definite (n.n.d.) weight matrix,

$$EL_2(\theta, X) = tr(Q),\ EL_2(\theta, \hat{\theta}_B) = (1-B)tr(Q); \tag{11}$$

$$EL_2(\theta, \hat{\theta}_{EB}^{(1)}) = (1-B)tr(Q) + B\ tr(Qp^{-1}J_p). \tag{12}$$

**Proof.** Note that $\hat{\theta}_B = (1-B)X + B\mu 1_p$. It is immediate that $EL_1(\theta, X) = E[(X-\theta)(X-\theta)^T] = E(I_p) = I_p$ and $EL_1(\theta, \hat{\theta}_B) = E[V(\theta|X)] = E[(1-B)I_p] = (1-B)I_p$. Also, since marginally $\bar{X} \sim N(\mu, (Bp)^{-1})$

$$EL_1(\theta, \hat{\theta}_{EB}^{(1)}) = EL_1(\theta, \hat{\theta}_B) + E\left[(\hat{\theta}_B-\hat{\theta}_{EB}^{(1)})(\hat{\theta}_B-\hat{\theta}_{EB}^{(1)})^T\right]$$

$$= (1-B)I_p + B^2E\left[(\bar{X}-\mu)^2 1_p 1_p^T\right]$$

$$= (1-B)I_p + Bp^{-1}J_p.$$

This completes the proof of (10). To prove (11) and (12), write $L_2(\theta, a) = (\theta-a)^T Q(\theta-a) = tr(QL_1(\theta, a))$ and use (10).

**Remark 1.** It follows from (10)–(12) that $E[L_i(\theta, X) - L_i(\theta, \hat{\theta}_{EB}^{(1)})]$ is nonnegative definite for each $i = 1, 2$. Accordingly, $\hat{\theta}_{EB}^{(1)}$ has smaller Bayes risk than that of $X$ both under the matrix loss $L_1$, and *a fortiori* the quadratic loss $L_2$. To our knowledge, this particular optimality of the Lindley-Smith estimator has not been pointed out before.

The perfect agreement between the EB and the HB point estimators of $\theta$ in Case I is an exception rather than the rule. We now consider cases II and III which reveal that the point estimators of $\underset{\sim}{\theta}$ can also differ under the two approaches.

**Case II.** Assume that $\underset{\sim}{\mu}$ is known, but its components need not be all equal. Moreover, this time $A$ is unknown. The marginal distribution of $\underset{\sim}{X}$ is $N(\underset{\sim}{\mu}, B^{-1}\underset{\sim}{I}_p)$. Then $||\underset{\sim}{X}-\underset{\sim}{\mu}||^2$ is complete sufficient, and is distributed as $B^{-1}\chi_p^2$. Accordingly, for $p \geq 3$, the UMVUE of $B$ is given by $(p-2)/||\underset{\sim}{X}-\underset{\sim}{\mu}||^2$. Substituting this estimator of B in (1), an EB estimator of $\underset{\sim}{\theta}$ is given by

$$\hat{\underset{\sim}{\theta}}_{EB}^{(2)} = \left(1 - \frac{p-2}{||\underset{\sim}{X} - \underset{\sim}{\mu}||^2}\right) \underset{\sim}{X} + \frac{p-2}{||\underset{\sim}{X} - \underset{\sim}{\mu}||^2} \underset{\sim}{\mu}$$

$$= \underset{\sim}{X} - \frac{p-2}{||\underset{\sim}{X} - \underset{\sim}{\mu}||^2} (\underset{\sim}{X}-\underset{\sim}{\mu}). \tag{13}$$

This is the celebrated James-Stein estimator (James and Stein, 1961). The EB interpretation of this estimator was given in a series of articles by Efron and Morris (1972, 1973, 1975). The most popular version of this estimator takes $\underset{\sim}{\mu} = \underset{\sim}{0}$.

It is shown in James and Stein that for $p \geq 3$, the risk of $\hat{\underset{\sim}{\theta}}_{EB}^{(2)}$ is smaller than that of $\underset{\sim}{X}$ under the squared error loss. However, if the loss is changed to the arbitrary quadratic loss $L_2$ of Theorem 1, then the risk dominance of $\hat{\underset{\sim}{\theta}}_{EB}^{(2)}$ over $\underset{\sim}{X}$ does not necessarily hold. Indeed, it is well-known that (see, e.g., Bock, 1975, or Berger, 1975) that under the loss $L_2$, $\hat{\underset{\sim}{\theta}}_{EB}^{(2)}$ dominates $\underset{\sim}{X}$ if (i) $tr(\underset{\sim}{Q}) > 2ch_1(\underset{\sim}{Q})$ and (ii) $0 < p-2 < 2[tr(\underset{\sim}{Q})/ch_1(\underset{\sim}{Q}) - 2]$, where $ch_1(\underset{\sim}{Q})$ denotes the largest eigen-value of $\underset{\sim}{Q}$.

The Bayes risk of $\hat{\underset{\sim}{\theta}}_{EB}^{(2)}$ is, however, smaller than that of $\underset{\sim}{X}$ under the losses $L_1$ and $L_2$, the model given in Theorem 1, and the prior $N_p(\underset{\sim}{\mu}, A\underset{\sim}{I}_p)$. As before, let $E$ denote expectation over the joint distribution of $\underset{\sim}{X}$ and $\underset{\sim}{\theta}$. The following theorem is proved.

**Theorem 2**

Let $\underset{\sim}{X}|\underset{\sim}{\theta} \sim N(\underset{\sim}{\theta}, \underset{\sim}{I}_p)$ and $\underset{\sim}{\theta} \sim N_p(\underset{\sim}{\mu}, A\underset{\sim}{I}_p)$. Then for $p \geq 3$,

$$E[L_1(\underset{\sim}{\theta}, \hat{\underset{\sim}{\theta}}_{EB}^{(2)})] = \underset{\sim}{I}_p - B(p-2)p^{-1}\underset{\sim}{I}_p; \tag{14}$$

$$E[L_2(\underset{\sim}{\theta}, \hat{\underset{\sim}{\theta}}_{EB}^{(2)})] = tr(\underset{\sim}{Q}) - B(p-2)p^{-1}tr(\underset{\sim}{Q}). \tag{15}$$

**Proof.** To prove (14), use the identity

$$E[L_1(\underset{\sim}{\theta}, \hat{\underset{\sim}{\theta}}_{EB}^{(2)})] = E[L_1(\underset{\sim}{\theta}, \hat{\underset{\sim}{\theta}}_B)] + E[(\hat{\underset{\sim}{\theta}}_B - \hat{\underset{\sim}{\theta}}_{EB}^{(2)})(\hat{\underset{\sim}{\theta}}_B - \hat{\underset{\sim}{\theta}}_{EB}^{(2)})^T]. \tag{16}$$

Next write

$$E[(\hat{\underset{\sim}{\theta}}_B - \hat{\underset{\sim}{\theta}}_{EB}^{(2)})(\hat{\underset{\sim}{\theta}}_B - \hat{\underset{\sim}{\theta}}_{EB}^{(2)})^T] = E\left[\left(B - \frac{p-2}{||\underset{\sim}{X} - \underset{\sim}{\mu}||^2}\right)^2 (\underset{\sim}{X} - \underset{\sim}{\mu})(\underset{\sim}{X} - \underset{\sim}{\mu})^T\right]. \tag{17}$$

Marginally, $\underset{\sim}{X} \sim N(\underset{\sim}{\mu}, B^{-1}\underset{\sim}{I}_p)$. Hence, $||\underset{\sim}{X} - \underset{\sim}{\mu}||^2$ is complete sufficient, while $(\underset{\sim}{X} - \underset{\sim}{\mu})(\underset{\sim}{X} - \underset{\sim}{\mu})^T/||\underset{\sim}{X} - \underset{\sim}{\mu}||^2 = B^{-1}(\underset{\sim}{X} - \underset{\sim}{\mu})(\underset{\sim}{X} - \underset{\sim}{\mu})^T/(B^{-1}||\underset{\sim}{X} - \underset{\sim}{\mu}||^2)$ is ancillary. Hence, using Basu's Theorem (see Basu, 1955), $(\underset{\sim}{X} - \underset{\sim}{\mu})(\underset{\sim}{X} - \underset{\sim}{\mu})^T/||\underset{\sim}{X} - \underset{\sim}{\mu}||^2$ is distributed independently of $||\underset{\sim}{X} - \underset{\sim}{\mu}||^2$. Hence,

$$\begin{aligned} B^{-1}\underset{\sim}{I}_p &= E[(\underset{\sim}{X} - \underset{\sim}{\mu})(\underset{\sim}{X} - \underset{\sim}{\mu})^T] \\ &= E[||\underset{\sim}{X} - \underset{\sim}{\mu}||^2\{(\underset{\sim}{X} - \underset{\sim}{\mu})(\underset{\sim}{X} - \underset{\sim}{\mu})^T/||\underset{\sim}{X} - \underset{\sim}{\mu}||^2\}] \\ &= E(||\underset{\sim}{X} - \underset{\sim}{\mu}||^2)E[(\underset{\sim}{X} - \underset{\sim}{\mu})(\underset{\sim}{X} - \underset{\sim}{\mu})^T/||\underset{\sim}{X} - \underset{\sim}{\mu}||^2]. \end{aligned}$$

Now using $E||\underset{\sim}{X} - \underset{\sim}{\mu}||^2 = B^{-1}p$, $E||\underset{\sim}{X} - \underset{\sim}{\mu}||^{-2} = B(p-2)^{-1}$ for $p \geq 3$, one gets

$$E[(\underset{\sim}{X} - \underset{\sim}{\mu})(\underset{\sim}{X} - \underset{\sim}{\mu})^T/||\underset{\sim}{X} - \underset{\sim}{\mu}||^2] = p^{-1}\underset{\sim}{I}_p; \tag{18}$$

$$\begin{aligned} &E[(\underset{\sim}{X} - \underset{\sim}{\mu})(\underset{\sim}{X} - \underset{\sim}{\mu})^T/||\underset{\sim}{X} - \underset{\sim}{\mu}||^4] \\ &\quad = E[(\underset{\sim}{X} - \underset{\sim}{\mu})(\underset{\sim}{X} - \underset{\sim}{\mu})^T/||\underset{\sim}{X} - \underset{\sim}{\mu}||^2]E(||\underset{\sim}{X} - \underset{\sim}{\mu}||^{-2}) \\ &\quad = (p^{-1}\underset{\sim}{I}_p)B(p-2)^{-1}. \end{aligned} \tag{19}$$

It follows from (17)–(19) that

$$\begin{aligned} &E[(\hat{\underset{\sim}{\theta}}_B - \hat{\underset{\sim}{\theta}}_{EB}^{(2)})(\hat{\underset{\sim}{\theta}}_B - \hat{\underset{\sim}{\theta}}_{EB}^{(2)})^T] \\ &\quad = B^2(B^{-1}\underset{\sim}{I}_p) - 2B(p-2)p^{-1}\underset{\sim}{I}_p + B(p-2)p^{-1}\underset{\sim}{I}_p \\ &\quad = B\underset{\sim}{I}_p - B(p-2)p^{-1}\underset{\sim}{I}_p. \end{aligned} \tag{20}$$

Combining (10), (16) and (20), one gets (14). The proof of (15) is immediate from (14) by writing $L_2(\underset{\sim}{\theta}, \underset{\sim}{a}) = tr[\underset{\sim}{Q}L_1(\underset{\sim}{\theta}, \underset{\sim}{a})]$.

**Remark 2.** Taking $\underset{\sim}{Q}$ as a matrix with its $(i, i)^{th}$ element equal to 1 and the rest zeroes, it follows that the $i^{th}$ component of $\hat{\underset{\sim}{\theta}}_{EB}^{(2)}$ dominates $X_i$ when one compares their Bayes risks. This co-ordinatewise Bayes risk dominance of $\hat{\underset{\sim}{\theta}}_{EB}^{(2)}$ over $\underset{\sim}{X}$ appears in Efron and Morris (1973). One can derive (15) from their work by using an orthogonal transformation. The dominance of $\hat{\underset{\sim}{\theta}}_{EB}^{(2)}$ over $\underset{\sim}{X}$ under the matrix loss $L_1$ has not been pointed out before, but the approach appears in Reinsel (1985) for a more complex EB problem.

**Remark 3.** Efron and Morris (1973) found it convenient to define the concept of relative savings loss (RSL). Denote the given prior by $\xi$ and the Bayes risk of an estimator $\underset{\sim}{e}$ of $\underset{\sim}{\theta}$ under the prior $\xi$ and the loss $L_2$ by $r(\xi, \underset{\sim}{e})$. The RSL of $\hat{\underset{\sim}{\theta}}_{EB}^{(2)}$ with respect to $\underset{\sim}{X}$ is defined by

$$\begin{aligned} RSL(\hat{\underset{\sim}{\theta}}_{EB}^{(2)}; \underset{\sim}{X}) &= [r(\xi, \hat{\underset{\sim}{\theta}}_{EB}^{(2)}) - r(\xi, \hat{\underset{\sim}{\theta}}_B)]/[r(\xi, \underset{\sim}{X}) - r(\xi, \hat{\underset{\sim}{\theta}}_B)] \\ &= 1 - [r(\xi, \underset{\sim}{X}) - r(\xi, \hat{\underset{\sim}{\theta}}_{EB}^{(2)})]/[r(\xi, \underset{\sim}{X}) - r(\xi, \hat{\underset{\sim}{\theta}}_B)]. \end{aligned} \quad (21)$$

This is the proportion of the possible Bayes risk improvement over $\underset{\sim}{X}$ that is sacrificed by using $\hat{\underset{\sim}{\theta}}_{EB}^{(2)}$ rather than the ideal estimator $\hat{\underset{\sim}{\theta}}_B$ under the prior $\xi$. It follows from (11), (15) and (16) that $RSL(\hat{\underset{\sim}{\theta}}_{EB}^{(2)}; \underset{\sim}{X}) = 2/p$ for an arbitrary n.n.d. non-null matrix $\underset{\sim}{Q}$. Efron and Morris (1973) proved the result when $\underset{\sim}{Q} = \underset{\sim}{I}_p$ as well as when the $(i, i)^{th}$ element of $\underset{\sim}{Q}$ is 1 and the rest zeroes ($i = 1,\ldots,p$). For the matrix loss $L_1$, the RSL concept of Efron and Morris (1973) can be generalized to get

$$\begin{aligned} \mathrm{RSL}(\hat{\underset{\sim}{\theta}}_{EB}^{(2)}; \underset{\sim}{X}) &= [\underset{\sim}{r}(\xi, \underset{\sim}{X}) - \underset{\sim}{r}(\xi, \hat{\underset{\sim}{\theta}}_B)]^{-1}[\underset{\sim}{r}(\xi, \hat{\underset{\sim}{\theta}}_{EB}^{(2)}) - \underset{\sim}{r}(\xi, \hat{\underset{\sim}{\theta}}_B)] \\ &= (B\underset{\sim}{I}_p)^{-1}(B(2/p))\underset{\sim}{I}_p = (2/p)\underset{\sim}{I}_p. \end{aligned} \quad (22)$$

Suppose now we consider a HB approach in this case, where conditional on $\underset{\sim}{\theta}$ and $A$, $\underset{\sim}{X} \sim N(\underset{\sim}{\theta}, \underset{\sim}{I}_p)$, and conditional on $A$, $\underset{\sim}{\theta} \sim N(\underset{\sim}{\mu}, A\underset{\sim}{I}_p)$. Also, let $A$ have marginal pdf $g_0(A)$. Then, the joint pdf of $\underset{\sim}{X}$, $\underset{\sim}{\theta}$ and $A$ is

$$f(\underset{\sim}{x}, \underset{\sim}{\theta}, A) \propto exp\left[-\tfrac{1}{2}||\underset{\sim}{x} - \underset{\sim}{\theta}||^2\right] A^{-\frac{1}{2}p}\, exp\left[-\tfrac{1}{2}A^{-1}||\underset{\sim}{\theta} - \underset{\sim}{\mu}||^2\right] g_0(A). \quad (23)$$

As before, the conditional distribution of $\underset{\sim}{\theta}$ given $\underset{\sim}{x}$ and $A$ is $N((1-B)\underset{\sim}{x} + B\underset{\sim}{\mu}, (1-B)\underset{\sim}{I}_p)$, where $B = (A+1)^{-1}$. But integrating with respect to $\underset{\sim}{\theta}$, the joint pdf of $\underset{\sim}{X}$ and $A$ is

$$f(\underset{\sim}{x}, A) \propto (A+1)^{-\frac{1}{2}p}\, exp\left[-\frac{1}{2(A+1)}||\underset{\sim}{x} - \underset{\sim}{\mu}||^2\right] g_0(A). \quad (24)$$

Since $B = (A+1)^{-1}$, the joint pdf of $\underset{\sim}{X}$ and $B$ is of the form

$$f(\underset{\sim}{x}, B) \propto B^{\frac{1}{2}p}\, exp\left[-\tfrac{1}{2} B \,||\underset{\sim}{x} - \underset{\sim}{\mu}||^2\right] g(B). \tag{25}$$

The HB approach of the above type was first proposed by Strawderman (1971), and was later generalized by Faith (1978). Assuming the Type II Beta density for $A$, namely $g_0(A) \propto A^{m-1}(1+A)^{-(m+n)}$, where $m\ (>0)$ and $n\ (>0)$, it is easy to see that

$$f(\underset{\sim}{x}, B) \propto B^{\frac{1}{2}p+n-1}(1-B)^{m-1}\, exp\left[-\tfrac{1}{2} B \,||\underset{\sim}{x} - \underset{\sim}{\mu}||^2\right]. \tag{26}$$

Now, using the iterated formula for conditional expectations,

$$\hat{\underset{\sim}{\theta}}_{HB}^{(2)} \equiv E(\underset{\sim}{\theta}|\underset{\sim}{x}) = E(E[\underset{\sim}{\theta}|B, \underset{\sim}{x}] \mid \underset{\sim}{x}) = (1 - \hat{\hat{B}})\underset{\sim}{x} + \hat{\hat{B}}\underset{\sim}{\mu}, \tag{27}$$

where

$$\hat{\hat{B}} = E(B|\underset{\sim}{x}) = \int_0^1 B^{\frac{1}{2}p+n}(1-B)^{m-1}\, exp\left[-\tfrac{1}{2} B||\underset{\sim}{x} - \underset{\sim}{\mu}||^2\right] dB$$

$$\div \int_0^1 B^{\frac{1}{2}p+n-1}(1-B)^{m-1}\, exp\left[-\tfrac{1}{2} B||\underset{\sim}{x} - \underset{\sim}{\mu}||^2\right] dB. \tag{28}$$

Strawderman (1971) considered the case $m = 1$, and found sufficient conditions on $n$ under which the risk of $\hat{\underset{\sim}{\theta}}_{HB}^{(2)}$ is smaller than that of $\underset{\sim}{X}$. His results were generalized to a certain extent by Faith (1978).

We consider also the case $m = 1$, and interpreting (26) as the posterior pdf of $B$ given $\underset{\sim}{x}$, find the posterior mode of $B$ as

$$\hat{B}_{MO} = min((p+2n-2)/||\underset{\sim}{x} - \underset{\sim}{\mu}||^2, 1). \tag{29}$$

Substituting this estimator of $B$ in (1), one gets the estimator

$$\hat{\underset{\sim}{\theta}}_{HB}^{(3)} = (1 - \hat{B}_{MO})\underset{\sim}{X} + \hat{B}_{MO}\underset{\sim}{\mu} = \underset{\sim}{X} - \hat{B}_{MO}(\underset{\sim}{X} - \underset{\sim}{\mu}) \tag{30}$$

of $\underset{\sim}{\theta}$. The special choice $n = 0$ leads to the positive part James-Stein estimator which is known to dominate the usual James-Stein estimator (see Lehmann, 1983, p. 302). This is intuitively very clear since the usual James-Stein estimator substitutes the UMVUE of $B$ in (1), and this UMVUE can take values exceeding 1 with positive probability while $0 < B < 1$. This deficiency is rectified by $\hat{B}_{MO}$.

**Case III.** The model is similar to the one in Case I, except that now $\mu$ (real) and $A$ ($>0$) are both unknown. Recall that marginally $\underset{\sim}{X} \sim N(\mu \underset{\sim}{1}_p, B^{-1} \underset{\sim}{I}_p)$ where $B = (A+1)^{-1}$. Hence, $(\bar{X}, \sum_1^p (X_i - \bar{X})^2)$ is complete sufficient, so that the UMVUE's of $\mu$ and $B$ are given respectively by $\bar{X}$ and $(p-3)/\sum_1^p (X_i-\bar{X})^2$. Substituting these estimators of $\mu$ and $B$ in (1), the EB estimator of $\underset{\sim}{\theta}$ is given by

$$\hat{\underset{\sim}{\theta}}_{EB}^{(3)} = \left(1 - \frac{p-3}{\sum_1^p (X_i-\bar{X})^2}\right) \underset{\sim}{X} + \frac{p-3}{\sum_1^p (X_i-\bar{X})^2} \bar{X} \underset{\sim}{1}_p$$

$$= \underset{\sim}{X} - \frac{p-3}{\sum_1^p (X_i-\bar{X})^2} (\underset{\sim}{X} - \bar{X}\underset{\sim}{1}_p). \tag{31}$$

This modification of the James-Stein estimator was proposed by Lindley (1962). Whereas, the original James-Stein estimator shrinks $\underset{\sim}{X}$ towards a specified point, the modified estimator given in (31) shrinks $\underset{\sim}{X}$ towards a hyperplane spanned by $\underset{\sim}{1}_p$.

The estimator $\hat{\underset{\sim}{\theta}}_{EB}^{(3)}$ is known to dominate $\underset{\sim}{X}$ for $p \geq 4$. Its Bayes risk under the $L_1$ and $L_2$ losses are not known however. We now prove a theorem to this effect quite in the spirit of Theorems 1 and 2.

**Theorem 3**

Assume the model and the prior given in Theorem 1. Then, for $p \geq 4$,

$$E[L_1(\underset{\sim}{\theta}, \hat{\underset{\sim}{\theta}}_{EB}^{(3)})] = \underset{\sim}{I}_p - B(p-3)(p-1)^{-1}(\underset{\sim}{I}_p - p^{-1}\underset{\sim}{J}_p); \tag{32}$$

$$E[L_2(\underset{\sim}{\theta}, \hat{\underset{\sim}{\theta}}_{EB}^{(3)})] = tr(\underset{\sim}{Q}) - B(p-3)(p-1)^{-1} tr[\underset{\sim}{Q}(\underset{\sim}{I}_p - p^{-1}\underset{\sim}{J}_p)]. \tag{33}$$

**Proof.** First write

$$E[L_1(\underset{\sim}{\theta}, \hat{\underset{\sim}{\theta}}_{EB}^{(3)})] = E[L_1(\underset{\sim}{\theta}, \hat{\underset{\sim}{\theta}}_B)] + E(\hat{\underset{\sim}{\theta}}_{EB}^{(3)} - \hat{\underset{\sim}{\theta}}_B)(\hat{\underset{\sim}{\theta}}_{EB}^{(3)} - \hat{\underset{\sim}{\theta}}_B)^T. \tag{34}$$

We write

$$\hat{\underset{\sim}{\theta}}_{EB}^{(3)} - \hat{\underset{\sim}{\theta}}_B = \left(B - \frac{p-3}{\sum_1^p (X_i-\bar{X})^2}\right)(\underset{\sim}{X} - \bar{X}\underset{\sim}{1}_p) + B(\bar{X} - \mu)\underset{\sim}{1}_p. \tag{35}$$

Now using the independence of $\underset{\sim}{X} - \bar{X}\underset{\sim}{1}_p$ and $\bar{X}$, and using the fact that $\bar{X} \sim N(\mu, (Bp)^{-1})$, one gets from (35),

$$E\left[\left(\hat{\underset{\sim}{\theta}}_{EB}^{(3)} - \hat{\underset{\sim}{\theta}}_B\right)\left(\hat{\underset{\sim}{\theta}}_{EB}^{(3)} - \hat{\underset{\sim}{\theta}}_B\right)^T\right]$$

$$= E\left[\left(B - \frac{p-3}{\sum (X_i - \bar{X})^2}\right)^2 (\underset{\sim}{X} - \bar{X}\underset{\sim}{1}_p)(\underset{\sim}{X} - \bar{X}\underset{\sim}{1}_p)^T\right] + B^2(Bp)^{-1}\underset{\sim}{J}_p. \qquad (36)$$

Next using the independence of $(\underset{\sim}{X} - \bar{X}\underset{\sim}{1}_p)(\underset{\sim}{X} - \bar{X}\underset{\sim}{1}_p)^T \Big/ \sum_1^p (X_i - \bar{X})^2$ with $\sum_1^p (X_i - \bar{X})^2$ (again by applying Basu's Theorem) and the facts that $E\,[(\underset{\sim}{X} - \bar{X}\underset{\sim}{1}_p)(\underset{\sim}{X} - \bar{X}\underset{\sim}{1}_p)^T] = B^{-1}(\underset{\sim}{I}_p - p^{-1}\underset{\sim}{J}_p)$, while $\sum_1^p (X_i - \bar{X})^2 \sim B^{-1}\chi^2_{p-1}$, it follows from (36) that for $p \geq 4$,

$$E\left[\left(\hat{\underset{\sim}{\theta}}_{EB}^{(3)} - \hat{\underset{\sim}{\theta}}_B\right)\left(\hat{\underset{\sim}{\theta}}_{EB}^{(3)} - \hat{\underset{\sim}{\theta}}_B\right)^T\right]$$

$$= B^2 B^{-1}\left(\underset{\sim}{I}_p - p^{-1}\underset{\sim}{J}_p\right) - 2B(p-3)(p-1)^{-1}\left(\underset{\sim}{I}_p - p^{-1}\underset{\sim}{J}_p\right)$$

$$+ (p-3)^2 B(p-3)^{-1}(p-1)^{-1}\left(\underset{\sim}{I}_p - p^{-1}\underset{\sim}{J}_p\right) + Bp^{-1}\underset{\sim}{J}_p$$

$$= B\underset{\sim}{I}_p - B(p-3)(p-1)^{-1}\left(\underset{\sim}{I}_p - p^{-1}\underset{\sim}{J}_p\right). \qquad (37)$$

Combining (10), (34) and (37), one gets (32). The proof of (33) is immediate from (32).

We now proceed to find the HB estimator of $\underset{\sim}{\theta}$. Consider the model where (i) conditional on $\underset{\sim}{\theta}$, $\mu$ and $A$, $\underset{\sim}{X} \sim N(\underset{\sim}{\theta}, \underset{\sim}{I}_p)$; (ii) conditional on $\mu$ and $A$, $\underset{\sim}{\theta} \sim N(\mu\underset{\sim}{1}_p, A\underset{\sim}{I}_p)$; (iii) marginally $\mu$ and $A$ are independently distributed with $\mu$ uniform on $(-\infty, \infty)$, and $A$ has uniform improper pdf on $(0, \infty)$. Then the joint (improper) pdf of $\underset{\sim}{X}$, $\underset{\sim}{\theta}$, $\mu$ and $A$ is given by

$$f(\underset{\sim}{x}, \underset{\sim}{\theta}, \mu, A) \propto exp\left[-\tfrac{1}{2}\left\|\underset{\sim}{x} - \underset{\sim}{\theta}\right\|^2\right] A^{-\frac{1}{2}p}\, exp\left[-\frac{1}{2A}\left\|\underset{\sim}{\theta} - \mu\underset{\sim}{1}_p\right\|^2\right]. \qquad (38)$$

Now integrating with respect to $\mu$, it follows from (38) that the joint (improper) pdf of $\underset{\sim}{X}$, $\underset{\sim}{\theta}$ and $A$ is

$$f(\underset{\sim}{x}, \underset{\sim}{\theta}, A) \propto A^{-\frac{1}{2}(p-1)} exp\left[-\tfrac{1}{2}(\underset{\sim}{\theta} - \underset{\sim}{D}^{-1}\underset{\sim}{x})^T \underset{\sim}{D}(\underset{\sim}{\theta} - \underset{\sim}{D}^{-1}\underset{\sim}{x}) - \frac{1}{2(A+1)}\sum_1^p (x_i - \bar{x})^2\right], \qquad (39)$$

where $\underset{\sim}{D}$ is defined after (6). Recall $\underset{\sim}{D}^{-1} = (1-B)\underset{\sim}{I}_p + Bp^{-1}\underset{\sim}{J}_p$. Hence, conditional on $\underset{\sim}{x}$ and $A$, $\underset{\sim}{\theta} \sim N[(1-B)\underset{\sim}{x} + B\bar{x}\underset{\sim}{1}_p, (1-B)\underset{\sim}{I}_p + Bp^{-1}\underset{\sim}{J}_p]$. Also, integrating with respect to $\underset{\sim}{\theta}$ in (39), one gets the joint pdf of $\underset{\sim}{X}$ and $A$ given by

$$f(\underset{\sim}{x}, A) \propto (A+1)^{-\frac{1}{2}(p-1)} \, exp\left[-\frac{1}{2(A+1)} \sum_1^p (x_i-\bar{x})^2\right]. \tag{40}$$

Since $B = (A+1)^{-1}$, it follows from (40) that the joint pdf of $\underset{\sim}{X}$ and $B$ is given by

$$\begin{aligned} f(\underset{\sim}{x}, B) &\propto B^{\frac{1}{2}(p-1)} \, exp\left[-\frac{1}{2} B \sum_1^p (x_i-\bar{x})^2\right] B^{-2} \\ &= B^{\frac{1}{2}(p-5)} \, exp\left[-\frac{1}{2} B \sum_1^p (x_i-\bar{x})^2\right]. \end{aligned} \tag{41}$$

It follows from (41) that

$$\begin{aligned} E(B|\underset{\sim}{x}) = &\int_0^1 B^{\frac{1}{2}(p-3)} exp\left[-\frac{1}{2} B \sum_1^p (x_i-\bar{x})^2\right] dB \\ &\div \int_0^1 B^{\frac{1}{2}(p-5)} exp\left[-\frac{1}{2} B \sum_1^p (x_i-\bar{x})^2\right] dB; \end{aligned} \tag{42}$$

$$\begin{aligned} E(B^2|\underset{\sim}{x}) = &\int_0^1 B^{\frac{1}{2}(p-1)} exp\left[-\frac{1}{2} B \sum_1^p (x_i-\bar{x})^2\right] dB \\ &\div \int_0^1 B^{\frac{1}{2}(p-5)} exp\left[-\frac{1}{2} B \sum_1^p (x_i-\bar{x})^2\right] dB. \end{aligned} \tag{43}$$

One can obtain $V(B|\underset{\sim}{x})$ from (42) and (43), and use these to obtain

$$E(\underset{\sim}{\theta}|\underset{\sim}{x}) = \underset{\sim}{x} - E(B|\underset{\sim}{x})(\underset{\sim}{x}-\bar{x}\underset{\sim}{1}_p); \tag{44}$$

$$\begin{aligned} V(\underset{\sim}{\theta}|\underset{\sim}{x}) &= V[E(\underset{\sim}{\theta}|B, \underset{\sim}{x})| \underset{\sim}{x}] + E[V(\underset{\sim}{\theta}|B, \underset{\sim}{x})| \underset{\sim}{x}] \\ &= V[\underset{\sim}{x} - B(\underset{\sim}{x}-\bar{x}\underset{\sim}{1}_p)| \underset{\sim}{x}] + E[(1-B)\underset{\sim}{I}_p + Bp^{-1}\underset{\sim}{J}_p| \underset{\sim}{x}] \\ &= V(B| \underset{\sim}{x})(\underset{\sim}{x}-\bar{x}\underset{\sim}{1}_p)(\underset{\sim}{x}-\bar{x}\underset{\sim}{1}_p)^T + \underset{\sim}{I}_p - E(B|\underset{\sim}{x})(\underset{\sim}{I}_p - p^{-1}\underset{\sim}{J}_p). \end{aligned} \tag{45}$$

Also, one can obtain a positive-part version of Lindley's estimator by substituting the posterior mode of $B$ namely $min\left((p-5)\Big/\sum_1^p (X_i-\bar{X})^2,\ 1\right)$ in (1). Morris (1981) suggests approximations to $E(B|\underset{\sim}{x})$ and $E(B^2|\underset{\sim}{x})$ involving replacement of $\int_0^1$ by $\int_0^\infty$ both in the numerator as well in the denominator of (42) and (43). The resulting approximations turn out to be $E(B|\underset{\sim}{x}) \doteq (p-3)\Big/\sum_1^p (x_i-\bar{x})^2$ and, $E(B^2|\underset{\sim}{x}) \doteq (p-1)(p-3)\Big/\left\{\sum_1^p (x_i-\bar{x})^2\right\}^2$, so that $V(B|\underset{\sim}{x}) \doteq 2(p-3)\Big/\left\{\sum_1^p (x_i-\bar{x})^2\right\}^2$. Morris (1981) points out that the above approximations amount to putting a uniform prior to $A$ on $(-1, \infty)$ rather than on $(0, \infty)$. Note that with Morris's approximations

$$E(\underset{\sim}{\theta}|\underset{\sim}{X}) \doteq \underset{\sim}{X} - \frac{p-3}{\sum_1^p (X_i-\bar{X})^2}(\underset{\sim}{X} - \bar{X}\underset{\sim}{1}_p) = \hat{\theta}^{(3)}_{EB}, \tag{46}$$

which is Lindley's modification of the James-Stein estimator, while

$$V(\underset{\sim}{\theta}|\underset{\sim}{X}) \doteq \frac{2(p-3)}{\left(\sum_1^p (X_i-\bar{X})^2\right)^2}(\underset{\sim}{X} - \bar{X}\underset{\sim}{1}_p)(\underset{\sim}{X} - \bar{X}\underset{\sim}{1}_p)^T$$
$$+ \underset{\sim}{I}_p - \frac{p-3}{\sum_1^p (X_i-\bar{X})^2}(\underset{\sim}{I}_p - p^{-1}\underset{\sim}{J}_p). \tag{47}$$

Morris (1981) considered a slightly more general version of the model where conditional on $\underset{\sim}{\theta}$, $\mu$ and $A$, $\underset{\sim}{X} \sim N(\underset{\sim}{\theta}, \sigma^2\underset{\sim}{I}_p)$, while the distributions of $\underset{\sim}{\theta}$, $\mu$ and $A$ remain the same. If one redefines $B = \sigma^2/(\sigma^2+A)$, the only change that is needed in the calculations is that conditional on $\underset{\sim}{x}$ and $A$, $\underset{\sim}{\theta} \sim N((1-B)\underset{\sim}{x} + B\bar{x}\underset{\sim}{1}_p,\ \sigma^2[(1-B)\underset{\sim}{I}_p + Bp^{-1}\underset{\sim}{J}_p])$, while the conditional pdf of $B$ given $\underset{\sim}{x}$, and accordingly $E(B|\underset{\sim}{x})$ and $V(B|\underset{\sim}{x})$ are modified by putting $B/\sigma^2$ in place of $B$ in the exponents.

We now revisit the famous baseball data of Efron and Morris (1975). They considered the batting averages of 18 baseball players in 1970 after each had batted 45 times. Based on these batting averages, they estimated (in fact, predicted) the players' batting averages for the remainder of the season. We used formulas (42) and (43) with $B/\sigma^2$ replacing $B$ in the exponents to get the exact expressions for $E(\theta_i|\underset{\sim}{x})$ and $V(\theta_i|\underset{\sim}{x})$. Also, we used Morris's approximations which are obtained by modifying (46) and (47). The results are given in Table 1. In

TABLE 1. The True Values ($\theta_i$), the Maximum Likelihood Estimates ($Y_i$), the Hierarchical Bayes Estimates ($\hat{\theta}_{i,HB}$), the Hierarchical Bayes S.D.'s ($s_{i,HB}$), Morris's Approximate Estimates ($\hat{\theta}_{i,M}$), and Morris's Approximate S.D.'s ($s_{i,M}$).

| $i$ | $\theta_i$ | $Y_i$ | $\hat{\theta}_{i,HB}$ | $s_{i,HB}$ | $[\hat{\theta}_{i,HB}-2s_{i,HB}, \hat{\theta}_{i,HB}+2s_{i,HB}]$ | $\hat{\theta}_{i,M}$ | $s_{i,M}$ | $[\hat{\theta}_{i,M}-2s_{i,M}, \hat{\theta}_{i,M}+2s_{i,M}]$ |
|---|---|---|---|---|---|---|---|---|
| 1 | 0.346 | 0.395 | 0.308 | 0.046 | [0.216,0.400] | 0.293 | 0.073 | [0.147,0.439] |
| 2 | 0.300 | 0.375 | 0.301 | 0.044 | [0.213,0.389] | 0.288 | 0.071 | [0.142,0.430] |
| 3 | 0.279 | 0.355 | 0.295 | 0.043 | [0.209,0.381] | 0.284 | 0.069 | [0.146,0.422] |
| 4 | 0.223 | 0.334 | 0.288 | 0.042 | [0.204,0.372] | 0.280 | 0.067 | [0.146,0.414] |
| 5 | 0.276 | 0.313 | 0.281 | 0.041 | [0.199,0.363] | 0.275 | 0.066 | [0.143,0.407] |
| 6 | 0.273 | 0.291 | 0.281 | 0.041 | [0.199,0.363] | 0.275 | 0.066 | [0.143,0.407] |
| 7 | 0.266 | 0.269 | 0.274 | 0.040 | [0.194,0.354] | 0.271 | 0.066 | [0.139,0.405] |
| 8 | 0.211 | 0.247 | 0.267 | 0.040 | [0.187,0.347] | 0.266 | 0.066 | [0.134,0.398] |
| 9 | 0.271 | 0.247 | 0.260 | 0.040 | [0.180,0.340] | 0.262 | 0.067 | [0.128,0.396] |
| 10 | 0.232 | 0.247 | 0.260 | 0.040 | [0.180,0.340] | 0.262 | 0.067 | [0.128,0.396] |
| 11 | 0.266 | 0.224 | 0.252 | 0.040 | [0.172,0.332] | 0.257 | 0.068 | [0.121,0.393] |
| 12 | 0.258 | 0.224 | 0.252 | 0.040 | [0.172,0.332] | 0.257 | 0.068 | [0.121,0.393] |
| 13 | 0.306 | 0.224 | 0.252 | 0.040 | [0.172,0.332] | 0.257 | 0.068 | [0.121,0.393] |
| 14 | 0.267 | 0.224 | 0.252 | 0.040 | [0.172,0.332] | 0.257 | 0.068 | [0.121,0.393] |
| 15 | 0.228 | 0.224 | 0.252 | 0.040 | [0.172,0.332] | 0.257 | 0.068 | [0.121,0.393] |
| 16 | 0.288 | 0.200 | 0.244 | 0.041 | [0.162,0.326] | 0.252 | 0.070 | [0.112,0.392] |
| 17 | 0.318 | 0.175 | 0.236 | 0.043 | [0.150,0.322] | 0.247 | 0.073 | [0.101,0.393] |
| 18 | 0.200 | 0.148 | 0.227 | 0.045 | [0.137,0.317] | 0.241 | 0.077 | [0.087,0.395] |

what follows the true values $\theta_i$'s refer to the baseball players' actual batting averages for the remainder of the season. Also, $\hat{\theta}_{i,HB}$ and $\hat{\theta}_{i,M}$ denote respectively the HB estimate of $\theta_i$ and Morris's approximate estimate of $\theta_i$. The standard errors associated with $\hat{\theta}_{i,HB}$ and $\hat{\theta}_{i,M}$ are denoted respectively by $s_{i,HB}$ and $s_{i,M}$. It turns out that

$$(18\sigma^2)^{-1}\sum_{i=1}^{18}(X_i-\theta_i)^2 = 0.976,$$

$$(18\sigma^2)^{-1}\sum_{i=1}^{18}(\hat{\theta}_{i,HB}-\theta_i)^2 = 0.299,$$

and

$$(18\sigma^2)^{-1}\sum_{i=1}^{18}(\hat{\theta}_{i,M}-\theta_i)^2 = 0.286$$

so that Morris's approximations serve well as point estimates. However, Morris's (1981) approximations to the s.d.'s are consistently larger than the actual ones, leading thereby to wider confidence intervals. It appears that Morris (1981) has reported that $\hat{\theta}_{i,HB}$'s and $s_{i,HB}$'s in his Table 1, p. 31, but his notations seem to suggest that these are $\hat{\theta}_{i,M}$'s and $s_{i,M}$'s.

So far we have considered only the case when the sampling variance $\sigma^2$ is known. In a more realistic set up, $\sigma^2$ is unknown. In such instances, one approach is to first find the Bayes estimator of $\underset{\sim}{\theta}$ assuming $\sigma^2$ to be known. Next find an estimator of $\sigma^2$, and substitute this estimator in the Bayes estimator found earlier. Berger (1985) discusses this approach. A slightly different classical EB approach can be found in Ghosh and Meeden (1986) or Ghosh and Lahiri (1987). These methods do not take into account the uncertainty involved in estimating $\sigma^2$. This deficiency can be rectified by putting a prior distribution (often non-informative) on $\sigma^2$ as well.

One important example is the unbalanced one-way ANOVA model. We propose a HB analysis with an unknown $\sigma^2$ as well as unknown parameters involved in the prior distribution of $\underset{\sim}{\theta}$. We find it convenient to reparametrize into $\sigma^2 = r^{-1}$ and $A = (\lambda r)^{-1}$. The remainder of this section is an adaptation of the arguments of Ghosh and Lahiri (1988).

Assume that

(a) conditional on $\underset{\sim}{\theta}$, $m$, $\lambda$ and $r$, the random variables $X_1,\ldots,X_p$ and $U$ are mutually independent with $X_i \sim N\left(\theta_i, (rn_i)^{-1}\right)$ $(i = 1,\ldots,p)$, while $U \sim r^{-1}\chi^2_{N-p}$ $(N = \sum_{i=1}^{p} n_i)$;

(b) conditional on m, $\lambda$ and $r$, $\underset{\sim}{\theta} \sim N\left(m\underset{\sim}{1}_p, (\lambda r)^{-1}\underset{\sim}{I}_p\right)$;

(c) marginally, $M$, $\Lambda$ and $R$ are independently distributed with $M \sim$ *uniform*$(-\infty, \infty)$, $R$ has pdf $g(r) \propto r^{-2}$, while $\Lambda$ has pdf $h(\lambda) \propto \lambda^{-2}$.

**Remark 3.** Note that we have changed the notation from $\mu$ to $m$. If one assigns the noninformative prior $g(A, \sigma^2) \propto (\sigma^2)^{-1}$, then noting that $r = (\sigma^2)^{-1}$ and $\lambda r = A^{-1}$, one gets the prior on $R$ and $\Lambda$ as given in (c). It is possible to assign gamma priors (informative or noninformative) on $R$ and $\Lambda R$ as in Ghosh and Lahiri (1988), but we have decided to sacrifice that generality.

To identify the above model with an unbalanced one-way random effects ANOVA model, write $Y_{ij} = m + \tau_i + e_{ij}$ ($j = 1,\ldots,n_i$; $i = 1,\ldots,p$). Here, $\tau_i$'s and $e_{ij}$'s are mutually independent with $\tau_i$'s iid $N(0, (\lambda r)^{-1})$ and $e_{ij}$'s iid $N(0, r^{-1})$. Write $\theta_i = m + \tau_i$, $X_i = \bar{Y}_i = n_i^{-1}\sum_{j=1}^{n_i} Y_{ij}$ ($i = 1,\ldots,p$) and $U = \sum_{i=1}^{p}\sum_{j=1}^{n_i}(Y_{ij} - \bar{Y}_i)^2$. Clearly, $(X_1,\ldots,X_p, U)$ is minimal sufficient with joint distribution given in (a).

Under the above model, the joint pdf of $X_1,\ldots,X_p$, $U$, $\underset{\sim}{\theta}$, $M$, $R$ and $\Lambda$ is given by

$$f(\underset{\sim}{x}, u, \underset{\sim}{\theta}, m, r, \lambda) \propto r^{\frac{1}{2}p} exp\left[-\frac{1}{2}r\sum_{i=1}^{p} n_i(x_i - \theta_i)^2\right] r^{\frac{1}{2}(N-p)}$$

$$\times\ exp\left(-\frac{1}{2}ru\right)u^{\frac{1}{2}(N-p)-1}$$

$$\times\ (\lambda r)^{\frac{1}{2}p} exp\left[-\frac{1}{2}\lambda r\sum_{i=1}^{p}(\theta_i - m)^2\right](\lambda r)^{-2}. \tag{48}$$

Integrating with respect to $m$ in (48), one gets the joint pdf of $\underset{\sim}{X}$, U, $\underset{\sim}{\theta}$, $R$ and $\Lambda$ given by

$$f(\underset{\sim}{x}, \underset{\sim}{u}, \underset{\sim}{\theta}, r, \lambda) \propto r^{\frac{1}{2}(N+p-1)-2} exp\left[-\frac{1}{2}r(\underset{\sim}{\theta}^T \underset{\sim}{D}\underset{\sim}{\theta} - 2\underset{\sim}{\theta}^T \underset{\sim}{G}\underset{\sim}{x} + \underset{\sim}{x}^T \underset{\sim}{G}\underset{\sim}{x} + u)\right]$$

$$\times\ u^{\frac{1}{2}(N-p)-1}\ \lambda^{\frac{1}{2}(p-1)-2}, \tag{49}$$

where $\underset{\sim}{G} = Diag(n_1,\ldots,n_p)$, $\underset{\sim}{D} = \underset{\sim}{G} + \lambda(\underset{\sim}{I}_p - p^{-1}\underset{\sim}{J}_p)$. Next integrating with respect to $r$ in (49), it follows that the joint pdf of $\underset{\sim}{X}$, $U$, $\underset{\sim}{\theta}$ and $\underset{\sim}{\Lambda}$ is

$$f(\underset{\sim}{x}, u, \underset{\sim}{\theta}, \lambda)$$

$$\propto\ u^{\frac{1}{2}(N-p)-1}\lambda^{\frac{1}{2}(p-1)-2}(\underset{\sim}{\theta}^T \underset{\sim}{D}\underset{\sim}{\theta} - 2\underset{\sim}{\theta}^T \underset{\sim}{G}\underset{\sim}{x} + \underset{\sim}{x}^T \underset{\sim}{G}\underset{\sim}{x} + u)^{-\frac{1}{2}(N+p-1)-1}$$

$$= u^{\frac{1}{2}(N-p)-1}\lambda^{\frac{1}{2}(p-1)-2}$$

$$\times \left[(\underset{\sim}{\theta} - \underset{\sim}{D}^{-1}\underset{\sim}{G}\underset{\sim}{x})^T\underset{\sim}{D}(\underset{\sim}{\theta} - \underset{\sim}{D}^{-1}\underset{\sim}{G}\underset{\sim}{x}) + \underset{\sim}{x}^T(\underset{\sim}{G} - \underset{\sim}{G}\underset{\sim}{D}^{-1}\underset{\sim}{G})\underset{\sim}{x} + u\right]^{-\frac{1}{2}(N+p-3)}. \tag{50}$$

It is clear from (50) that the conditional pdf of $\underset{\sim}{\theta}$ given $\underset{\sim}{x}$, $u$ and $\lambda$ is multivariate-$t$ with location parameter $\underset{\sim}{D}^{-1}\underset{\sim}{G}\underset{\sim}{x}$, scale parameter $(N-1)^{-1}[\underset{\sim}{x}^T(\underset{\sim}{G} - \underset{\sim}{G}\underset{\sim}{D}^{-1}\underset{\sim}{G})\underset{\sim}{x} + u]\underset{\sim}{D}^{-1}$ and degrees of freedom $N-1$. On simplification, one gets

$$\underset{\sim}{D}^{-1} = \underset{\sim}{K} + \lambda\left(\sum_{i=1}^{p} n_i(n_i + \lambda)^{-1}\right)^{-1}\underset{\sim}{K}\underset{\sim}{J}_p\underset{\sim}{K}$$

$$\times \left(\underset{\sim}{K} = Diag[(n_1 + \lambda)^{-1},\ldots,(n_p + \lambda)^{-1}]\right); \tag{51}$$

$$\underset{\sim}{D}^{-1}\underset{\sim}{G} = Diag\left(n_1(n_1 + \lambda)^{-1},\ldots,n_p(n_p + \lambda)^{-1}\right)$$

$$+ \lambda\left(\sum_{i=1}^{p} n_i(n_i + \lambda)^{-1}\right)^{-1}\begin{bmatrix}(n_1+\lambda)^{-1}\\ \vdots \\ (n_p+\lambda)^{-1}\end{bmatrix}$$

$$\times\ [n_1(n_1 + \lambda)^{-1},\ldots,n_p(n_p + \lambda)^{-1}]; \tag{52}$$

$$\underset{\sim}{D}^{-1}\underset{\sim}{G}\underset{\sim}{x} = [n_1(n_1+\lambda)^{-1}x_1 + \lambda(n_1+\lambda)^{-1}\bar{x}.,\ldots,n_p(n_p+\lambda)^{-1}x_p + \lambda(n_p+\lambda)^{-1}\bar{x}.]^{\mathrm{T}}, \tag{53}$$

where $\bar{x}. = \left(\sum_{i=1}^{p} n_i(n_i + \lambda)^{-1}\right)^{-1}\left(\sum_{i=1}^{p} n_i(n_i + \lambda)^{-1}x_i\right)$. Further after much simplifications, one can write

$$\underset{\sim}{x}^T(\underset{\sim}{G} - \underset{\sim}{G}\underset{\sim}{D}^{-1}\underset{\sim}{G})\underset{\sim}{x}$$

$$= \lambda\left\{\sum_{i=1}^{p} n_i(n_i + \lambda)^{-1}x_i^2 - \left(\sum_{i=1}^{p} n_i(n_i + \lambda)^{-1}\right)^{-1}\left(\sum_{i=1}^{p} n_i(n_i + \lambda)^{-1}x_i\right)^2\right\}$$

$$= Q_\lambda(\underset{\sim}{x})\ \text{(say)}. \tag{54}$$

Integrating with respect to $\underset{\sim}{\theta}$ in (50), one finds the joint pdf of $\underset{\sim}{X}$, U and $\Lambda$ given by

$$f(\underset{\sim}{x}, u, \lambda) \propto \lambda^{\frac{1}{2}(p-1)-2} u^{\frac{1}{2}(N-p)-1} \times \left(Q_\lambda(\underset{\sim}{x}) + u\right)^{-\frac{1}{2}(N+p-3)} \left|\left(Q_\lambda(\underset{\sim}{x}) + u\right)\underset{\sim}{D}^{-1}\right|^{\frac{1}{2}}$$

$$= \lambda^{\frac{1}{2}(p-1)-2} u^{\frac{1}{2}(N-p)-1}\left(Q_\lambda(\underset{\sim}{x}) + u\right)^{-\frac{1}{2}(N-3)} |\underset{\sim}{D}|^{-\frac{1}{2}}. \tag{55}$$

Using $|\underset{\sim}{D}| \propto \left\{\prod_1^p (n_i + \lambda)\right\}\left(\sum_{i=1}^p n_i(n_i + \lambda)^{-1}\right)$, it follows from (55) that the conditional pdf of $\underset{\sim}{\Lambda}$ given $\underset{\sim}{x}$ and $u$ is

$$f(\lambda|\underset{\sim}{x}, u) \propto \lambda^{\frac{1}{2}(p-3)-1}\left\{\prod_1^p (n_i + \lambda)^{-\frac{1}{2}}\right\} \times \left\{\sum_{i=1}^p n_i(n_i + \lambda)^{-1}\right\}^{-\frac{1}{2}} [Q_\lambda(\underset{\sim}{x}) + u]^{-\frac{1}{2}(N-3)}. \tag{56}$$

From the properties of multivariate-$t$, it follows that $E(\underset{\sim}{\theta}|\underset{\sim}{x}, u, \lambda) = \underset{\sim}{D}^{-1}\underset{\sim}{G}\underset{\sim}{x}$, given in (53), and $V(\underset{\sim}{\theta}|\underset{\sim}{x}, u, \lambda) = (N-3)^{-1}[Q_\lambda(\underset{\sim}{x}) + u]\underset{\sim}{D}^{-1}$. One obtains now $E(\underset{\sim}{\theta}|\underset{\sim}{x}, u)$ and $V(\underset{\sim}{\theta}|\underset{\sim}{x}, u)$ by using (56) and the formulas

$$\begin{aligned} E(\underset{\sim}{\theta}|\underset{\sim}{x}, u) &= E[E(\underset{\sim}{\theta}|\underset{\sim}{x}, u, \Lambda)|\underset{\sim}{x}, u]; \\ V(\underset{\sim}{\theta}|\underset{\sim}{x}, u) &= V[E(\underset{\sim}{\theta}|\underset{\sim}{x}, u, \Lambda)|\underset{\sim}{x}, u] + E[V(\underset{\sim}{\theta}|\underset{\sim}{x}, u, \Lambda)|\underset{\sim}{x}, u]. \end{aligned} \tag{57}$$

As noted already, the posterior mean of $\underset{\sim}{\theta}$ is given by (53) for known $\lambda$. Ghosh and Meeden (1986) used a classical EB procedure to estimate $\lambda$ and used this estimator of $\lambda$ in (53) to obtain an estimator of $\underset{\sim}{\theta}$. Although, the resulting estimator of $\underset{\sim}{\theta}$ was quite satisfactory for point estimation purposes (see Ghosh and Lahiri, 1987), the method suffered from the earlier criticism of not modelling the uncertainty in $\lambda$. The Ghosh-Meeden procedure was not particularly suitable for the construction of credible intervals or sets.

**Shrinking Towards Regression Surfaces**

In the preceding section, the sample mean was either shrunk towards a specified point or a subspace spanned by the vector $\underset{\sim}{1}_p$. The present section generalizes the ideas of the preceding section by shrinking the sample mean towards an arbitrary regression surface. This can be achieved by using either an EB or a HB approach. The HB approach is discussed in detail in Lindley and Smith (1972) with known variance components. Morris (1983) provides a thorough discussion of the EB procedure. We attempt a synthesis between the two, and argue that Morris's EB procedure is indeed an attempt to approximate a bonafide HB procedure, and is clearly superior to a naive EB procedure.

We begin with Morris's set up, except that we assign distributions on the unknown hyperparameters, rather than estimate them on the basis of the marginal distributions of the observations. The following model is proposed.

(A) Conditional on $\underset{\sim}{\theta}$, $\underset{\sim}{b}$ and $a$, let $X_1,\ldots,X_p$ be independently distributed with $X_i \sim N(\theta_i, V_i)$, $i = 1,\ldots,p$, where the $V_i$'s are known positive constants;

(B) Conditional on $\underset{\sim}{b}$ and $a$, $\Theta_1,\ldots,\Theta_p$ are in- dependently distributed with $\Theta_i \sim N(\underset{\sim}{z}_i^T\underset{\sim}{b},\ a)$ $(i = 1,\ldots,p)$, where $\underset{\sim}{z}_1,\ldots,\underset{\sim}{z}_p$ are known regression vectors of dimension $r$ and $\underset{\sim}{b}$ is $r \times 1$.

(C) $\underset{\sim}{B}$ and $A$ are marginally independent with $\underset{\sim}{B} \sim uniform(R^r)$ and $A \sim uniform(0, \infty)$. We assume that $p \geq r+3$. Also, we write $\underset{\sim}{Z}^T = (\underset{\sim}{z}_1,\ldots,\underset{\sim}{z}_p)$; $\underset{\sim}{G} = Diag(V_1,\ldots,V_p)$ and assume rank $(\underset{\sim}{Z}) = r$.

Now the joint (improper) pdf of $\underset{\sim}{X} = (X_1,\ldots,X_p)^T$, $\underset{\sim}{\Theta} = (\Theta_1,\ldots,\Theta_p)^T$, $\underset{\sim}{B}$ and $A$ is given by

$$f(\underset{\sim}{x},\underset{\sim}{\theta},\underset{\sim}{b},a) \propto exp\left[-\frac{1}{2}(\underset{\sim}{x}-\underset{\sim}{\theta})^T\underset{\sim}{G}^{-1}(\underset{\sim}{x}-\underset{\sim}{\theta})\right] a^{-\frac{1}{2}p} exp\left[-\frac{1}{2a}\|\underset{\sim}{\theta} - \underset{\sim}{Z}\underset{\sim}{b}\|^2\right]. \tag{58}$$

Integrating with respect to $\underset{\sim}{b}$ in (58), one finds the joint (improper) pdf of $\underset{\sim}{X}$, $\underset{\sim}{\Theta}$ and $A$ given by

$$f(\underset{\sim}{x},\underset{\sim}{\theta},a)$$
$$\propto a^{-\frac{1}{2}(p-r)} exp\left(-\frac{1}{2}(\underset{\sim}{x}-\underset{\sim}{\theta})^T\underset{\sim}{G}^{-1}(\underset{\sim}{x}-\underset{\sim}{\theta}) - \frac{1}{2a}\underset{\sim}{\theta}^T[\underset{\sim}{I}_p - \underset{\sim}{Z}(\underset{\sim}{Z}^T\underset{\sim}{Z})^{-1}\underset{\sim}{Z}^T]\underset{\sim}{\theta}\right). \tag{59}$$

Write $\underset{\sim}{E}^{-1} = \underset{\sim}{G}^{-1} + a^{-1}\left(\underset{\sim}{I}_p - \underset{\sim}{Z}(\underset{\sim}{Z}^T\underset{\sim}{Z})^{-1}\underset{\sim}{Z}^T\right)$. Then, one can write

$$\begin{aligned}(\underset{\sim}{x}-\underset{\sim}{\theta})^T\underset{\sim}{G}^{-1}(\underset{\sim}{x}-\underset{\sim}{\theta}) &+ a^{-1}\underset{\sim}{\theta}^T\left(\underset{\sim}{I}_p - \underset{\sim}{Z}(\underset{\sim}{Z}^T\underset{\sim}{Z})^{-1}\underset{\sim}{Z}^T\right)\underset{\sim}{\theta}\\ &= \underset{\sim}{\theta}^T\underset{\sim}{E}^{-1}\underset{\sim}{\theta} - 2\underset{\sim}{\theta}^T\underset{\sim}{G}^{-1}\underset{\sim}{x} + \underset{\sim}{x}^T\underset{\sim}{G}^{-1}\underset{\sim}{x}\\ &= (\underset{\sim}{\theta} - \underset{\sim}{E}\underset{\sim}{G}^{-1}\underset{\sim}{x})^T\underset{\sim}{E}^{-1}(\underset{\sim}{\theta} - \underset{\sim}{E}\underset{\sim}{G}^{-1}\underset{\sim}{x}) + \underset{\sim}{x}^T(\underset{\sim}{G}^{-1} - \underset{\sim}{G}^{-1}\underset{\sim}{E}\underset{\sim}{G}^{-1})\underset{\sim}{x}.\end{aligned} \tag{60}$$

From (59) and (60) it follows that

$$E(\underset{\sim}{\Theta}|\underset{\sim}{x},a) = \underset{\sim}{E}\underset{\sim}{G}^{-1}\underset{\sim}{x}\ ;\ V(\underset{\sim}{\Theta}|\underset{\sim}{x},a) = \underset{\sim}{E}. \tag{61}$$

Write $u_i = V_i/(a+V_i)$ $(i = 1,\ldots,p)$, and $\underset{\sim}{D} = Diag(1-u_1,\ldots,1-u_p)$. Then, on simplification, it follows that

$$\underset{\sim}{E} = a(\underset{\sim}{I}_p - D) + (\underset{\sim}{I}_p - \underset{\sim}{D})\underset{\sim}{Z}(\underset{\sim}{Z}^T\underset{\sim}{D}\underset{\sim}{Z})^{-1}\underset{\sim}{Z}^T\left(a(\underset{\sim}{I}_p - \underset{\sim}{D})\right); \tag{62}$$

$$\underset{\sim}{E}\underset{\sim}{G}^{-1} = \underset{\sim}{D} + (\underset{\sim}{I}_p - \underset{\sim}{D})\underset{\sim}{Z}(\underset{\sim}{Z}^T\underset{\sim}{D}\underset{\sim}{Z})^{-1}\underset{\sim}{Z}^T\underset{\sim}{D}; \tag{63}$$

$$\underset{\sim}{E}\underset{\sim}{G}^{-1}\underset{\sim}{x} = [(1-u_1)x_1 + u_1\underset{\sim}{z}_1^T\hat{\underset{\sim}{b}},\ldots,(1-u_p)x_p + u_p\underset{\sim}{z}_p^T\hat{\underset{\sim}{b}}]^T, \tag{64}$$

where $\hat{\underset{\sim}{b}} = (\underset{\sim}{Z}^T\underset{\sim}{D}\underset{\sim}{Z})^{-1}(\underset{\sim}{Z}^T\underset{\sim}{D}\underset{\sim}{x})$. Then,

$$\underset{\sim}{G}^{-1} - \underset{\sim}{G}^{-1}\underset{\sim}{E}\underset{\sim}{G}^{-1} = a^{-1}[\underset{\sim}{D} - \underset{\sim}{D}\underset{\sim}{Z}(\underset{\sim}{Z}^T\underset{\sim}{D}\underset{\sim}{Z})^{-1}\underset{\sim}{Z}^T\underset{\sim}{D}]. \tag{65}$$

Hence,

$$\underset{\sim}{x}^T(\underset{\sim}{G}^{-1} - \underset{\sim}{G}^{-1}\underset{\sim}{E}\underset{\sim}{G}^{-1})\underset{\sim}{x}$$

$$= a^{-1}\left[\sum_{i=1}^{p}(1-u_i)x_i^2 - \left(\sum_{i=1}^{p}(1-u_i)x_i\underset{\sim}{z}_i\right)^T(\underset{\sim}{Z}^T\underset{\sim}{D}\underset{\sim}{Z})^{-1}\left(\sum_{i=1}^{p}(1-u_i)x_i\underset{\sim}{z}_i\right)\right]$$

$$= Q_a(\underset{\sim}{x}) \text{ (say)}. \tag{66}$$

Combining (59), (60) and (66), the joint pdf of $\underset{\sim}{X}$ and $A$ is given by

$$f(\underset{\sim}{x}, a) \propto |\underset{\sim}{E}|^{\frac{1}{2}} a^{-\frac{1}{2}(p-r)} exp\left[-\frac{1}{2}Q_a(\underset{\sim}{x})\right]. \tag{67}$$

Writing $\underset{\sim}{F} = \underset{\sim}{G}^{-1} + a^{-1}\underset{\sim}{I}_p$, and using Exercise 2.4, p. 32 of Rao (1973), one gets

$$|\underset{\sim}{E}^{-1}| = \begin{vmatrix} \underset{\sim}{F} & \underset{\sim}{Z} \\ \underset{\sim}{Z}^T & a(\underset{\sim}{Z}^T\underset{\sim}{Z}) \end{vmatrix} \div |a(\underset{\sim}{Z}^T\underset{\sim}{Z})|$$

$$= |\underset{\sim}{F}|\,|a(\underset{\sim}{Z}^T\underset{\sim}{Z}) - \underset{\sim}{Z}^T\underset{\sim}{F}^{-1}\underset{\sim}{Z}| \div |a(\underset{\sim}{Z}^T\underset{\sim}{Z})|$$

$$\propto a^{-p}\left\{\prod_1^p(a + V_i)\right\}|\underset{\sim}{Z}^T\underset{\sim}{D}\underset{\sim}{Z}|. \tag{68}$$

It is clear from (67) and (68) that

$$f(a|\underset{\sim}{x}) \propto a^{\frac{1}{2}r}\left\{\prod_{1}^{p}(a+V_i)^{-\frac{1}{2}}\right\}\left|\underset{\sim}{Z}^T\underset{\sim}{D}\underset{\sim}{Z}\right|^{-\frac{1}{2}}exp\left[-\frac{1}{2}Q_a(\underset{\sim}{x})\right]. \tag{69}$$

Now writing $U_i = V_i/(A + V_i)$ $(i = 1,\ldots,p)$, using (69), and the iterated formulas for conditional expectations and variances, one gets

$$E[\Theta_i|\underset{\sim}{x}] = E[E(\Theta_i|\underset{\sim}{x}, A)|\underset{\sim}{x}] = E[(1-U_i)x_i + U_i\underset{\sim}{z}_i^T\hat{\underset{\sim}{b}}|\underset{\sim}{x}]; \tag{70}$$

$$\begin{aligned} V[\Theta_i|\underset{\sim}{x}] &= V[E(\Theta_i|\underset{\sim}{x}, A)|\underset{\sim}{x}] + E[V(\Theta_i|\underset{\sim}{x}, A)|\underset{\sim}{x}] \\ &= V[(1-U_i)x_i + U_i\underset{\sim}{z}_i^T\hat{\underset{\sim}{b}}|\underset{\sim}{x}] + E[AU_i + AU_i^2\underset{\sim}{z}_i^T(\underset{\sim}{Z}^T\underset{\sim}{D}\underset{\sim}{Z})^{-1}\underset{\sim}{z}_i|\underset{\sim}{x}] \\ &= V[U_i(x_i - \underset{\sim}{z}_i^T\hat{\underset{\sim}{b}})|\underset{\sim}{x}] + E[V_i(1-U_i) + V_iU_i(1-U_i)\underset{\sim}{z}_i^T(\underset{\sim}{Z}^T\underset{\sim}{D}\underset{\sim}{Z})^{-1}\underset{\sim}{z}_i|\underset{\sim}{x}]; \end{aligned} \tag{71}$$

$$\begin{aligned} &Cov[\Theta_i, \Theta_j|x_i] \\ &= Cov[U_i(x_i - \underset{\sim}{z}_i^T\hat{\underset{\sim}{b}}), U_j(x_j - \underset{\sim}{z}_j^T\hat{\underset{\sim}{b}})|\underset{\sim}{x}] + E[AU_iU_j\underset{\sim}{z}_i^T(\underset{\sim}{Z}^T\underset{\sim}{D}\underset{\sim}{Z})^{-1}\underset{\sim}{z}_j|\underset{\sim}{x}]. \end{aligned} \tag{72}$$

Morris (1983) provides approximations for $E(\Theta_i|\underset{\sim}{x})$ and $V(\Theta_i|\underset{\sim}{x})$, $i = 1,\ldots,p$. He estimates the parameter $a$ from the marginal distribution of $X_1,\ldots,X_p$ by employing some non-Bayesian method, and substitutes this estimate in the expressions for $E[\Theta_i|\underset{\sim}{x}, a]$ and $V[\Theta_i|\underset{\sim}{x}, a]$ instead of finding posterior expectations and variances of functions involving $A$. Thus, using Morris's method, $E[\Theta_i|\underset{\sim}{x}]$ is approximated by $(1 - \hat{u}_i)x_i + \hat{u}_i\underset{\sim}{z}_i^T\hat{\hat{\underset{\sim}{b}}} = x_i - \hat{u}_i(x_i - \underset{\sim}{z}_i^T\hat{\hat{\underset{\sim}{b}}})$, while $V(\Theta_i|\underset{\sim}{x})$ is approximated by $v_i(x_i - \underset{\sim}{z}_i^T\hat{\hat{\underset{\sim}{b}}})^2 + V_i(1 - \hat{u}_i)[1 + \hat{u}_i\underset{\sim}{z}_i^T(\underset{\sim}{Z}^T\hat{\underset{\sim}{D}}\underset{\sim}{Z})^{-1}\underset{\sim}{z}_i]$, $i = 1,\ldots,p$. In the above $v_i = [2/(p-r-2)]\hat{u}_i^2(\tilde{V}+\hat{a}) \div (V_i+\hat{a})$, $i = 1,\ldots,p$, $\tilde{V} = \sum_{i=1}^{p} V_i(V_i+\hat{a})^{-1} \div \sum_{i=1}^{p}(V_i+\hat{a})^{-1}$, $\hat{D} = Diag(1-\hat{u}_1,\ldots,1-\hat{u}_p)$, and $\hat{\hat{\underset{\sim}{b}}}$ is obtained from $\hat{\underset{\sim}{b}}$ by substituting the estimator of $a$. The $v_i$'s are purported to estimate $V(U_i|\underset{\sim}{x})$'s. It is not clear whether such an approximation can be justified very rigorously since $\hat{\hat{\underset{\sim}{b}}}$ also involves the $\hat{u}_i$'s and $\hat{u}_i$ is *not* distributed independently of the $x_i - \underset{\sim}{z}_i^T\hat{\hat{\underset{\sim}{b}}}$.

We examine now how formulas (70) and (71) work in estimating the batting averages of Ty Cobb during 1905-1928. Morris (1983) took a similar undertaking except that his major emphasis was to examine whether Ty Cobb was "ever a true .400 hitter". To make our results comparable to those of Morris (1983), we fit a quadratic to Ty Cobb's batting averages, that is we take $\underset{\sim}{b} = (b_1, b_2, b_3)^T$, $\underset{\sim}{x}_i = (1, i, i^2)^T$, $i = 1,\ldots,24$. In the average year 1 refers to 1905, and year 24 refers to 1928. We provide in Table 2 the actual batting averages

TABLE 2. The Actual Batting Averages of Ty Cobb ($Y_i$), the Number of Times He Was at Bat ($n_i$), the HB Estimates ($\hat{\theta}_{i,HB}$), the Corresponding S.D.'s ($s_{i,HB}$), Morris's Approximate Estimates ($\hat{\theta}_{i,M}$), and the Corresponding S.D.'s ($s_{i,M}$).

| $i$ | $n_i$ | $Y_i$ | $\hat{\theta}_{i,HB}$ | $s_{i,HB}$ | $[\hat{\theta}_{i,HB}-2s_{i,HB}, \hat{\theta}_{i,HB}+2s_{i,HB}]$ | $\hat{\theta}_{i,M}$ | $s_{i,M}$ | $[\hat{\theta}_{i,M}-2s_{i,M}, \hat{\theta}_{i,M}+2s_{i,M}]$ |
|---|---|---|---|---|---|---|---|---|
| 1 | 150 | .240 | .298 | .020 | [.258, .338] | .303 | .026 | [.251, .355] |
| 2 | 350 | .320 | .325 | .015 | [.295, .355] | .327 | .018 | [.293, .363] |
| 3 | 605 | .350 | .344 | .013 | [.318, .370] | .345 | .015 | [.315, .375] |
| 4 | 581 | .324 | .337 | .014 | [.309, .365] | .339 | .015 | [.309, .369] |
| 5 | 573 | .377 | .366 | .014 | [.338, .394] | .366 | .015 | [.336, .396] |
| 6 | 509 | .385 | .373 | .014 | [.345, .401] | .373 | .015 | [.343, .403] |
| 7 | 591 | .420 | .393 | .016 | [.361, .425] | .393 | .017 | [.359, .427] |
| 8 | 553 | .410 | .391 | .015 | [.361, .421] | .391 | .015 | [.361, .421] |
| 9 | 428 | .390 | .384 | .015 | [.354, .414] | .385 | .015 | [.355, .415] |
| 10 | 345 | .368 | .379 | .015 | [.349, .409] | .379 | .016 | [.347, .411] |
| 11 | 563 | .369 | .379 | .014 | [.351, .407] | .380 | .014 | [.352, .408] |
| 12 | 542 | .371 | .381 | .014 | [.353, .409] | .381 | .015 | [.351, .411] |
| 13 | 588 | .383 | .386 | .013 | [.350, .412] | .386 | .014 | [.358, .414] |
| 14 | 421 | .382 | .385 | .015 | [.355, .415] | .385 | .015 | [.355, .415] |
| 15 | 497 | .384 | .385 | .014 | [.357, .413] | .385 | .014 | [.357, .413] |
| 16 | 428 | .334 | .364 | .016 | [.332, .396] | .365 | .018 | [.329, .400] |
| 17 | 507 | .389 | .383 | .014 | [.355, .411] | .383 | .014 | [.355, .411] |
| 18 | 526 | .401 | .385 | .015 | [.355, .415] | .384 | .015 | [.354, .414] |
| 19 | 556 | .340 | .355 | .014 | [.327, .383] | .356 | .015 | [.326, .386] |
| 20 | 625 | .338 | .350 | .014 | [.322, .378] | .351 | .014 | [.323, .379] |
| 21 | 415 | .378 | .362 | .015 | [.332, .392] | .361 | .016 | [.329, .393] |
| 22 | 233 | .339 | .342 | .016 | [.310, .374] | .342 | .018 | [.306, .378] |
| 23 | 490 | .357 | .342 | .015 | [.312, .372] | .342 | .017 | [.308, .376] |
| 24 | 353 | .323 | .322 | .015 | [.290, .352] | .322 | .019 | [.284, .360] |

$(Y_i)$ of Ty Cobb, the number of times he was at bat $(n_i)$, our estimated batting averages $(\hat{\theta}_{i,HB})$, the corresponding standard errors $(s_{i,HB})$, Morris's approximations $(\hat{\theta}_{i,M})$ for these batting averages, and the corresponding approximate standard errors $(s_{i,M})$. Following Morris, we took $V_i = (.367)(.633)/n_i$, $i = 1,\ldots,24$.

It follows from Table 2 that $\sum_{i=1}^{24}(\hat{\theta}_{i,HB} - Y_i)^2 = .007377$ and $\sum_{i=1}^{24}(Y_i - \hat{\theta}_{i,M})^2 = .008244$. Thus, Morris's approximations lead to about a 11.0% increase in the overall mean squared error. Also, the $s_{i,M}$'s though mostly very close to $s_{i,HB}$'s can lead upto a 30% increase. More important, our two standard deviation confidence intervals around the posterior means are usually mugh tighter than the corresponding ones given in Morris (1983). However, as mentioned earlier, Morris's EB procedure is much superior to a naive EB procedure, since the latter can seriously underestimate the actual standard errors. This is evidenced in our actual calculations which are not reported here. We should also point out that both $[\hat{\theta}_{i,HB} \pm 2s_{i,HB}]$'s and $[\hat{\theta}_{i,M} \pm 2s_{i,M}]$'s cover the true $Y_i$'s 23 out of 24 times which is approximately 95.8%. Also, $[\hat{\theta}_{i,HB} \pm s_{i,HB}]$'s and $[\hat{\theta}_{i,M} \pm s_{i,M}]$'s cover the true $Y_i$'s 17 out of 24 times which is approximately 70.8%. Thus a normal approximation to the posterior distribution is not totally out of the way.

One of Cobb's greatest claims to fame is that he has the highest lifetime batting average of any baseball player in the modern era. Ty Cobb's actual overall batting average in 1905-1928 is .367. Also, $\hat{\theta}_{HB} = \sum_{i=1}^{24}n_i\hat{\theta}_{i,HB}/\sum_{i=1}^{24}n_i = .366$ and $\hat{\theta}_M = \sum_{i=1}^{24}n_i\hat{\theta}_{i,M}/\sum_{i=1}^{24}n_i = .366$. This shows that both the HB and EB estimates of the overall batting average of Ty Cobb essentially match the reality.

It is instructive to look at the special case of *equal variances*, that is, when $V_1 = \ldots = V_p = V$. Then $u_1 = \ldots = u_p = V/(V+a) = u$ (say). In this case $\underset{\sim}{D} = (1-u)\underset{\sim}{I}_p$, $\underset{\sim}{Z}^T\underset{\sim}{D}\underset{\sim}{Z} = (1-u)\underset{\sim}{Z}^T\underset{\sim}{Z}$, $\hat{\hat{\underset{\sim}{b}}} = (\underset{\sim}{Z}^T\underset{\sim}{Z})^{-1}\underset{\sim}{Z}^T\underset{\sim}{x} = \hat{\underset{\sim}{b}}$, the usual least squares estimate of $\underset{\sim}{b}$. Moreover, $a + V = Vu^{-1}$ so that $a = V(1-u)/u$, $Q_a(\underset{\sim}{x}) = a^{-1}(1-u)SSE$, where $SSE = \sum_{i=1}^{p} x_i^2 - \left(\sum_{i=1}^{p} x_i\underset{\sim}{z}_i\right)^T(\underset{\sim}{Z}^T\underset{\sim}{Z})^{-1}\left(\sum_{i=1}^{p} x_i\underset{\sim}{z}_i\right)$, the usual error $SS$. Since $|da/du| = Vu^{-2}$, it follows from (69) that the conditional pdf of $U$ given $\underset{\sim}{x}$ is

$$f(u|\underset{\sim}{x}) \propto \left((1-u)/u\right)^{\frac{1}{2}r} u^{\frac{1}{2}p-2}(1-u)^{-\frac{1}{2}r} exp\left(-\frac{u}{2V}SSE\right)$$

$$= u^{\frac{1}{2}(p-r-4)} exp\left(-\frac{u}{2V}SSE\right). \tag{73}$$

It follows from (70) and (71) that

$$E(\Theta_i|\underset{\sim}{x}) = x_i - E(U|\underset{\sim}{x})(x_i - \underset{\sim}{z}_i^T\hat{\underset{\sim}{b}}); \tag{74}$$

$$V(\Theta_i|\underset{\sim}{x}) = V(U|\underset{\sim}{x})(x_i - \underset{\sim}{z}_i^T\hat{\underset{\sim}{b}})^2 + V - VE(U|\underset{\sim}{x})\left(1 - \underset{\sim}{z}_i^T(\underset{\sim}{Z}^T\underset{\sim}{Z})^{-1}\underset{\sim}{z}_i\right). \tag{75}$$

If one adopts Morris's approximations as in the second section, then one estimates $E(U|\underset{\sim}{x})$ by

$$\hat{\hat{U}} = \int_0^\infty u^{\frac{1}{2}(p-r-2)} exp\left(-\frac{u}{2V}SSE\right)du \Big/ \int_0^\infty u^{\frac{1}{2}(p-r-2)} exp\left(-\frac{u}{2V}SSE\right)du$$

$$= V(p-r-2)/SSE$$

and $E(U^2|\underset{\sim}{x})$ by

$$\int_0^\infty u^{\frac{1}{2}(p-r)} exp\left(-\frac{u}{2V}SSE\right)du \Big/ \int_0^\infty u^{\frac{1}{2}(p-r-4)} exp\left(-\frac{u}{2V}SSE\right)du$$

$$= V^2(p-r)(p-r-2)/(SSE)^2.$$

Accordingly, $V(U|x)$ is approximated by $2V^2(p-r-2) \div (SSE)^2 = [2/(p-r-2)]\hat{\hat{U}}^2$. These calculations suggest that $E(\Theta_i|\underset{\sim}{x})$ should be approximated by $x_i - \hat{\hat{U}}(x_i - \underset{\sim}{z}_i^T\hat{\underset{\sim}{b}})$ and $V(\Theta_i|\underset{\sim}{x})$ should be approximated by

$$s_{i,G}^2 = \left(2/(p-r-2)\right)\hat{\hat{U}}^2(x_i - \underset{\sim}{z}_i^T\hat{\underset{\sim}{b}})^2 + V\left[1 - \hat{\hat{U}}\left(1 - \underset{\sim}{z}_i^T(\underset{\sim}{Z}^T\underset{\sim}{Z})^{-1}\underset{\sim}{z}_i\right)\right]. \tag{76}$$

The expression $s_{i,G}^2$ *does not* agree with the expression $s_i^2$ given in (4.1) of Morris (1983) (with the obvious changes in his notations). It seems to us that Morris's (4.1) uses his (1.17) which involves a slight oversight. We shall discuss this point now.

Morris (1983) starts with an EB approach, where he assumes conditions (A) and (B) with $V_1 = \ldots = V_p = V$ (say). With this formula for known $\underset{\sim}{b}$ and $a$, the Bayes estimator of $\underset{\sim}{\theta}$ is given by

$$\hat{\underset{\sim}{\theta}}_B = (1-u)\underset{\sim}{X} + u\underset{\sim}{Z}\underset{\sim}{b}, \; u = V/(V+a). \tag{77}$$

If $\underset{\sim}{b}$ and $u$ are unknown, Morris (1983) estimates them by $\hat{\underset{\sim}{b}}$ and $\hat{a}$ respectively, where $\hat{b} = (Z^TZ)^{-1}Z^TX$, the least squares estimator of $b$ and $\hat{u} = (p-r-2)V/SSE$, $SSE = \sum_{i=1}^{p}(X_i - z_i^T\hat{b})^2$, the error $SS$. Note that $\hat{u}$ is the UMVUE of $u$ since marginally $SSE \sim Vu^{-1}\chi^2_{p-r}$.

Morris (1983) proposes the EB estimator $\hat{\theta}_{EB} = (\hat{\theta}_{1,EB},\ldots,\hat{\theta}_{p,EB})^T$ of $\theta$, where

$$\hat{\theta}_{i,EB} = (1-\hat{u})X_i + \hat{u}z_i^T\hat{b} \quad (i = 1,\ldots,p). \tag{78}$$

Then,

$$\begin{aligned} E(\hat{\theta}_{i,EB} - \theta_i)^2 &= E(\theta_i - \hat{\theta}_{i,B})^2 + E(\hat{\theta}_{i,B} - \hat{\theta}_{i,EB})^2 \\ &= V(1-u) + E\Big[(u-\hat{u})(X_i - z_i^T\hat{b}) + uz_i^T(\hat{b}-b)\Big]^2. \end{aligned} \tag{79}$$

Using the marginal independence of $X_i - z_i^T\hat{b}$ and $z_i^T\hat{b}$, and noting that $V(\hat{b}) = Vu^{-1}(Z^TZ)^{-1}$, it follows from (79) that

$$E(\hat{\theta}_{i,EB} - \theta_i)^2 = V(1-u) + E\Big[(u-\hat{u})^2(X_i - z_i^T\hat{b})^2\Big] + Vuz_i^T(Z^TZ)^{-1}z_i. \tag{80}$$

Since $(\hat{b}, SSE)$ is complete sufficient for $(b, u)$ and $(X_i - z_i^T\hat{b})^2/SSE$ is ancillary, they are independently distributed by Basu's (1955) theorem. Now using $E(X_i - z_i^T\hat{b})^2 = Vu^{-1}\Big(1 - z_i^T(Z^TZ)^{-1}z_i\Big)$ and $SSE \sim Vu^{-1}\chi^2_{p-r}$, it follows on simplification that

$$\begin{aligned} &E\Big[(u-\hat{u})^2(X_i - z_i^T\hat{b})^2\Big] \\ &\quad = Vu\Big(1 - z_i^T(Z^TZ)^{-1}z_i\Big) - Vu\,\frac{p-r-2}{p-r}\Big(1 - z_i^T(Z^TZ)^{-1}z_i\Big). \end{aligned} \tag{81}$$

Combining (80) and (81), it follows that

$$E(\hat{\theta}_{i,EB} - \theta_i)^2 = V - V\,\frac{p-r-2}{p-r}\Big(1 - z_i^T(Z^TZ)^{-1}z_i\Big)\,u. \tag{82}$$

In Morris's (1.17), $z_i^T(Z^TZ)^{-1}z_i = r/p$ for every $i$ which does not seem to be the case.

**Acknowledgements**

I am indebted to the referee and Mr. Gauri Sankar Datta for their constructive comments which led to a substantial improvement over the original version of the paper. I am also indebted to Dr. Li Chu Lee and Mr. Gauri Sankar Datta for carrying out the numerical computations, to Mr. Gwowen Shieh for catching a number of typos, and to Professor Carl Morris and Professor Ron Christensen for supplying me with the full batting average data of Ty Cobb.

**References**

Basu, D. (1955): On statistics independent of a complete sufficient statistic, *Sankhyā* 15, 377-380.

Berger, J. (1975): Minimax estimation of location vectors for a wide class of densities, *Ann. Statist.* 3, 1318-1328.

Berger, J. (1985): *Statistical Decision Theory and Bayesian Analysis*, 2nd Edition, Springer-Verlag, New York.

Bock, M. E. (1975): Minimax estimators of the mean of a multivariate normal distribution, *Ann. Statist.* 3, 209-218.

Casella, G. (1985): An introduction to empirical Bayes data analysis, *American Statistician* 39, 83-87.

Deely, J. J., and Lindley, D. V. (1981): Bayes empirical Bayes, *J. Amer. Statist. Assoc.* 76, 833-841.

Efron, B., and Morris, C. (1972): Limiting the risk of Bayes and empirical Bayes estimators – Part II: The empirical Bayes case, *J. Amer. Statist. Assoc.* 67, 130-139.

Efron, B., and Morris, C. (1973): Stein's estimation rule and its competitors – an empirical Bayes approach, *J. Amer. Statist. Assoc.* 68, 117-130.

Efron, B., and Morris, C. (1975): Data analysis using Stein's estimator and its generalizations, *J. Amer. Statist. Assoc.* 70, 311-319.

Faith, R. E. (1978): Minimax Bayes and point estimations of a multivariate normal mean, *J. Mult. Anal.* 8, 372-379.

Ghosh, M., and Lahiri, P. (1987): Robust empirical Bayes estimation of means from stratified samples, *J. Amer. Statist. Assoc.* 82, 1153-1162.

Ghosh, M., and Lahiri, P. (1988): A hierarchical Bayes approach to small area estimation with auxiliary information, presented at the Joint Indo. U.S. Workshop on Bayesian Analysis in Statistics and Econometrics held in Bangalore, India, December 19-23, 1988.

Ghosh, M., and Meeden, G. (1986): Empirical Bayes estimation in finite population sampling, *J. Amer. Statist. Assoc.* 81, 1058-1062.

Good, I.J. (1965): *The Estimation of Probabilities, An Essay on Modern Bayesian Methods*, M.I.T. Press, Cambridge, Massachusetts.

James, W., and Stein, C. (1961): Estimation with quadratic loss, *Proc. 4th Berk. Symp. Math. Statist. Prob.* 1, 361-380, Univ. California Press, Berkeley.

Lehmann, E. L. (1983): *Theory of Point Estimation*, Wiley, New York.

Lindley, D. (1962): Discussion of Professor Stein's paper, *J. R. Statist. Soc. B* 24, 265-296.

Lindley, D., and Smith, A. F. M. (1972): Bayes estimates for the linear model, *J. R. Statist. Soc. B* 34, 1-41.

Morris, C. (1981): Parametric empirical Bayes confidence intervals, *Scientific Inference, Data Analysis, and Robustness*, eds. G.E.P. Box, T. Leonard and C.F. Jeff Wu, Academic Press, 25-50.

Morris, C. (1983): Parametric empirical Bayes inference and applications, *J. Amer. Statist. Assoc.* 78, 47-65.

Rao, C. R. (1973): *Linear Statistical Inference and Its Applications*, 2nd Edition, Wiley, New York.

Reinsel, G. (1985): Mean squared error properties of empirical Bayes estimators in a multivariate random effects general linear model, *J. Amer. Statist. Assoc.* 80, 642-651.

Robbins, H. (1951): Asymptotically subminimax solutions of compound statistical decision problems, *Proc. 2nd Berk. Symp. Math. Statist. Prob.* V1, University of California Press, Berkeley, 131-184.

Robbins, H. (1955): An empirical Bayes approach to statistics, *Proc. 3rd Berk. Symp. Math. Statist. Prob.* V1, 157-164.

Strawderman, W. E. (1971): Proper Bayes minimax estimators of the multivariate normal mean, *Ann. Math. Statist.* 42, 385-388.

# BASU'S CONTRIBUTIONS TO THE FOUNDATIONS OF SAMPLE SURVEY

Glen Meeden, Department of Statistics, Iowa State University, Ames

## Introduction

Whenever I read a paper by Dev I am impressed with the clarity of his writing and thinking. He is able to distill the essence of the topic at hand and present it in such a way that it seems almost obvious to me. This is particularly true in the foundations of sample survey where he has elegantly demonstrated the proper role of the sufficiency and likelihood principles. Because these principles fail to justify much of the current design based practice and because he has presented his arguments in a Bayesian context some survey samplers have chosen to either ignore or attempted to modify the consequence of these principles. This coldness to Bayesian ideas in survey sampling could be considered surprising since it is the one area in statistics where everyone agrees prior information should be used.

In the next section, the results of Basu and Ghosh (1967), which characterize the minimal sufficiency partition for discrete models, will be briefly summarized. In the third section, the results of Basu (1969) will be summarized. Here he demonstrated the role of the sufficiency and likelihood principles in sample survey, from which it follows, that once the sample has been drawn the inference should not depend in any way on the sampling design. In the fourth section, some of the implications of these results will be noted. In particular, the famous *Jumbo* example of Basu (1971) will be discussed. It will be shown how Basu's argument there suggests a pseudo-Bayesian approach to survey sampling. This approach is quite flexible in that one can incorporate various levels of prior information without specifying a prior distribution. Finally, the role of random sampling in survey sampling will be discussed briefly. It should be noted that Basu (1978) contains some further reflections on his earlier work.

## Sufficiency in Discrete Models

For many years, in statistical decision theory, it has been an accepted convention, to begin by assuming the existence of a nonempty set $X$, equipped with a $\sigma$-algebra of subsets of $X$, say $\beta$, along with $\mathbb{P} = \{P_\theta | \theta \varepsilon \Omega\}$ a family of probability measures on $(X, \beta)$. One of the consequences of Basu's work (along with others) was to fit survey sampling into this scheme. For such a model it is of interest to find the minimal sufficient statistic, assuming it exists. Now, in general, for such models a minimal sufficient statistic need not exist. However, for discrete models, which includes the sample survey model, a minimal sufficient statistic always exists and is easy to find.

The triple $(X, \beta, \mathbb{P})$ is said to be a discrete model if i) $\beta$ is the class of all subsets of $X$ and ii) each $P_\theta$ is a discrete probability measure. (We are also assuming that for each $x \varepsilon X$, there exists a $\theta \varepsilon \Omega$, such that $P_\theta(\{x\}) = P_\theta(x) > 0$.) Note that a discrete model is undominated if and only if $X$ is uncountable.

Now a statistic is just a function, $T$, defined on $X$. By our choice of $\beta$ every function $T$ is measurable. Every statistic $T$ defines an equivalence relation ($x \sim x'$ if $T(x) = T(x')$) on the space $X$. This leads to a partition of $X$ into equivalent classes of points. Since we need not distinguish between statistics that induce the same partition of $X$, we may think of a statistic $T$ as a partition $\{\pi\}$ of $X$ into a family of mutually exclusive and collectively exhaustive parts $\pi$.

Using the usual measure theoretic definition of sufficiency one can prove the following factorization theorem for discrete models:

**Theorem (Basu and Ghosh, 1967).**

If $(X, \beta, \mathbb{P})$ is a discrete model, then a necessary and sufficient condition for a statistic (partition) $T = \{\pi\}$ to be sufficient is that there exists a real valued function $g$ on $X$ such that, for all $\theta \varepsilon \Omega$ and $x \varepsilon X$

$$P_\theta(x) = g(x) P_\theta(\pi_x)$$

where $\pi_x$ is the part of the partition $\{\pi\}$ that contains $x$.

Using this theorem, it is easy to find the minimal sufficient partition for a discrete model. For each $x \varepsilon X$ let

$$\Omega_x = \{\theta | \; P_\theta(x) > 0\}.$$

Consider the binary relation on $X$: "$x \sim x'$ if $\Omega_x = \Omega_{x'}$, and $P_\theta(x)/P_\theta(x')$ is a constant in $\theta$ for all $\theta \varepsilon \Omega_x = \Omega_{x'}$." This is an equivalence relationship on $X$ and defines the minimal sufficient partition.

The minimal sufficient statistic has an alternative characterization. For each $x \varepsilon X$ let $L_x(\theta)$ be the likelihood function, i.e.

$$\begin{aligned} L_x(\theta) &= P_x(\theta) && \text{for } \theta \varepsilon \Omega_x \\ &= 0 && \text{for } \theta \notin \Omega_x \end{aligned}$$

and

$$\bar{L}_x(\theta) = L_x(\theta) / \sup_\theta L_x(\theta)$$

be the standardized likelihood function. Consider the mapping

$$x \rightarrow \bar{L}_x(\cdot)$$

a mapping of $X$ into a class of real-valued functions on $\Omega$. This mapping is a minimal sufficient statistic, i.e. induces the minimal sufficient partition given above.

**The Sufficiency and Likelihood Principles in Survey Sampling**

The sufficiency and likelihood principles were widely used in other areas of statistics before their role in survey sampling was properly understood. The sufficiency principle states that if $T$ is a sufficient statistic and $T(x) = T(x')$ then the inference about $\theta$ should be the same whether the sample is $x$ or $x'$. This principle has gained wide acceptance. In discrete models since the mapping $x \to \bar{L}_x(\theta)$ is a minimal sufficient statistic, according to the sufficiency principle two sample points $x$ and $x'$ are equally informative if

$$\bar{L}_x(\theta) = \bar{L}_{x'}(\theta) \qquad \text{for all } \theta.$$

Note the sufficiency principle does not say anything about the nature of the information supplied by $x$. For this we need the likelihood principle which states that the information supplied by $x$ is just the standardized likelihood function $\bar{L}_x(\theta)$.

To see the implications of these principles in survey sampling we consider a simple survey model. Let $U$ denote a finite population of $N$ units labeled 1, $2,\ldots,N$. Attached to unit $i$ let $y_i$ be the unknown value of some characteristic of interest. For this problem

$$\theta = (y_1,\ldots,y_N)$$

is the unknown state of nature. $\theta$ is assumed to belong to $\Omega$ a subset of $N$-dimensional Euclidean space, $\mathbb{R}^N$. The statistician usually has some prior information about $\underset{\sim}{y}$ and this could influence the choice of $\Omega$. Often it is assumed that $\Omega = \mathbb{R}^N$ but this need not be so. We will assume that, in addition, associated with each unit $i$ is $m_i$, a possible vector of other characteristics all of which are known to the statistician. We assume that the $m_i$'s and their possible relationship to the $y_i$'s summarize the statisticians prior information about $\underset{\sim}{y}$.

A subset $s$ of $\{1, 2,\ldots,N\}$ is called a sample. Let $n(s)$ denote the number of elements belong to $s$. Let $S$ denote the set of all possible samples. A (nonsequential) sampling design is a function $\Delta$ defined on $S$ such that $\Delta(s)\varepsilon[0, 1]$ and $\sum_{s\varepsilon S}\Delta(s) = 1$. Given $\theta\varepsilon\Omega$ and $s = \{i_1,\ldots,i_{n(s)}\}$ where $1 \le i_1 < \ldots < i_{n(s)} \le N$ let $\theta(s) = (y_{i_1},\ldots,y_{i_{n(s)}})$. Suppose we wish to estimate the population total

$$\gamma(\theta) = \sum_{i=1}^{N} y_i$$

with squared error loss. Note $e(s, \theta)$ will denote an estimator of $\gamma(\theta)$ where

$e(s, \theta)$ depends on $\theta$ only through $\theta(s)$. If the design $\Delta$ is used in conjunction with the estimator $e$, then the risk function is

$$r(\theta; \Delta, e) = \sum_s \left[e(s, \theta) - \gamma(\theta)\right]^2 \Delta(s).$$

Typically a frequentist sampler uses the prior information summarized in the $m_i$'s to choose some design $\Delta$ and then looks for estimators which are unbiased for estimating $\gamma(\theta)$. For such an unbiased estimator the risk function is just its variance.

For such a problem a typical sample point is the set of labels of the units contained in the observed sample along with their values of the characteristic of interest. We will denote such a point by

$$\begin{aligned} x &= (s, x_s) \\ &= \left(s, (x_{i_1}, \ldots, x_{i_{n(s)}})\right) \end{aligned}$$

when $s = \{i_1, \ldots, i_{n(s)}\}$ is the observed sample.

Hence for a given design $\Delta$ the sample space is given by

$$X = \{(s, x_s) | \Delta(s) > 0 \text{ and } x_s = \theta(s) \text{ for some } \theta \varepsilon \Omega\}.$$

So for a fixed $\theta \varepsilon \Omega$ the probability function over $X$ is given by

$$\begin{aligned} P_\theta(x) = P_\theta(s, x_s) &= \Delta(s) && \text{if } x_s = \theta(s) \\ &= 0 && \text{otherwise.} \end{aligned}$$

This defines a discrete model. Note that

$$\begin{aligned} \Omega_x = \Omega_{(s, x_s)} &= \{\theta | P_\theta(x) > 0\} \\ &= \{\theta | \theta(s) = x_s\} \end{aligned}$$

from which it follows that

$$\begin{aligned} P_\theta(x) = P_\theta(s, x_s) &= \Delta(s) && \text{if } \theta \varepsilon \Omega_x \\ &= 0 && \text{elsewhere.} \end{aligned}$$

If as before, $\bar{L}_x(\cdot)$ denotes the standardized likelihood function, we see that

$$\bar{L}_x(\theta) = \bar{L}_{(s,\, x_s)}(\theta) = 1 \qquad \text{if } \theta \varepsilon \Omega_x$$
$$= 0 \qquad \text{elsewhere.}$$

Since the mapping $x \to \bar{L}_x(\cdot)$ is a minimal sufficient statistic and the likelihood function is constant over $\Omega_x$, all we learn from the observed data $x = (s,\, x_s)$ are the values of the characteristic for the units in sample and that the *true* $\theta$ must be consistent with these observed values.

Note that this observation is independent of the sampling design. That is, after the sample $x = (s,\, x_s)$ is observed the minimal sufficient statistic does not depend in any way on the value of $\Delta(s)$. (In fact, Basu demonstrated that this is true even for sequential sampling plans where the choice of a population unit at any stage is allowed to depend on the observed $y$-values of the previously selected units.) Furthermore, the principle of maximizing the likelihood function cannot be invoked to find an estimate of the population total since the standardized likelihood function is constant over $\Omega_x$.

In the next section some implications of these results will be discussed.

**Some Implications**

For most statisticians, perhaps the most unsettling aspect of Basu's argument is his demonstration that the likelihood principle implies that the design probability should not be considered in analyzing the data, after the sample has been observed. In particular, choosing an estimator which is unbiased for a given design violates the likelihood principle. But from a naive point of view this is not surprising when one recalls the *strange* way probability is used in survey sampling. Since the characteristic $y_i$ is assumed to be measured without error the only way probability enters the model is through the design $\Delta$. That is the phenomenon of randomness is not inherent within the problem but is artificially injected into it by the statistician. In other areas of statistics the statistician uses probability theory to model uncontrollable randomness while in survey sampling the whole analysis is based on a controlled *randomness* introduced by the statistician.

Godambe (1966) had noted before Basu (1969) that the application of the likelihood principle to survey sampling would mean that the sampling design is irrelevant for data analysis. But he, as many other non-Bayesian statisticians since then, has chosen to ignore the likelihood principle and tried to justify a role for the design when analyzing the data.

Scott (1977) and Sugden and Smith (1984) considered situations where some information available to the person who designed the sample is not available to the one who must analyze the data. They argued that in such situations the design may become informative. Although such examples are interesting I do not feel that they lessen the force of Basu's argument.

Recall that the likelihood principle in survey sampling justifies a very intuitively appealing notion, that is, given the observed data $x = (s,\, x_s)$ one just learns the $y_i$'s for $i \in s$ and that the unsampled $y_j$'s for $j \notin s$ must come from a $\theta$ which is consistent with $x$. So the basic question of survey sampling is how can one relate the unseen, $\theta(s') = \{y_j: j \notin s\}$, to the seen, $\theta(s) = \{y_i: i \in s\}$. Without some assumptions about how these two sets are related, knowing $\theta(s)$ does not tell one anything at all about $\theta(s')$. Presumably, for a frequentist, the design $\Delta$ along with the unbiasedness requirement is a way to relate the unseen to the seen. But I have never understood the underlying logic of the relationship.

On the other hand, the Bayesian paradigm allows one to relate the unseen to the seen in a straightforward way which does not violate the likelihood principle. Let $q(\theta)$ denote the Bayesians' prior density over $\Omega$. $q$ would be chosen to represent and summarize the statisticians prior beliefs about $\theta$. Given the sample $x = (s, x_s)$ one then computes the conditional density of $\theta$ given $x$, say $q(\theta| x)$. This is concentrated on the set $\Omega_x$ and is just $q$ with the seen, $\theta(s) = \{y_i: i \in s\}$, inserted in their appropriate places and normalized, so it integrates to one over $\Omega_x$. Then the Bayes estimator against $q$ for the populational total is

$$\sum_{i \in s} y_i + \sum_{j \notin s} E_q(y_j| x)$$

where for $j \notin s$, $E_q(y_j|x)$ is the conditional expectation of $y_j$ with respect to $q(\theta|x)$.

The form of the Bayes estimator emphasizes that estimation in survey sampling can be thought of as a prediction problem, i.e. of predicting the unseen from the seen. That is, in these problems one should argue conditionally from the seen to the unseen.

As was to be expected the Bayes estimator does not depend on the design. In most of the standard statistical decision problems an estimator is admissible if and only if it is a Bayes estimator or limit of Bayes estimators. This suggests that in survey sampling the admissibility of an estimator should not depend on a particular design. This was demonstrated in Scott (1975). Let $\Delta_1$ and $\Delta_2$ be two designs with $\Delta_1$ dominating $\Delta_2$, i.e. if $s$ is such that $\Delta_1(s) = 0$ then $\Delta_2(s) = 0$ as well. Then Scott proved if the estimator $e$ is admissible for design $\Delta_1$ then it is also admissible for design $\Delta_2$.

From the Bayesian point of view the statistician should use a design which minimizes the overall Bayes risk. In practice such designs are very difficult to find but often such minimizers are purposeful designs, i.e., designs which put probability one on a single set. Hence Basu has elegantly outlined a coherent theory of survey sampling in which random sampling or more generally the sampling design has little or no role to play. Ericson (1969) is one example of a Bayesian approach to survey sampling very much in the spirit of Basu. However, one serious difficulty in using a Bayesian approach to survey sampling is specifying a realistic prior distribution. Even for those who are somewhat sympathetic to Bayesian ideas, choosing a prior in survey sampling is almost impossible because of the larger number of parameters. Hence, it would be of interest to have an approach to survey sampling which did not violate the likelihood principle, allowed one to think conditionally given the sample, and allowed one to incorporate various levels of prior information relating the unseen to the seen without actually specifying a prior distribution. Such an approach is suggested in Basu's famous *Jumbo* example in Basu (1971).

Here Basu was discussing the Horvitz-Thompson estimator and other estimators which were suggested for some unequal probability designs. The Jumbo example dealt with estimating the total weight of a group of elephants where Jumbo was the largest.

Following Basu, let $N$ be the size of the herd and $y_i$ the weight of the $i^{th}$ elephant. Let $m_i$ be our best prior guess, before the sample is observed, of the weight of elephant $i$, that is, the $m_i$'s incorporate all our prior information about the herd. We begin by assuming that the herd is reasonably homogeneous (in contrast to Basu, there is no Jumbo). Suppose a sample $s$ with $n(s) = n > 1$ is chosen and the corresponding $y_i$'s observed. Suppose we believe that these $n$ observed ratios $\{y_i/m_i\colon i\epsilon s\}$ are *representative* of the $N - n$ unobserved ratios $\{y_j/m_j\colon j\not\epsilon s\}$. Although we may not be able to define *representative* we have an intuitive idea of what it means. Furthermore, if in practice we obtained a sample which we believed was not representative then we would be foolish to act as if it were.

Assuming the sample is representative then Basu suggested that $\bar{r} = \frac{1}{n}\sum_{i\epsilon s}(y_i/m_i)$ should be a good guess for $y_j/m_j$ when $j\epsilon s'$. Hence, for a typical unsample unit $j$, a reasonable estimate of $y_j$ is $m_j\bar{r}$. This suggests a sensible estimate of the population total is

$$\sum_{i\epsilon s} y_i + \left[\frac{1}{n}\sum_{i\epsilon s}(y_i/m_i)\right]\sum_{j\not\epsilon s} m_j. \tag{1}$$

This estimator can be given a pseudo Bayesian justification by creating a *posterior distribution* for the unseen given the seen which is appropriate when one believes the sample is representative. Suppose in the sample of $n$ observations there are $r$ distinct values of these ratios, say $\alpha_1,\ldots,\alpha_r$. Let $k_j$ be the number of observed $y_i/m_i$'s which are $\alpha_j$ for $j = 1,\ldots,r$. Construct an urn which contains $n$ balls where $k_j$ are labeled $\alpha_j$ for $j = 1,\ldots,r$. Then take as the pseudo posterior distribution for the $N - n$ unobserved ratios the distribution generated by simple Polya sampling from the urn. To begin, a ball is chosen at random from the urn and the observed value is given to the unobserved ratio with the smallest label. This ball and an additional ball with the same value are returned to the urn. Another ball is chosen from the urn and its value is given to the unobserved ratio with the next smallest label. This ball and another with the same value are returned to the urn. The process is continued until all $N - n$ unobserved ratios are given a value. We will call this pseudo posterior the *Polya posterior* for the unseen given the seen. The *Polya posterior* is a pseudo posterior because it does not arise from any single prior distribution over the parameter space. This is intuitively clear since it is data dependent. On the other hand, it does reflect the belief that the unseen are like the seen. Finally it is easy to check that the Bayes estimate of the population total using the *Polya posterior* is just the estimate given in (1).

Note in the special case when little is known about the herd, i.e. all the $m_i$'s are equal, then the estimator in (1) reduces to $(N/n)\sum_{i\varepsilon s} y_i$ which is the classical estimator of the population total.

In Meeden and Ghosh (1983) the estimator given in (1) was shown to be admissible. The proof used the stepwise Bayes technique. In the proof the *Polya posterior* played a crucial role. Hence, Basu's argument not only gives an intuitive justification for the estimator (1) but suggests a method for proving its admissibility. This approach can be extended to prove the admissibility of a variety of other estimators. (See Vardeman and Meeden (1984) for details.)

For example, suppose the population can be stratified into various strata each of which is relatively homogeneous. If the sample contains units from each stratum then the estimator in (1) can be used within each stratum, where within each stratum the $m_i$'s are assumed to be equal, to produce an estimate of the population total. If in a given stratum, say $k$, we decide to sample $n_k$ units then the stepwise Bayes argument shows that any set of $n_k$ units within the stratum is optimal. That is, we may choose our $n_k$ units by simple random sampling without replacement. This type of argument gives an noninformative Bayesian justification for a variety of the usual estimators in survey sampling along with a justification for choosing the sample at random.

One can argue that it is a relatively weak justification since it justifies any method of selecting the sample. In spite of Basu's arguments even some through going Bayesians, still admit to being attracted to the notion of randomization even though they do not know any intellectual justification for it. I however find Basu's statement on page 594 of Basu (1980), in slightly different context, quite compelling.

"I have no objection to prerandomization as such. Indeed, I think that the scientist ought to prerandomize and have the physical art of randomization properly witnessed and notarized. In this crooked world, how else can he avoid the charge of doctoring his own data?"

**References**

Basu, D. (1969): Role of sufficiency and likelihood principles in sample survey theory, *Sankhyā A* 31, 441-454.

Basu, D. (1971): An essay on the logical foundations of survey sampling, part one, *Foundations of Statistical Inference*, Holt, Reinhart and Winston, Toronto, 203-242.

Basu, D. (1978): *On the Relevance of Randomization in Data Analysis in Survey Sampling and Measurement*, N. K. Namboodiri, ed., Academic Press, New York, 267-292.

Basu, D. (1980): Randomization analysis of experimental data: The Fisher randomization test (with comments and rejoinder), *Journal of the American Statistical Association* 75, 575-595.

Basu, D. and Ghosh, J. K. (1967): Sufficient statistics in sampling from a finite universe, *Bull. Int. Stat. Inst.* 42, BK. 2, 850-859.

Godambe, V. P. (1966): A new approach to sampling from finite populations, I: Sufficiency and linear estimation, *Journal of the Royal Statistical Society B* 28, 310-319.

Meeden, G. and Ghosh, M. (1983): Choosing between experiments: Application to finite population sampling, *Annals of Statistics* 11, 296-305.

Scott, A. J. (1975): On admissibility and uniform admissibility in finite population sampling, *Annals of Statistics* 3, 489-491.

Scott, A. J. (1977): On the problem of randomization in survey sampling, *Sankhyā A* 39, 1-9.

Sugden, R. A. and Smith, T. M. F. (1984): Ignorable and informative designs in survey sampling inference, *Biometrika* 71, 495-506.

Vardeman, S. and Meeden, G. (1984): Admissible estimators for the total of a stratified population that employ prior information, *Annals of Statistics* 12, 675-684.

# SURVEY SAMPLING – AS I UNDERSTAND IT
## (A Development of Optimality Criterion)

V. P. Godambe, University of Waterloo

This was the Gold Medalist Presentation at the Statistical Society of Canada meetings held in Victoria, 6th June 1988.

> For since the fabric of the universe is most perfect and the work of a most wise Creator, nothing at all takes place in the universe in which some rule of maximum or minimum does not appear.
> – Leonhard Euler

**Introduction**

This is a brief overview of the historical development of the optimality criterion in survey sampling theory and practice. The presentation here has been considerably simplified for it takes for granted a fundamental result. In survey sampling set-up the entire data can be effectively summarized by the set of observed units (or individual labels) together with the corresponding variate values as in (1) to follow. This is a basic discovery due to Basu. He (1958) proved that in survey-sampling set-up (1) constitutes a *minimal sufficient statistic.*

**Definitions, Notation and the Problem**

Survey Population $P$ is a finite collection of individuals (houses, blocks, farms, households, etc.), each bearing a distinctive label $i$; we may write

$$P = \{i{:}1,\ldots,N\},$$

where $N$ is the size of $P$. Variate under study such as income, size, produce, etc. is denoted by $y$. The value of $y$ associated with the individual $i$ is $y_i$, $i = 1,\ldots,N$.

We want to estimate some *unknown* characteristic, say the mean

$$\bar{Y} = \sum_1^N y_i \Big/ N$$

of the population $P$. For this purpose a sample $s$ of size $n$ is drawn from $P$ ($s \subset P$), using a *sampling design* (simple random sampling or stratified sampling, etc.) and the values $y_i$, $i \in s$ are ascertained through a survey.

Problem I: To estimate $\bar{Y}$ given the data

$$d = \{(i, y_i) : i \in s\} \tag{1}$$

and the sampling design. (A related problem is, how to use the pre-survey knowledge about $P$, particularly in the choice of a sampling design?)

For historical reasons the above problem remained confused, until recently, with the following quite different problem.

A treatment is tried $n$ times with the following results

$$y_1, y_2, \ldots, y_n. \tag{2}$$

Problem II: To estimate the average treatment effect $\theta$ on the basis of the data (2).

## Fundamental Distinction

The fundamental distinction between the two problems above becomes at once clear by the fact that while in problem (II), the sample mean $\sum_1^n y_i/n$ is the *unbiased minimum variance* (UMV) estimate for "$\theta$", the corresponding mean $\sum_{i \in s} y_i/n$ in problem (I) is *not* UMV for $\bar{Y}$, even for a simple random sampling design.

The above phenomenon, as is now well understood, is due to the existence of individual labels "$i$" in the data (1), unlike in data (2). "$\bar{Y}$" in problem (I) is the mean of the *actual* (survey) population. In contrast "$\theta$" in problem (II), is the mean of a *hypothetical* population generated by repeated (independent) trials of the treatment.

Why was problem (I) confused with problem (II) for a long time?

Answer: When the survey sampler arrived on the statistical stage (at about the beginning of this century), there already was a statistical theory developed by Galton, Pearson and others (to study primarily biological phenomena) which essentially dealt with problems akin to (II) of *hypothetical* populations. The confusion arose out of the attempts of the early survey samplers to use the then existing statistical theory to solve problem (I) concerning the *actual* (survey) population.

## Historical Comments

Today's popular understanding of statistics consists of probabilistic estimates, say for instance, of country's average income, based on some random samples. But essentially this meshing of probability calculus with actual social statistics, historically proved to be far more formidable than establishing central limit theorem or Bayes theorem and the like. Actually both social statistics (Graunt) and probability theory (Pascal & Fermat) originated around 1660, but the meshing of the two occurred only in this century. Even in earlier history (for instance Jewish & Jain literature) one can find discussions of *uncertain* (probabilistic) inference; almost none relate to survey sampling. One exception I have temptation to quote. This is from Mahabharat, the old Indian epic (Vana-Parva; Nala-Damayanti Akhyan).

The God Kali has his eye on a beautiful princess and is dismayed when Nala wins her hand. In revenge an evil spirit enters the body of the virtuous prince. Crazed with frenzy for gambling, Nala loses his kingdom, and wanders demented for many years. Nala's change of fortune is described in a remarkable anecdote.

In an alien form, he has been travelling with another king, Bhangasuri. This latter, wanting to flaunt his skill in numbers, estimates the number of leaves, and the number of fruit, on two great branches of a spreading tree. There are, he avers, 2,095 fruits. Nala counts all night and is duly amazed by morning. Bhangasuri accepts his due:

I of dice possess the science, and in numbers thus am skilled.

He agrees to teach this science to Nala in exchange for some classes in horsemanship, in which, despite his exile, Nala still excels. At the end of this sensational course in survey-sampling Nala vomits out the poison of Kali, and is restored his normal form. Kali, exorcised by mathematics, retires to the tree. Nala returns to his kingdom, offers his still faithful bride as his final stake and quickly recoups all his losses, and lives happily ever after.

(Reproduced from History and Philosophy of Science Seminar by Ian Hacking)

**Neyman's UMV-Criterion**

The first well publicized attempt to solve the survey sampling problem, Problem I, using the then available statistical theory developed by Galton, Pearson, Fisher and others was due to Neyman, 1934. Actually this theory, as said before, was meant for hypothetical populations of Problem II. Following this theory, for simple random sampling (with or without replacement) Neyman considered the class of unbiased estimates (for the population mean $\bar{Y}$) of the form

$$\sum_{r=1}^{n} a_r y_r$$

where $a_r$ is the coefficient associated with the $r^{th}$ draw and $y_r$ is the observed value of $y$ at the $r^{th}$ draw. The variance of this estimate is minimized, Neyman argued, using Gauss-Markov theorem, for $a_r = 1/n$, $r = 1,\ldots,n$. In this sense, Neyman demonstrated the UMV-ness of the sample mean. Similarly for stratified sampling he established UMV-ness of the corresponding weighted mean. (Similar previous, but little known results are due to Tchuprow (1923); see

Bellhouse, 1987.) In retrospect it appears Neyman obtained UMV estimates by restricting himself to the class of estimates which depended on individual labels $i$, *only* to the extent they determined the stratum to which the individual belonged. That is, *labels were ignored within each stratum.*

For several years, following Neyman, survey samplers investigated UMV estimation for more sophisticated designs than stratification. For reducing variance of estimates Hansen and Hurwitz (1943) introduced unequal probability sampling. Here however individual labels ($i$) were used not just for stratification but also were used even *within strata.* That is, in a stratum, two individuals could be selected with different probabilities.

What happened to Neyman's UMV-estimation here? Using individual labels $i$, Horwitz and Thompson (1952) constructed three different classes of estimates and investigated UMV estimation in each class. Though these latter investigations were inconclusive, the work clearly established that wider classes of estimates, than those considered by Neyman, could be constructed, using individual labels.

Neyman's introduction of UMV estimation in survey sampling led to an improved practice of stratified sampling, a better understanding of randomization and finally suggested the innovation of unequal probability sampling and general sampling designs.

Here however, the UMV-criterion appeared to have reached its limits of usefulness.

During 1935-1955 and even afterwards, while comparing variances of different estimates, possibly under different designs, proved to be rewarding, a search for UMV estimation led to futile confusion mentioned earlier; for such estimation was generally nonexistent!

Godambe (1955) introduced a general class of label dependent estimates of which all the known estimates were special cases. For this class, he demonstrated that UMV estimation was nonexistent, for any sampling designs (trivial exceptions apart). Particularly the sample mean was *not* UMV for the simple random sampling design.

Looking back, it would appear that survey samplers made considerable progress in sampling practice and theory, in their search for the nonexistent UMV estimation! But such things can happen in Science. Or one may say, survey samplers, in their investigation of UMV estimation, *informally* restricted themselves to the use of labels only to the extent they *intuitively* looked useful. This was the case with Neyman (1934). For a general development of this approach we refer to Hartley and Rao (1968).

## A New Criterion: UMSV

Godambe (1955) also showed that in the class of all (linear) label dependent unbiased estimates, for the population mean $\bar{Y}$, the $HT$-estimate

$$e_{HT} = \frac{1}{N} \sum_{i \in s} y_i/\pi_i, \tag{3}$$

(due to Horwitz and Thompson, 1952), where $\pi_i$ is the *probability of including the individual* $i$ in the sample $s$ drawn by the specified sampling design, has minimum *expected* variance. Here expectation is w.r.t. any distribution belonging to a *class* of distributions on the variate values $(y_1,\ldots,y_i,\ldots,y_N)$ under study. This class of distributions, called a *Superpopulation Model* (SPM), is supposed to be a formalization of our pre-survey knowledge of the survey-population $P$ (see next section for illustration). Thus w.r.t. the SPM the $HT$-estimate is UM$\mathcal{E}$V: $U \equiv$ unbiased, $V \equiv$ variance, w.r.t. sampling design and $\mathcal{E} \equiv$ expected w.r.t. the SPM. Note that many estimates in common use, such as the sample mean for simple random sampling and the appropriately weighted mean for stratified sampling are but special cases of the estimate $e_{HT}$ in (3). Hence they are UM$\mathcal{E}$V w.r.t. suitable SPMs.

Actually, since much earlier than 1955, variances had been compared in terms of their expectations w.r.t. the SPM (Cochran, 1939). Thus in the absence of UMV-estimation its replacement by UM$\mathcal{E}$V-estimation seemed natural. By now UM$\mathcal{E}$V-criterion seems to have received a general acceptance in theory as well as in practice. It is also used, somewhat reluctantly though, by *Model Theorists* in sampling.

The discovery that in survey-sampling the likelihood function is independent of the sampling design and hence according to the "*Likelihood Principle*" (LP) the inference must be independent of the design (randomization) probabilities (Godambe, 1966), gave impetus to the development of the model theory (Royall, 1970). This theory, to implement the above conclusion of LP, *restricts* inference/estimation exclusively to the probabilities given by SPM. (Such restriction was previously proposed by Brewer (1963), but he did not tie it to the LP. For this reason, possibly, Brewer's work was not effective in the development of model theory. By this time due to the works of Barnard, Birnbaum and Savage, LP became respectable.) With this restriction, the model theory estimation, using the notation above, proceeds as follows.

For a *given (fixed)* sample $s$, in terms of $y_i$: $i \in s$ construct the *class* of all linear estimates which are SPM-unbiased for the survey-population characteristic, say $\bar{Y}$. From this class, the minimum variance estimate (SPM-UMV) is recommended, for practical use, by the model theory.

Now for any sample $s$, the SPM-UMV estimate exists for rather restrictive SPMs. On the other hand when design and model probabilities are combined one can obtain UM$\mathcal{E}$V estimates (or close approximations) for far more flexible SPMs incorporating nuisance parameters of high dimension (Godambe, 1982, 1983).

Anyway even model theorists, in an attempt to make their estimation robust (to departures from the assumed SPM), have relied on the UM$\mathcal{E}$V-criterion (Brewer, 1979). Actually, from the model theorists ideological criticism and rejection, randomization emerged with new meaning, vigor and applications.

As I mentioned before, the UMV-criterion led to better understanding and practice of stratified sampling; the same thing can be said to have been achieved by the UMSV-criterion for unequal probability sampling beyond stratification.

Yet, the UMSV-criterion is rather restrictive. It is generally non-vacuous only for fixed sample size designs. As mentioned before the HT-estimator is UMSV—but generally only for fixed sample size designs. It is absurd for the following (rather *extreme*) random sample size design: with probability 1/2, a random sample of size "1" is drawn, and with remaining probability 1/2, the *whole population* is sampled. Now when the whole population is sampled, the *HT*-estimate (3) of the population mean $\bar{Y}$ is approximately $2\bar{Y}$!. Yet *random sample size* designs do occur in practice. For instance in surveys having *non-respondents* the (effective) sample size is essentially a random variate. The same thing happens for domain estimation.

Just as the extension of the UMV criterion to the UMSV criterion was necessary to cover label dependent estimates, a further *extension* of the UMSV itself is necessary to cover *random sample size designs.* This is achieved by the UMSV-f criterion introduced in the next section. With this introduction, we can use even more flexible/broader SPMs than was possible under the UMSV criterion. This will be clear soon.

## UMSV-f Criterion

Here we present the work of Godambe and Thompson (1986a). In addition to the notation above we denote by $x_i$ the *covariate* value associated with the individual $i$, $i = 1,\ldots,N$. We *assume* $\boldsymbol{x} = (x_1,\ldots,x_i,\ldots,x_N)$ *known* and the SPM to be a *class* of distributions on $(y_1,\ldots,y_N)$ satisfying the following conditions:

(I) Given the covariate $\boldsymbol{x}$, $\boldsymbol{y} = (y_1,\ldots,y_i,\ldots,y_N)$ are distributed mutually independently.

(II) With respect to any distribution in the class the expectation $\mathcal{E}(y_i-\theta x_i) = 0$, $i = 1,\ldots,N$.

That is, under the SPM, $\theta$ is the regression parameter, intercept terms being ignored for simplicity. We define

$$\tilde{g} = \sum_1^N (y_i-\theta x_i); \tag{4}$$

$\tilde{g}$ is said to be a population or $\boldsymbol{y}$-based *unbiased estimating function*, since $\mathcal{E}(\tilde{g}) = 0$. If $[\tilde{g} = 0] \Rightarrow [\theta = \theta_N]$, $\theta_N$ is a $\boldsymbol{y}$-based estimate of the SPM-parameter $\theta$. Further $\theta_N = \left(\sum_1^N y_i/\sum_1^N x_i\right)$ is itself a Survey Population parameter. Godambe and Thompson (1986a) theory provides *optimal* (sample based) estimation for $\theta_N$ as follows: Let $h(d, \theta)$ be any function of the parameter $\theta$ and the data $d = \{(i, y_i) : i \in s\}$ in (1), with

$$E(h - \tilde{g}) = 0, \tag{5}$$

"$E$" being the expectation under the sampling design, holding $\boldsymbol{y}$ and $\tilde{g}$ in (I) and (4) above fixed. The function $h$ satisfying (5) is called a (design) *unbiased* estimating function; a solution of the equation $h(d, \theta) = 0$, provides an (data $d$ based) estimate of both the parameters $\theta_N$ and $\theta$. Now the function $h^*(d, \theta)$ satisfying (5) is said to be UM$\mathcal{E}$V-f Optimum (f for estimating function), if for any $h$ satisfying (5)

$$\mathcal{E}E(h^* - \tilde{g})^2 \leq \mathcal{E}E(h - \tilde{g})^2 \tag{6}$$

where $\mathcal{E}$ as before is the expectation w.r.t. the SPM-I&II above.

**Theorem.** For SPM-I&II, and any sampling design with $\pi_i > 0$, $i = 1,\ldots,N$, UM$\mathcal{E}$V-f $h^*$ is given by

$$h^* = \sum_{i \in s} (y_i - \theta x_i)/\pi_i. \tag{7}$$

Solving the equation $h^* = 0$, we get for $\theta$ and $\theta_N$, the optimum estimate

$$e = \frac{\sum_{i \in s}(y_i/\pi_i)}{\sum_{i \in s}(x_i/\pi_i)}. \tag{8}$$

As a special case for all $x_i \equiv 1$, in (8),

$$e = \frac{\sum_{i \in s}(y_i/\pi_i)}{\sum_{i \in s}(1/\pi_i)}. \tag{9}$$

The relationship between the estimates $e$ in (9) and the $HT$-estimate $e_{HT}$ in (3) is given by the fact that for any sampling design

$$E\left\{\sum_{i \in s}(1/\pi_i)\right\} = N.$$

Note, now, for the random sample size design, considered before in previous section, $\pi_i = (N+1)/2N$, $i = 1,\ldots,N$ and when the whole population is sampled $e$ in (9) unlike $e_{HT}$ in (3) equals $\bar{Y}$!

A generalization of the theorem just stated is obtained by replacing in (II) $y_i - \theta x_i$ by any function

$$\pi_i(y_i, \theta),$$

covering many practical situations including (optimal) estimation of quantiles. The appeal, to the practitioners, of this approach is evident from the fact that special cases of the function $\phi_i$ above were already in common use (Binder, 1983) before the present theory (Godambe & Thompson, 1986a) was developed. For further applications we refer to a later paper of Godambe and Thompson (1986b).

**References**

Basu, D. (1958): On sampling without replacement, *Sankhya* 20, 287-294.

Bellhouse, D. R. (1988): A brief history of random sampling methods, in *Handbook of Statistics*, Volume 6, P. R. Krishnaiah and C. R. Rao (eds.), Neth. Holland, 1-14.

Binder, D. A. (1983): On the variances of asymptotically normal estimators from complex surveys, *Int. Statist. Rev.* 51, 279-292.

Brewer, K. W. R. (1963): Ratio estimation and finite populations: Some results deducible from the assumption of an underlying stochastic process, *As. J. Statist.* 5, 93-105.

Brewer, K. W. R. (1979): A class of robust sampling designs for large-scale surveys, *J. Amer. Statist.* 74, 911-915.

Cochran, W. G. (1939): The use of analysis of variance in enumeration by sampling, *J. Amer. Statist. Ass.* 34, 492-510.

Godambe, V. P. (1955): A unified theory of sampling from finite populations, *J. R. Statist. Soc.* 28, 269-278.

Godambe, V. P. (1966): A new approach to sampling from finite populations, *J. R. Statist. Soc.* 3, 310-328.

Godambe, V. P. (1982): Estimation in survey sampling: Robustness and optimality, *J. Amer. Statist. Ass.* 77, 393-406.

Godambe, V. P. (1983): Survey-sampling: Modelling, randomization and robustness–a unified theory view, *Proc. Amer. Statist. Ass.* 26-29.

Godambe, V. P. and Thompson, M. E. (1986a): Parameters of superpopulation and survey population: Their relationships and estimation, *Int. Statist. Rev.* 54, 127-138.

Godambe, V. P. and Thompson, M. E. (1986b): Some optimality results in the presence of nonresponse, *Survey Methodology* 12, 29-36.

Hansen, M. H. and Hurwitz, W. N. (1943): On the theory of sampling from finite populations, *Ann. Math. Statist.* 14, 333-362.

Hartley, H. O. and Rao, J. N. K. (1968). A new estimation theory for sampling surveys, *Biometrika* 55, 547-557.

Horwitz, D. G. and Thompson, D. J. (1952). A generalization of sampling without replacement from a finite universe, *J. Amer. Statist. Ass.* 47, 663-685.

Neyman, J. (1934). On two different aspects of the representative method: The method of stratified sampling and the method of purposive selections, *J. Roy. Statist. Soc.* 97, 558-625.

Royall, R. M. (1970). On finite population sampling theory under certain linear regression models, *Biometrika* 57, 377-387.

Tchuprow, A. A. (1923). On the mathematical expectation of the moments of frequency distributions in the case of correlated observations, *Metron* 2, 461-493, 646-680.

# TWO BASIC PARTIAL ORDERINGS FOR DISTRIBUTIONS DERIVED FROM SCHUR FUNCTIONS AND MAJORIZATION

Kumar Joag-Dev, University of Illinois and
Florida State University

and

Jayaram Sethuraman, Florida State University

**Abstract**

Researchers in applied fields have long recognized the usefulness of inequalities when exact results are not available. The use of inequalities allows us to say that one estimate is better than another, that one maintenance policy is better than another or that a certain selection procedure is better than another etc., even though, we may not know the best estimator, the best maintenance policy or the best selection procedure. Such results are generally obtained from inequalities between two probability measures or random variables. Inequalities between random variables are in turn obtained from deterministic inequalities or deterministic partial orderings.

Hardy, Littlewood and Pólya (1952) in their classical book entitled *Inequalities* have discussed various partial orderings in $R^n$, one of which is known as majorization. Majorization is intimately related to Schur functions. This partial ordering was used to derive the partial orderings of stochastic majorization and DT ordering among distributions in a series of papers by Proschan and Sethuraman (1977); Nevius, Proschan and Sethuraman (1977); Hollander, Proschan and Sethuraman (1977); and Hollander, Proschan and Sethuraman (1981). Even though many more partial orderings of this type have been studied in recent papers and books by Marshall and Olkin (1979), Tong (1980), Boland, Tong and Proschan (1987, 1988), Abouammoh, El-Neweihi and Proschan (1989), the above two partial orderings remain the centerpiece in this type of research endeavor. In this expository paper, we describe the essentials of stochastic majorization and DT ordering and demonstrate some applications. A new proof of a slight generalization of earlier result on DT functions in Hollander et al., 1981 is given.

## Introduction

Researchers in applied fields have long recognized the usefulness of inequalities when exact results are not available. The use of inequalities allows

---

Research supported by the United States Army Research Office, Durham, under Grant No. DAAGLO3 86-K-0094. The United States Government is authorized to reproduce and distribute reprints for governmental purposes. FSU Technical Report Number M-814; USARO Technical Report Number D-109, September 1989.

us to say that one estimate is better than another, that one maintenance policy is better than another or that a certain selection procedure is better than another etc., even though, we may not know the best estimator, the best maintenance policy or the best selection procedure. Such results are generally obtained from inequalities between two probability measures or random variables. Inequalities between random variables are in turn obtained from deterministic inequalities or deterministic partial orderings.

Hardy, Littlewood and Pólya (1952) in their classical book entitled *Inequalities* have discussed various partial orderings in $R^n$, one of which is known as majorization. Majorization is intimately related to Schur functions. This partial ordering was used to derive the partial orderings of stochastic majorization and DT ordering among distributions in a series of papers by Proschan and Sethuraman (1977) [PS 77]; Nevius, Proschan and Sethuraman (1977) [NPS 77]; Hollander, Proschan and Sethuraman (1977) [HPS 77]; and Hollander, Proschan and Sethuraman (1981) [HPS 81]. Even though many more partial orderings of this type have been studied in recent papers and books by Marshall and Olkin (1979), Tong (1980), Boland, Tong and Proschan (1987, 1988), Abouammoh, El-Neweihi and Proschan (1989), the above two partial orderings remain the centerpiece in this type of research endeavor. In this expository paper, we describe the essentials of stochastic majorization and DT ordering and demonstrate some applications in the second and third sections. A new proof of a slight generalization of earlier result on DT functions is given in the third section.

**Schur Functions**

We begin by reviewing some basic concepts and results involving Schur functions. Given a vector $\boldsymbol{x} = (x_1, x_2, \ldots, x_n)$, let $x_{[1]}, x_{[2]}, \ldots, x_{[n]}$ be a permutation of its co-ordinates satisfying $x_{[1]} \geq x_{[2]} \geq \ldots \geq x_{[n]}$. A vector $\mathbf{x}$ is said to *majorize* a vector $\boldsymbol{y}$, $\boldsymbol{x} \overset{m}{\geq} \boldsymbol{y}$ in symbols, if

$$\sum_{i=1}^{j} x_{[i]} \geq \sum_{i=1}^{j} y_{[i]}, \qquad j = 1, 2, \ldots, n-1,$$

and

$$\sum_{i=1}^{n} x_{[i]} = \sum_{i=1}^{n} y_{[i]}.$$

Majorization is not a true partial ordering on $R^n$ since $\boldsymbol{x} \overset{m}{\geq} \boldsymbol{y}$ and $\boldsymbol{y} \overset{m}{\geq} \boldsymbol{x}$ implies only that the co-ordinate sequence of $\boldsymbol{x}$ is a permutation of the co-ordinate sequence of $\boldsymbol{y}$. However it is a partial ordering in the cone $\{\boldsymbol{x} : \boldsymbol{x} \in R^n, x_1 \geq x_2 \geq \ldots x_n\}$. In any case, $\boldsymbol{x} \overset{m}{\geq} \boldsymbol{y}$ means that the co-ordinates of $\boldsymbol{x}$ are more spread out than those of $\boldsymbol{y}$.

A measurable function $f$ defined on $R^n$ will be called a *Schur function* if it is either *Schur-convex*, that is, if $f(\boldsymbol{x}) \geq f(\boldsymbol{y})$ whenever $\boldsymbol{x} \overset{m}{\geq} \boldsymbol{y}$, or is *Schur-concave*, that is, if $f(\boldsymbol{x}) \leq f(\boldsymbol{y})$ whenever $\boldsymbol{x} \overset{m}{\geq} \boldsymbol{y}$. It is easy to construct Schur functions from the example below.

**Example 1**

Let $f(\boldsymbol{x}) = \sum_1^n g(x_i)$. Then $f(\boldsymbol{x})$ is Schur convex if and only if $g$ is Schur convex.

A subset A of $R^n$ is called *Schur increasing* if it satisfies:

$$\boldsymbol{x} \in \mathrm{A},\ \boldsymbol{y} \overset{m}{\geq} \boldsymbol{x} \Rightarrow \boldsymbol{y} \in \mathrm{A}.$$

Note that the indicator of a Schur increasing set is a Schur *convex* function and in fact such indicators are the building blocks of the class of Schur convex functions and act as their level sets.

A partial ordering for random vectors can be defined as follows using Schur increasing sets. Let $\boldsymbol{X}$ and $\boldsymbol{X}'$ be random $n$-vectors. Then $\boldsymbol{X}$ is said to *stochastically majorize* $\boldsymbol{X}'$ if for every Schur increasing set A in $R^n$, $P[\boldsymbol{X} \in \mathrm{A}] \geq P[\boldsymbol{X}' \in \mathrm{A}]$, or equivalently, $E[f(\boldsymbol{X})] \geq E[f(\boldsymbol{X}')]$, for every bounded Schur convex function $f$ on $R^n$. This is stated, in symbols, as $\boldsymbol{X} \overset{st.m.}{\geq} \boldsymbol{X}'$.

Stochastic majorization is a way of comparing distributions of random vectors in much the same way as the stochastic ordering is for comparing distributions functions of real random variables. In fact stochastic majorization can be equivalently defined as stochastic ordering between certain transformed random vectors. Recall that $Z$ is said to be *stochastically larger* than $Z'$ if for every bounded nondecreasing function $h$, $E[h(Z)] \geq E[h(Z')]$. Consider the transformation $\boldsymbol{y} = (y_1, y_2,\ldots,y_n) \overset{def}{=} T(\boldsymbol{x})$, where $y_i = \sum_{j=1}^{i} x_{[j]}$, $i = 1, 2,\ldots,n$. It is clear that $C \overset{def}{=} TR^n$ is a cone. Let $\boldsymbol{X}$ and $\boldsymbol{X}'$ be two random vectors and let $\boldsymbol{Y} = T\boldsymbol{X}$ and $\boldsymbol{Y}' = T\boldsymbol{X}'$. Then it is easy to see that $\boldsymbol{X} \overset{st.m.}{\geq} \boldsymbol{X}'$ if and only if $E[g(\boldsymbol{Y})] \geq E[g(\boldsymbol{Y}')]$ for all bounded measurable functions $g$ such that $g(\boldsymbol{y}) \geq g(\boldsymbol{y}')$ whenever $y_i \geq y_i'$, $i = 1, 2,\ldots,n-1$ and $y_n = y_n'$, that is, if and only if $\boldsymbol{Y} \overset{st}{\geq} \boldsymbol{Y}'$ and $Y_n \overset{st}{=} Y_n'$.

Oftentimes one shows that families of random variables are stochastically ordered by showing that they satisfy a stronger condition called $\mathrm{TP}_2$ defined below. A function $\phi$ defined on $R_2$ is said to be *totally positive of order* 2 ($\mathrm{TP}_2$) if it is nonnegative and satisfies

$$\phi(\lambda_1, x_1)\phi(\lambda_2, x_2) \geq \phi(\lambda_1, x_2)\phi(\lambda_2, x_1),$$

whenever $\lambda_1 < \lambda_2$, $x_1 < x_2$.

Let $\mu$ denote either the Lebesgue measure on $[0, \infty]$ or the counting measure on the set of non-negative integers. A function defined on $(0, \infty) \times [0, \infty)$ is said to possess a *semigroup property* in $\lambda$ if

$$\phi(\lambda_1 + \lambda_2, x) = \int_0^\infty \phi(\lambda_1, x - y)\phi(\lambda_2, y)\, d\mu(y).$$

A class of theorems generally known as preservation theorems allows us to construct new Schur functions and understand their structure. The following is one of the first preservation theorems for Schur functions. We will see later that by using the $TP_2$ and Schur properties with a variety of preservation theorems, several commonly used parametric families of distributions possess interesting Schur properties.

**Theorem 1**

Let $f(\boldsymbol{x})$ be a Schur convex (Schur concave) function and let $\phi(\lambda, x)$ defined on $(0, \infty) \times [0, \infty)$ possess the $TP_2$ property and the semigroup property in $\lambda$. Let $\mu$ be the Lebesgue measure or the counting measure. Let the integral

$$h(\lambda) = \int \prod_{i=1}^{n} \phi(\lambda_i, x_i) f(\boldsymbol{x})\, d\mu(\boldsymbol{x})$$

be well defined. Then $h(\lambda)$ is Schur convex (Schur concave).

This theorem appears as the main theorem in [PS 77]. In the principal application of this theorem, one takes $\phi$ to be a probability density function and shows that the operation of taking the expected value of a Schur convex function transfers the Schur convexity to the parameter vector.

**Theorem 2**

Let $X$ and $X'$ be a pair of $n$-vectors and define $S = \sum_{i=1}^{n} X_i$ and $S' = \sum_{i=1}^{n} X_i'$. Then $X \overset{st.m.}{\geq} X'$ if and only if (a) $S \overset{st}{=} S'$ and (b) for each bounded Schur convex function $f$, $E[f(X)|S = s] \geq E[f(X')|S' = s]$, for all $s \in A_f$, where the distribution of $S$ assigns probability one to $A_f$.

This theorem is one of the important tools to be found in [NPS 77]. The notion of a *Schur family* extends the concept of stochastic majorization to a family of random variables. Let $X_{\boldsymbol{\lambda}}$ be a family of random vectors with a distribution $P_{\boldsymbol{\lambda}}$ indexed by $\boldsymbol{\lambda}$ in $R^n$. The family $X_{\boldsymbol{\lambda}}$ and the family $P_{\boldsymbol{\lambda}}$ are said to be Schur families if $\boldsymbol{\lambda} \overset{m}{\geq} \boldsymbol{\lambda}'$ implies that $X_{\boldsymbol{\lambda}} \overset{st.m.}{\geq} X_{\boldsymbol{\lambda}'}$.

The following theorem shows that in Schur families, stochastic majorization is preserved among the posterior distributions when there is stochastic majorization among the prior distributions.

**Theorem 3**

Let $\{X_\lambda\}$ be a Schur family in $\lambda$. Let $G_1$ and $G_2$ be two prior distributions for $\lambda$, such that $G_1 \overset{st.m.}{\geq} G_2$. Then the posterior of $X_\lambda$ under $G_1$ stochastically majorizes the posterior of $X_\lambda$ under $G_2$.

**Example 2. Shock Models.**

Consider a system subject to a series of shocks and assume that the different types of shocks arrive in a Poissonian fashion. For example, suppose that $X_i(t)$ denote the number of shocks of the $i^{th}$ type arriving in the interval $[0, t]$. Let $\bar{P}(\boldsymbol{k})$, where $\boldsymbol{k} = (k_1, k_2,\ldots,k_n)$, be the survival probability of the system surviving $k_i$ shocks of the type $i$, $i = 1, 2,\ldots,n$. Suppose that for each $i$, the random variable $X_i(t)$ has a Poisson distribution with parameter $\lambda_i t$. Then it follows that the survival function of the system is given by

$$\bar{H}(t; \lambda) = E\left[\bar{P}\left(X_1(t), X_2(t),\ldots,X_n(t)\right)\right].$$

Assume further that $\bar{P}$ is Schur concave in $\boldsymbol{k}$. This assumption holds, for example, if the effects of shocks are independent and the $\bar{P}$ is the product of $n$ survival functions, each of which is logconcave. The $TP_2$ property of Poisson density functions and Theorem 1 show that the survival function $\bar{H}(t; \lambda)$ is Schur concave in $\lambda$. For details see [PS 77].

**Example 3. Schur Function of Partial Sums.**

Let $X_{ij}$, $i = 1, 2,\ldots,n$; $j = 1, 2,\ldots,k_i$ be independent identically distributed random variables with common logconcave density function $g$. Let $f$ be a Schur concave function and consider

$$h(\boldsymbol{k}) = E\left[f\left(\sum_{j_1=1}^{k_1} X_{1,j_1}, \sum_{j_2=1}^{k_2} X_{2,j_2},\ldots, \sum_{j_n=1}^{k_n} X_{1,j_n}\right)\right].$$

According to a result of Karlin and Proschan (1960), the $k$-fold convolution $g^{(k)}(x)$ is $TP_2$ in $k$ and $x$. Using this and Theorem 1, it follows that $h(\boldsymbol{k})$ is Schur concave in $\boldsymbol{k}$.

**Example 4. Schur Concavity of Moments.**

Let $g$ be a Schur concave density with the support $[0, \infty]^n$. Let $\alpha_i$, $i = 1, 2,\ldots,n$, be positive numbers and let

$$M(\boldsymbol{\alpha}) = \int \cdots \int \frac{\prod_{i=1}^{n} x_i^{\alpha_i - 1}}{\prod_{i=1}^{n} \Gamma(\alpha_i)}\, g(\boldsymbol{x})\, d\boldsymbol{x}$$

be a *multivariate normalized moment.* One can rewrite the integrand as $\left[\prod_{i=1}^{n}\left\{x_i^{\alpha_i-1}e^{-x_i}/\Gamma(\alpha_i)\right\}g(\boldsymbol{x})exp\{\sum x_i\}\right]$. Note that $g(\boldsymbol{x})exp\{\sum x_i\}$ is Schur concave and that $\{x^{\alpha-1}e^{-x}/\Gamma(\alpha)\}$ is $TP_2$ in $(\alpha, x)$ and is a semigroup on $(0, \infty)$. From Theorem 1 it follows that $M(\alpha)$ is Schur concave. Note that there are examples where $M(\alpha)$ is Schur convex if the normalizing constant $\Gamma(\alpha)$ is omitted in the integrand.

**Example 5. Schur Families.**

A number of parametric families found in standard textbooks can be shown to be Schur families. To name a few: multinomial, multivariate negative binomial, multivariate hypergeometric, Dirichlet. Furthermore, families of independent random variables such as Poisson, Gamma etc. also form Schur families. A host of such examples are listed and demonstrated in [NPS 77].

**Functions Decreasing in Transposition**

The partial ordering of majorization can sometimes be better understood by a standard partial ordering on the space of permutation on the set of $n$ integers $(1, 2,\ldots,n)$. This leads to the concept of functions which are *decreasing in transposition* (DT) which extends the concept of Schur functions.

Let $\boldsymbol{\pi} = (\pi_1, \pi_2, \ldots,\pi_n)$ denote a permutation of $(1, 2,\ldots,n)$. Let $S$ denote the group of such permutations $\boldsymbol{\pi}$. Suppose that $\boldsymbol{\pi}$ and $\boldsymbol{\pi}'$ differ only in two of their components, say the $i^{th}$ and $j^{th}$, where $i < j$, $\pi_i < \pi_j$ and that $\pi_i' = \pi_j$, $\pi_j' = \pi_i$. We say that $\boldsymbol{\pi}'$ is a *simple transposition* of $\boldsymbol{\pi}$. If a member of $S$, say $\boldsymbol{\pi}''$ is obtained from $\boldsymbol{\pi}$ by successive simple transpositions, we say that $\boldsymbol{\pi}$ dominates $\boldsymbol{\pi}''$ in transposition and write $\boldsymbol{\pi} \overset{t}{\geq} \boldsymbol{\pi}''$. Clearly this relation establishes a partial ordering in $S$.

Suppose that the components of $\boldsymbol{x}$ are such that $x_1 \leq x_2 \leq \ldots \leq x_n$. A permutation obtained by composing it with $\boldsymbol{\pi}$ is denoted by $\boldsymbol{\pi} \circ \boldsymbol{x}$ and defined by

$$\boldsymbol{\pi} \circ \boldsymbol{x} = (x_{\pi_1}, x_{\pi_2},\ldots,x_{\pi_n}).$$

The partial ordering defined above can be extended in an obvious way to the vectors obtained by permuting components of $\boldsymbol{x}$.

In many applications one considers two vectors, the first vector corresponding to a parameter and the second vector to an observed random variable. It is useful to describe in a mathematical fashion the fact that a random vector and its parameter vector increase and decrease together. Oftentimes one needs to study and compare the way in which two random vectors vary together. For instance, one use of rank correlation is to measure how similarly two random vectors vary together. We will see below that the partial ordering on permutations, defined above, provides a satisfactory way to compare how similarly two vectors, which may be random or deterministic, vary together.

Let $\Lambda$ and $\Xi$ be subsets of $R$. A function $g(\boldsymbol{\lambda}, \boldsymbol{x})$ is said to be *decreasing in transposition* (DT) on $\Lambda^n \times \Xi^n$ if $g(\boldsymbol{\lambda} \circ \boldsymbol{\pi}, \boldsymbol{x} \circ \boldsymbol{\pi}) = g(\boldsymbol{\lambda}, \boldsymbol{x})$, for every $\boldsymbol{\pi}$ (that is, $g$ is invariant under the same permutation on the two vectors) and $g(\boldsymbol{\lambda}, \boldsymbol{x} \circ \boldsymbol{\pi}) \geq g(\boldsymbol{\lambda}, \boldsymbol{x} \circ \boldsymbol{\pi}')$, where $\lambda_1 \leq \lambda_2 \leq \ldots \leq \lambda_n$; $x_1 \leq x_2 \leq \ldots \leq x_n$ and $\boldsymbol{\pi} \overset{t}{\geq} \boldsymbol{\pi}'$.

When $g(x, y)$ is DT the function $g(x, y)$ gives larger values when the ranking in the pair $(x, y)$ is more similarly ordered than when the ranking is less similarly ordered.

In certain applications there is only one vector and it is desirable to define functions of a single vector which exhibit a monotonicity under this partial ordering. Let $h$ be defined on $\Xi^n$ and suppose that the components of $\boldsymbol{x}$ are in increasing order. Then $h$ is said to be DT if $h(\boldsymbol{x} \circ \boldsymbol{\pi}) \geq h(\boldsymbol{x} \circ \boldsymbol{\pi}')$ whenever $\boldsymbol{\pi} \overset{t}{\geq} \boldsymbol{\pi}'$.

DT functions occur quite frequently in statistics. The book of Marshall and Olkin (1979) has popularized the notion of DT functions under the more positive sounding name of *Arrangement Increasing* (AI) functions. The following theorem shows the relation between DT, Schur and $TP_2$ functions.

**Theorem 4**

(a) Suppose $g(\boldsymbol{\lambda}, \boldsymbol{x}) = h(\boldsymbol{\lambda} - \boldsymbol{x})$. Then $g$ is DT on $R^{2n}$ if and only if $h$ is Schur concave.

(b) Suppose $g(\boldsymbol{\lambda}, \boldsymbol{x}) = h(\boldsymbol{\lambda} + \boldsymbol{x})$. Then $g$ is DT on $R^{2n}$ if and only if $h$ is Schur convex.

(c) Suppose $g(\boldsymbol{\lambda}, \boldsymbol{x}) = \prod h(\lambda_i, x_i)$, then $g$ is DT on $R^{2n}$ if and only if $h$ is $TP_2$.

The main result on DT functions is the following preservation theorem which states that the DT property is preserved under the operation of composition.

**Theorem 5**

Let $g_i$, $i = 1, 2$ be DT on $R^{2n}$ and $\sigma$ be a measure on $R^n$ such that for every Borel set $A$ in $R^n$, $\sigma(A) = \sigma(\boldsymbol{\pi} \circ A)$ for every $\boldsymbol{\pi}$. Suppose that

$$g(\boldsymbol{x}, z) = \int_A g_1(\boldsymbol{x}, \boldsymbol{y}) g_2(\boldsymbol{y}, \boldsymbol{z}) d\sigma(\boldsymbol{y}),$$

is well defined. Then $g$ is DT on $R^{2n}$.

The proofs of the above two theorems can be found in [HPS 77].

Theorem 1 can be derived as a consequence of Theorem 5 and Theorem 4(b). Furthermore, the following result of Marshall and Olkin (1974) can also be obtained from Theorem 5 and Theorem 4(a).

**Theorem 6**

The convolution of two Schur concave functions is Schur concave.

Most of the families considered in the second section can also be shown to have DT property. In some sense this provides a better tool than Schur concavity because of the connections seen earlier. One of the interesting applications is the problem in ranking. Suppose the vector $X$ has density $\phi(\boldsymbol{\lambda}, \boldsymbol{x})$ which is a DT function. Let $g(\boldsymbol{\lambda}, \boldsymbol{r})$ be the probability that the rank vector of $X$-observations is $\boldsymbol{r}$. By using Theorem 5 above, it can be shown that $g$ is DT. This has important consequences in nonparametric statistics. For details of this please see [HPS 77].

It should be noted that the concept of Schur concavity is closely related to that of unimodality. From the above discussion it can be seen that a function defined on $R^2$ is Schur concave if and only if it is permutation invariant and its graph is such that it is unimodal on every section perpendicular to the line of equality. This definition can be extended to $R_n$ by considering all bivariate sections obtained by fixing $(n-2)$ arguments and requiring Schur concavity for each section, in the sense just described.

The convolution of two symmetric univariate unimodal densities can be shown to be a symmetric unimodal density. This is known as Wintner's theorem. Using this result it follows that the convolution of two bivariate Schur concave densities is Schur concave. Again by considering sections, an alternative proof for Theorem 6 can be provided.

The condition that the set $\{\boldsymbol{x}\colon f(\boldsymbol{x}) \geq c\}$ be convex and permutation invariant, for every $c > 0$, is sufficient for all the required sections of an $n$-variate density $f(x)$ to be symmetric unimodal. Many results that follow from such basic unimodality have been explored in a book by Joag-Dev and Dharmadhikari (1988) which are useful in deriving various properties of Schur concave functions. For instance, consider a random vector whose density function is logconcave. The logconcavity implies that the set where the density exceeds a given constant is a convex set and hence it satisfies the condition state above. If the components of this random vector are also exchangeable, then the density function is Schur concave.

An important theorem for multivariate logconcave densities is due to Prékopa (1973) which is stated below.

**Theorem 7**

Let $Y = (Y_1, Y_2, \ldots, Y_m)$ have logconcave density. Then $Z = (Z_1, Z_2, \ldots, Z_k) \overset{def}{=} (\sum a_{1,i} Y_i, \sum a_{2,i} Y_i, \ldots, \sum a_{k,i} Y_i)$ also has a logconcave density. In particular all marginals have logconcave densities.

We will use this theorem to derive Schur concavity and DT properties of densities of some random vectors obtained as overlapping sums of random variables. We begin with a simple case before going to the general case because the notation can get quite complicated.

**Theorem 8**

Let $X_{12}$, $X_{23}$, $X_{13}$ and $X_{123}$ be random variables such that $X_{12}$, $X_{23}$, $X_{13}$ are exchangeable. Define

$$X_1^{(2)} = X_{12} + X_{13},$$

$$X_2^{(2)} = X_{12} + X_{23},$$

$$X_3^{(2)} = X_{13} + X_{23},$$

$$T_1 = X_1^{(2)} + X_{123},$$

$$T_2 = X_2^{(2)} + X_{123},$$

$$T_3 = X_3^{(2)} + X_{123}.$$

Then the density of $T \overset{def}{=} (T_1, T_2, T_3)$ is Schur concave under either one of the following conditions:

(A) the joint density of $X_{12}$, $X_{13}$, $X_{23}$, $X_{123}$ is log concave

(B) the random vector $(X_{12}, X_{13}, X_{23})$ has a logconcave density and is independent of the random variable $X_{123}$.

**Proof.** Note that $T$ consists of overlapping sums of random variables. A more general case of overlapping sums will be considered later.

From the definition of $T$ it is easy to see that it is exchangeable. The logconcavity of the density of $T$ follows readily from Prékopa's theorem (Theorem 7) under condition (A). This establishes the Schur concavity of the density of $T$ under (A). When condition (B) holds, Prékopa's theorem (Theorem 7) once again shows that the density $f(x_1, x_2, x_3)$ of $(X_1^{(2)}, X_2^{(2)}, X_3^{(2)})$ is Schur concave. The density function of $T$ is given by

$$\int f(x_1 - y, x_2 - y, x_3 - y) g(y) dy$$

where $g(y)$ is the density function of $X_{1,2,3}$. Since a positive mixture of Schur concave functions is Schur concave, it follows that the density of $T$ is Schur concave. □

We now generalize the above to random vectors in $R^n$. Let $J = \{1, 2, 3,\ldots,n\}$. For $k = 2,\ldots,n$, let

$$I_k = \{I : I \text{ is a subset of } J \text{ with cardinality } k\},$$

and

$$I^* = \bigcup_2^k \{I \in I_k\} \text{ and } I_{k,i} = \{I \in I_k : i \in I\}.$$

Let $\{X_i, i = 1, 2,\ldots,n\}$ and $X_I$, $I \in I^*$ be a collection of random variables. Let $W(k) = \{X_I : I \in I_k\}$, $X_i^{(k)} = \sum_{I \in I_{k,i}} X_I$ and $X^{(k)} = (X_1^{(k)}, X_2^{(k)},\ldots,X_n^{(k)})$ where $i = 1, 2,\ldots,n$ and $k = 2, 3,\ldots,n$. Thus $X_i^{(k)}$ is the sum of random variables, each having $k$ subscripts, one of which is $i$.

**Theorem 9**

Let $X^{(1)} = (X_1, X_2,\ldots,X_n)$ be a random vector with probability density function which is DT. Suppose that the set $\{X_I, I \in I^*\}$ is independent of $X^{(1)}$ and one of the following conditions holds.

(A) The set of all variables $\{X_I, I \in I^*\}$ is exchangeable and has a logconcave joint density function.

(B) The collection of random variables in $W(k)$ has a logconcave density and is permutation invariant for $k = 2, 3,\ldots,n-1$, and the collections $W(2), W(3),\ldots, W(n)$ are independent.

Then the joint distribution of $Z = (Z_1, Z_2,\ldots,Z_n)$ is DT, where

$$Z_i = X_i + \sum_{k \geq 2} X_i^{(k)}.$$

**Proof.** The argument is similar to the proof of Theorem 8. Let $T_i = \sum_{k \geq 2} X_i^{(k)}$ and $T = \{T_1, T_2,\ldots,T_n\}$.

The density function of $Z$ is the convolution of the density functions of $T$ and $X^{(1)}$, the second of which is DT by assumption. If we can show that the density function of $T$ is Schur concave, then it will follow that the density function of $Z$ is DT from Theorems 4 and 5(a).

We will now show that condition (A) or (B) implies the Schur concavity of the density of $T$.

When condition (A) holds it is easy to see that Prékopa's theorem implies that the joint density of $T$ is logconcave. The permutation invariance of this joint density follows from the exchangeability of $\{X_I, I \in I^*\}$. This establishes the joint density function satisfies the DT property.

When condition (B) holds, Prékopa's theorem once again shows that the density function of $X^{(k)}$ is log concave for $k = 2, 3,\ldots,n-1$ and is permutation invariant. From the independence of $W(k)$, $k = 2, 3,\ldots,n-1$ it follows that the density function of $T_1 - X_{\{1,\ldots,n\}},\ldots, T_n - X_{\{1,\ldots,n\}}$ is logconcave and permutation invariant and hence Schur concave. Notice that $W(n) = X_{\{1,\ldots,n\}}$ consists of a single random variable. From the same argument given in case (B) of Theorem 8, it follows that the density of $T$ is Schur concave.

This completes the proof of Theorem 9. □

Theorem 9 generalizes Theorem 2.1 of [HPS 81] and contains a new proof. As an application of this theorem it can be shown that the density function of a generalized compound multivariate Poisson is DT. See [HPS 81] for details.

**References**

Abouammoh, A. M., El-Neweihi and Proschan, F. (1989): Schur structure functions, *Prob. Engg. Inform. Sc.* 3, 581-591.

Boland, P. J., Proschan, F. and Tong, Y. L. (1987): Fault diversity in software reliability, *Prob. Engg. Inform. Sc.* 1, 175-188.

Boland, P.J., Proschan, F. and Tong, Y. L. (1988): Moment and geometric probability inequalities arising from arrangement increasing functions, *Ann. Prob.* 16, 407-413.

Dharmadhikari, S. W. and Joag-Dev, K. (1988): *Unimodality, Convexity, and Applications*, Academic Press, New York.

Hardy, G. H., Littlewood, J.E. and Pólya, G. (1952): *Inequalities*, 2nd ed., Cambridge University Press, New York.

Hollander, M., Proschan, F. and Sethuraman, J. (1977): Functions decreasing in transposition and their applications in ranking problems, *Ann. Stat.* 5, 722-733.

Hollander, M., Proschan, F. and Sethuraman, J. (1981): Decreasing in transposition property of overlapping sums, and applications, *Jour. Mult. Anal.* 11, 50-57.

Karlin, S. and Proschan, F. (1960): Pólya type distributions of convolutions, *Ann. Math. Statist.* 31, 721-736.

Marshall, A. and Olkin, I. (1974): Majorization in multivariate distributions, *Ann. Statist.* 2, 1189-1200.

Marshall, A and Olkin, I. (1979): *Inequalities: Theory of Majorization and Its Applications*, Academic Press, New York.

Nevius, S.E., Proschan, F. and Sethuraman, J. (1977): Schur functions in Statistics II. Stochastic majorization, *Ann. Stat.* 5, 263-273.

Prékopa, A. (1973): On logarithmic concave measures and functions, *Acta Sci. Mat.* 34, 335-343.

Proschan, F. and Sethuraman, J. (1977): Schur functions in Statistics I. The preservation theorem, *Ann. Stat.* 5, 253-262.

Tong, Y. L. (1980): *Probability Inequalities in Multivariate Distributions*, Academic Press, New York.

# OPTIMAL INTEGRATION OF SURVEYS

P. K. Pathak, Department of Mathematics and
Statistics, University of New Mexico, Albuquerque

and

M. Fahimi, Department of Mathematics and
Statistics, University of New Mexico, Albuquerque

The problem of integration of surveys is known to be of considerable practical as well as theoretical interest in the design of multi-purpose and continuing surveys. The object of this paper is to present a brief review of current developments in this area and to furnish a unified framework within which integration of surveys can be studied from various angles.

## Introduction

The problem of integration of surveys, i.e., the problem of designing a sampling program for two or more surveys which maximizes the overlap between observed samples is known to be of considerable practical as well as theoretical interest in the design of multi-purpose and continuing surveys (Keyfitz, 1951). Development of cost-efficient sampling programs of this kind is a problem which agencies such as the National Sample Survey of India, Statistics Canada, the U.S. Bureau of the Census, the U.S.D.A., and others worldwide, have been continuing to tackle on an ad hoc basis. And although the literature on it is now over 40 years old, basic research in it has reached a modest level of maturity only recent (cf. Arthanari and Dodge, 1981; Causey et al. 1985; Krishnamoorthy and Mitra, 1987; Maczynski and Pathak, 1980; and others). Nevertheless despite these recent gains, there remains a pressing need for a unified framework within which integration of surveys and other similar problems of this nature, such as controlled selection and controlled rounding (Goodman and Kish, 1950, and Causey et al., 1985), can be studied with all their ramifications. In due course, such an approach is bound to provide a powerful guide to the cost-efficient design of survey programs commonly encountered in practice. The primary object of this paper is to briefly review the contemporary work in this area from the theoretical as well as computational viewpoints.

In broad terms, integration of surveys can be referred to as the sampling program for two or more surveys. It has its origin in multipurpose surveys and sampling over successive occasions. In multipurpose surveys, the population characteristics under study are often subdivided into two or more groups of

This research has been supported by the National Science Foundation Grants DMS-8703798 and DMS-8844575.

positively correlated characteristics and different sampling schemes are employed for data collection for the different groups of characteristics. For example in multipurpose surveys, traditionally population size is made the basis for socio-economic surveys and geographical area for agricultural surveys. This gives rise to the multivariate problem of designing an overall sampling program which imbeds the different sampling schemes into a single multi-variate sampling scheme in a cost-effective manner. A problem of similar kind arises in sampling over successive occasions (Keyfitz, 1951) in which a given population is sampled by probabilities proportional to size over two or more successive time periods. Over time, sizes of population units change and this necessitates sampling the given population according to a new set of probabilities at each new time period in a cost-effective manner. In this case, it makes sense to design an overall joint sampling program which in some sense maximizes the overlap between samples over different occasions, or equivalently minimizes the number of distinct population units sampled over different occasions. In applications of this nature, population units to be sampled are typically primary sampling units (psu's) and their selection represents a considerable financial investment. A new independent selection amounts to selecting an almost new set of psu's each time and is not cost-effective. On the other hand, the use of the same psu's on succeeding occasions, much like in the paired $t$-test, leads to significant reductions in errors of comparisons between periodic surveys (Kish and Scott, 1971). Thus in applications of this nature, an integrated joint sampling program which minimizes the number of distinct psu's selected over successive time periods is cost-effective and highly desirable.

The problem of integration of surveys for surveys involving two with replacement sampling schemes was originally formulated and solved by Keyfitz (1951). Lahiri (1954) proposed a serpentine arrangement of geographically contiguous psu's for optimal integration of two surveys. Raj (1957) studied the problem of integration of two surveys as a transportation problem and established the optimality of Lahiri's algorithm under a one-dimensional metric. Felligi (1966) studied the problem of integration of two without replacement sampling schemes and noted that even the simplest case of the sample size $n = 2$ causes added complications. In the context of $k$ ($\geq 2$) with replacement sampling schemes, Maczynski and Pathak (1980) presented a general solution in a closed form under certain assumptions. More recently Krishnamoorthy and Mitra (1986), and Mitra and Pathak (1984) have presented sequential algorithms for 'optimal' integration of two or three surveys in the context of with replacement sampling (cf. Arthanari and Dodge, 1981; Keyfitz, 1951; Lahiri, 1954; Raj, 1957; Kish and Scott, 1971; and others). The recent upsurge of research in this area is both practically useful and theoretically interesting.

Although the connection between integration of surveys and the transportation problem is well-known (Aragon and Pathak, 1990; Arthanari and Dodge, 1981; Causey et al., 1985; Maczynski and Pathak, 1980, p. 137; and Raj, 1957), it is only in recent years that serious attempts have been made to solve the problem of optimal integration of surveys as a transportation problem. In general, integration of surveys is a transportation problem with an exponentially large number of variables, e.g. a simple problem of integration of two samples of size $n = 5$ each from a population of size $N = 50$ is equivalent to a transportation problem with approximately $4.5 \times 10^{12}$ variables. At the present time solvers of the transportation problems of this size are unavailable in the public domain. An despite the unique sparse structure of the underlying tableaus of integration of surveys, there are very few results in the literature on size reduction techniques as an alternative to solving these very large transportation problems. Much remains to be done in the area of size reduction techniques for integration of surveys.

**Formulation of the Problem**

For clarity in the exposition, we adopt the following terminology:

$Z$ : the set of the first $N$ natural numbers. We use the artifice of identifying the population under study by $Z$.

$k$ : number of surveys to be carried out, $k \geq 2$.

$S$ : collection of all possible samples from which a sample is to be drawn for each of the $k$ surveys; $S$ being a subset of the power set of $Z$.

$n_i$ : sample size for the $i^{th}$ survey, $1 \leq i \leq k$.

$x_i$ : the outcome of the $i^{th}$ survey.

$X$ : the joint outcome of $k$ surveys, i.e., $X = (x_1,\ldots,x_k)$.

$P_{ij}$ : the probability of selecting the $j^{th}$ sample $s_j$ on the $i^{th}$ survey, i.e., $P_{ij} = P(x_i = s_j)$, $x_i \in X$, $s_j \in S$, $1 \leq i \leq k$.

$d$ : a cost function defined on $S^k$, i.e., it is non-negative and sub-additive on $S^k$.

A survey or a sampling scheme on $Z$ is a given, but otherwise quite arbitrary, collection $S$ of samples from $Z$ endowed with a given probability distribution $P$. Thus a survey is expressed by the pair $(S, P) = \{(s, P(s)): s \in S\}$ in which $P(s)$ denotes the probability of selection of the sample $s$.

The problem of optimal integration of $k$ surveys can now be stated as follows:

Given $k$ individual surveys, $(S, P_i)$, $1 \leq i \leq k$, and a cost function $d$ on $S^k$, find a joint probability distribution $\mathbb{P}$ for $X$ on $S^k$, which for each $x_i$ realizes the preassigned marginal probabilities $P_i$ determined by the $i^{th}$ survey, i.e.,

$\mathbb{P}(x_i = s_j) = P_{ij}$, and at the same time minimizes the expectation of the cost function over the class of all surveys of this kind.

Raj (1957) was perhaps the first to paraphrase the problem of integration of surveys as a transportation problem. In terms of our terminology, it is as follows:

**Problem 1**

Given $k$ surveys $\{(S, P_i): 1 \leq i \leq k\}$ and a cost function $d$ on $S^k$,

minimize $\phi(X) = \sum_X P(X) \cdot d(X)$

subject to $\sum_{X_{ij}} P(X) = P_{ij}$, $X_{ij} = \{X \in S^k: x_i = s_j\}$,

$P(X) \geq 0, \forall X \in S^k$.

**Example 1**

Consider a population of four psu's and suppose that on two occasions a sample of size two ($wor$) is to be drawn from the population according to the sampling schemes given in Table 1.

Table 1: Sampling Schemes for Example 1

| Sample | 1st Survey | 2nd Survey |
|---|---|---|
| $s_j$ | $P_{1j}$ | $P_{2j}$ |
| (1, 2) | 0.04 | 0.22 |
| (1, 3) | 0.16 | 0.15 |
| (1, 4) | 0.21 | 0.29 |
| (2, 3) | 0.13 | 0.07 |
| (2, 4) | 0.32 | 0.10 |
| (3, 4) | 0.14 | 0.17 |

To maximize the expectation of the overlap between the two samples selected on the two occasions, the cost function $d$ is taken to be the number of distinct psu's in the two samples, i.e., $d(s_m \cup s_n) = \#(s_m \cup s_n)$. The transportation problem representation of this problem is as follows:

minimize $$\sum_{m=1}^{6} \sum_{n=1}^{6} P(x_1 = s_m, x_2 = s_n) d(s_m, s_n)$$

subject to $$\sum_{n=1}^{6} P(x_1 = s_m, x_2 = s_n) = P_{1m},$$

$$\sum_{m=1}^{6} P(x_1 = s_m, x_2 = s_n) = P_{2n},$$

$$P(x_1 = s_m, x_2 = s_n) \geq 0,$$

$$\forall \ m, n = 1, \ldots, 6.$$

To solve this problem, one can use a standard linear programming package, e.g., the use of the SPLO-program (1981) yields the results summarized in Table 2.

Table 2: A Solution to Example 1
$P(x_1 = s_m, x_2 = s_n)$, $m, n = 1, \ldots, 6$

Survey II/I

| | 1, 2 | 1, 3 | 1, 4 | 2, 3 | 2, 4 | 3, 4 | |
|---|---|---|---|---|---|---|---|
| 1, 2 | 0.04 | | | | | | 0.04 |
| 1, 3 | 0.01 | 0.15 | | | | | 0.16 |
| 1, 4 | | | 0.21 | | | | 0.21 |
| 2, 3 | 0.06 | | | 0.07 | | | 0.13 |
| 2, 4 | 0.11 | | 0.08 | | 0.10 | 0.03 | 0.32 |
| 3, 4 | | | | | | 0.14 | 0.14 |
| | 0.22 | 0.15 | 0.29 | 0.07 | 0.10 | 0.17 | 1.00 |

Based on this solution, we find that the minimum value of the objective function as defined in Problem 1 is 2.29. It is worth noting that this represents the maximum expected overlap between the two sampling schemes.

**Integration of Surveys with Replacement Sampling Schemes**

Unfortunately, most realistic problems of integration of surveys are not as tractable as the example in the preceding section seems to indicate. For example, it is easily seen that a straight forward 3-dimensional integration of surveys problem for a population of 20 psu's for without replacement samples of size $n = 3$ amounts to a transportation problem with over a billion variables. At the present time, hardware or software which can handle a problem of this magnitude is unavailable in the public domain. So it is not surprising at all that earlier attempts at solutions of these problems were largely directed towards finding closed form solutions under special circumstances. The first general result of this nature was obtained by Maczynski and Pathak (1980). Based on the following lemma, they provided closed form solutions for the special case of the sample size $n_i = 1$ and $k$ surveys, $k \geq 2$.

**Lemma 1**

Consider the general problem of integration of $k$ surveys with $n_i = 1$ and suppose that there exists a joint probability distribution $\mathbb{P}$ on $S^k$ such that for each $h$, $1 \leq h \leq k$, and $1 \leq i_1 < \ldots < i_h \leq k$,

$$\mathbb{P}(x_{i_1} = x_{i_2} = \ldots = x_{i_h} = j) = min(P_{i_1 j},\ldots,P_{i_h j}).$$

Then

$$\mathbb{P}(\bigcup_i \{x_i = j\}) = max_i P_{ij}.$$

Moreover such a $\mathbb{P}$ minimizes the expected number of distinct psu's selected in all the $k$ samples.

**The case $k = 2$.** An immediate corollary of Lemma 1 is that for $k = 2$ surveys with $n_i = 1$, the following closed-form solution is optimal:

$$\mathbb{P}(x_1 = h,\ x_2 = j) = f(h), \qquad h = j$$

$$\mathbb{P}(x_1 = h,\ x_2 = j) = f_{12}(h, j), \qquad h <> j$$

where

$$f(h) = min(P_{1h}, P_{2h}),$$

$$f_{12}(h, j) = (P_{1h} - f(h))(P_{2j} - f(h))(1 - \Sigma_r f(r))^{-1}.$$

Algorithmic representations of the above solution have been provided by Keyfitz (1951) and Mitra and Pathak (1984).

**The case $k = 3$.** The problem of optimal integration of three surveys with $n_i = 1$ is essentially solved now. In this case Lemma 1 forms the basis of all closed-form solutions which minimize the expected number of distinct sample units over the three occasions. In a series of fundamental papers Krishnamoorthy and Mitra (1986, 1987) have established the optimality of the Mitra-Pathak type algorithmic approach (1984) for the integration of three with replacement sampling schemes. For further details in this connection we refer the reader to the elegant work of Krishnamoorthy and Mitra (1987). For completeness, it would be worthwhile indeed to investigate extensions of these results to the general case of $k$ surveys with $k > 3$.

**Size Reduction Techniques**

In this section we present a brief review of a technique which has the potential of significantly reducing the size of the induced transportation problem in the context of integration of two surveys. To illustrate this technique, consider the induced transportation problem of Example 1 and observe that the cost function $d$ of this example is in fact a metric. This simple observation allows one to establish the following interesting result (Aragon and Pathak, 1990):

**Theorem 1**

Consider the problem of integration of two surveys of equal size in which the cost function $d$ is induced by a metric. Then there is an optimal feasible solution $\mathbb{P}$ such that for each sample $s_m$

$$\mathbb{P}(x_1 = s_m,\, x_2 = s_m) = min(P_{1m},\, P_{2m}).$$

The theorem implies that by setting these diagonal probabilities equal to their largest admissible values, namely the minimum of the corresponding row and column marginal probabilities, at least half of the restrictions of the problem are satisfied. What is left then is the optimal determination of the remaining unknown nondiagonal probabilities, i.e. $\mathbb{P}(x_1 = s_m,\, x_2 = s_n)$, for only some of $m < > n$. This reduced problem is at most one-fourth the size of the original problem. For example, application of this technique to Example 1 reduces the original problem with 36 variables to a smaller problem with only 9 variables. The reduced problem is stated below and an optimal solution summarized in Table 3.

$$\text{minimize} \qquad \sum_{m=1}^{3} \sum_{n=1}^{3} P(x_1 = s_m,\, x_2 = s_n) \cdot d(s_m,\, s_n)$$

$$\text{subject to} \qquad \sum_{n=1}^{3} P(x_1 = s_m,\, x_2 = s_n) = P_{1m},$$

$$\sum_{m=1}^{3} P(x_1 = s_m,\, x_2 = s_n) = P_{2n},$$

$$P(x_1 = s_m,\, x_2 = s_n) \geq 0,$$

$$\forall\ m,\, n = 1,\ldots,3.$$

Table 3: A Solution to the Reduced Version of Example 1
$P(x_1 = s_m, x_2 = s_n)$, $m$, $n = 1,\ldots,3$

Survey II

| | | 1, 2 | 1, 4 | 3, 4 | |
|---|---|---|---|---|---|
| Survey I | 1, 3 | 0.01 | | | 0.01 |
| | 2, 3 | 0.06 | | | 0.06 |
| | 2, 4 | 0.11 | 0.08 | 0.03 | 0.22 |
| | | 0.18 | 0.08 | 0.03 | 0.29 |

Note that the marginal probabilities $P_{1m}$ and $P_{2n}$ are given by the marginal entries in Table 3. And that Table 3 was obtained from Table 1 after the assignment of the main diagonal probabilities $\mathbb{P}(s, s)$. Thus at the first stage of this size reduction technique, we set $\mathbb{P}(s_1, s_1) = .04$, $\mathbb{P}(s_1, s_2) = \ldots = \mathbb{P}(s_1, s_6) = 0$, $\mathbb{P}(s_2, s_1) = 0$, $\mathbb{P}(s_2, s_2) = .15,\ldots,\mathbb{P}(s_6, s_6) = .14$. Then at the second stage, the reduced problem is solved by using a standard transportation problem solver. The two solutions when combined together furnish a complete solution to the original problem of Example 1.

In order to present this size reduction technique in greater generality and scope, a slight digression from the main theme of the paper is necessary. We turn now to the following so-called Hitchock transportation problem (Chvatal, 1983, p. 345):

**Problem 2**

$$\text{minimize} \quad \sum_i \sum_j c_{ij} x_{ij}$$

$$\text{subject to} \quad \sum_j x_{ij} = p_i, \qquad (i = 1,\ldots,m),$$

$$\sum_i x_{ij} = q_j, \qquad (j = 1,\ldots,n),$$

$$x_{ij} \geq 0, \qquad \forall\ i, j.$$

Moreover, suppose that in the preceding transportation problem, there are cells in the cost matrix $C = \{c_{ij}\}$ with the following property of negative variation:

**Definition 1**

A cell $(i, j)$ of the cost matrix $C = \{c_{ij}\}$ is said to have *negative variation* if for all $k < > i$ and $l < > j$, the following inequality holds:

$$(c_{ij} + c_{lk}) - (c_{il} + c_{kl}) \leq 0.$$

Similarly, the cell $(i, j)$ is said to have *positive variation* if the above difference is always non-negative.

If the cost matrix of a given transportation problem has cells with negative variation, then the given problem can be reduced to a new problem of a smaller size. Specifically, if the original problem is of size $m \times n$ and has $c$ cells with negative variation, then the reduced problem is of size $(m-a) \times (n-b)$ with $a + b \geq c$. This size reduction is a consequence of the following theorem:

**Theorem 2**

Suppose that a given cell, say (1, 1), of the cost matrix of Problem 2 has negative variation. Then there exists an optimal feasible solution $X = \{x_{ij}\}$ with $x_{11} = min(p_1, q_1)$.

In a different guise, this theorem can be found hidden in the seminal work of the noted French mathematician Monsieur Monge (1781). In a totally different context in operations research, it has been used by A.J. Hoffman (1963). We independently discovered it in the context of integration of surveys and controlled selection. We take the liberty of referring to this theorem as the Monge-Hoffman size reduction theorem.

**Corollary 1**

If the objective of Problem 2 is maximization instead of minimization, then the above theorem goes through provided the cell (1, 1) has positive variation.

Now consider the Transportation Problem 2 and assume that the cell (1,1) of its cost matrix has negative variation. Also, without loss of generality assume that $p_1 < q_1$. Then the preceding theorem implies that the original problem can be replaced by the following smaller problem:

minimize $\quad \sum_i \sum_j c_{ij}x_{ij},\ 2 \leq i \leq m,\ 1 \leq j \leq n$

subject to $\quad \sum_j x_{ij} = p_i,\ (i = 2,\ldots,m),$

$$\sum_i x_{ij} = q_j,\ (j = 1,\ldots,n),$$

$$x_{ij} \geq 0,\ \forall\ i, j.$$

Clearly any optimal solution of this problem, along with $x_{11} = p_1$, $x_{12} = \ldots = x_{1n} = 0$, will provide an optimal solution to the original problem. This effectively reduces the number of variables from $m \times n$ to $(m-1) \times n$. If instead of $p_1 < q_1$, we have $q_1 < p_1$, a similar consideration will show that an optimal solution now is given by $x_{11} = q_{11}$, $x_{21} = \ldots = x_{m1} = 0$ and the values of the remaining variables are obtained by solving an analogous reduced problem involving $m \times (n-1)$ variables. Finally if $p_1 = q_1$, then an optimal solution is given by $x_{11} = p_1 = q_1$, $x_{12} = \ldots = x_{1n} = x_{21} = \ldots = x_{m1} = 0$, and the values of the remaining variables are obtained by solving another analogous reduced problem of $(m-1) \times (n-1)$ variables.

Note that if the cost matrix in the transportation problem has multiple cells with negative variation then the preceding size reduction algorithm can be carried out sequentially until all the cells with negative variation have been removed. In the special case of $c_{ij} = |j - i|$, this size reduction procedure can be carried out to the very end and an optimal solution can be obtained without ever having to invoke any solvers of the transportation problem. This last result is a consequence of the following theorem.

**Theorem 3**

Consider the $m \times n$ matrix $D = \{d_{ij}\}$ in which $d_{ij} = |j - i|$, and suppose that the first $r$ rows and the first $c$ columns of $D$ have been removed. Let $s = max(r, c)$ and $t = min(m, n)$. Then the following holds:

a) All the cells on the shortest path joining the cells $(1, 1)$, $(s-r+1, s-c+1)$, $(t-r, t-c)$ and $(m-r, n-c)$ have negative variation.

b) All the cells $(i, j)$ with $i \geq t - r$, $j \leq s - c + 1$, and all cells $(k, l)$ with $k \leq s - r + 1$, $l \geq t - c$ have positive variation.

**Corollary 1**

The above theorem also holds if some of the very last rows and columns of the matrix are removed as well. (This should be self-evident since the removal of rows and columns from the end leaves original structure of the matrix $D$ intact.)

**Corollary 2**

Suppose that $d(x, y)$ has the following property of a distribution function in two dimensions:

$$d(x+h, y+k) - d(x, y+k) - d(x+h, y) + d(x, y) \geq 0.$$

for all $h, k \Rightarrow 0$. Then the northeast and the southwest cells of the matrix $D$ have positive variation, while the northwest and the southeast cells have negative variation.

An immediate consequence of the above corollary is that for distance functions such as $d(x, y) = xy$, the conventional northwest (greedy) algorithm provides an optimum solution for the Hitchock Transportation Problem 2.

**Example 2**

The purpose of this example is to graphically illustrate the statement of Theorem 3. Consider a 9 × 7 matrix D for Theorem 3. Tables 4 through 6 summarize the variations of $D$ when certain initial rows and columns are removed from it.

Table 4: Variations of the Matrix $D$
(0 rows and 0 columns are removed)

| | 1 | 2 | 3 | 4 | 5 | 6 | 7 |
|---|---|---|---|---|---|---|---|
| 1 | — | | | | | | + |
| 2 | | — | | | | | |
| 3 | | | — | | | | |
| 4 | | | | — | | | |
| 5 | | | | | — | | |
| 6 | | | | | | — | |
| 7 | + | | | | | | — |
| 8 | + | | | | | | — |
| 9 | + | | | | | | — |

Table 5: Variations of the Matrix $D$
(3 rows and 2 columns are removed)

| | 3 | 4 | 5 | 6 | 7 |
|---|---|---|---|---|---|
| 4 | — | — | | | + |
| 5 | | | — | | |
| 6 | | | | — | |
| 7 | + | + | | | — |
| 8 | + | + | | | — |
| 9 | + | + | | | — |

Table 6: Variations of the Matrix $D$
(2 rows and 3 columns are removed)

| | 4 | 5 | 6 | 7 |
|---|---|---|---|---|
| 3 | — | | | + |
| 4 | — | | | + |
| 5 | | — | | |
| 6 | | | — | |
| 7 | + | | | — |
| 8 | + | | | — |
| 9 | + | | | — |

When the underlying cost coefficients $c_{ij}$'s of Problem 2 are given by the matrix $D$, the northeast algorithm (Algorithm I) provides a complete solution to the Problem 2. To determine the variations of the cells of an arbitrary matrix, the algorithms similar to the positive variation algorithm (Algorithm II) can be used. Both of these algorithms are given at the end of this paper.

**Example 3**

This example is taken from the paper by Causey, Cox, and Ernst (1985) on the problem of maximizing the overlap between two surveys. The sampling scheme is summarized in Table 7. The cost matrix $C = \{c_{ij}\}$, where $c_{ij} = \#(s_i \cap s_j)$, along with its variations are summarized in Table 8. The object is to:

maximize $\sum_i \sum_j c_{ij} p_{ij}$, $1 \leq i \leq 12$, $1 \leq j \leq 5$

subject to $\sum_j p_{ij} = p_i$, $1 \leq i \leq 12$,

$\sum_i p_{ij} = q_j$, $1 \leq j \leq 5$,

$p_{ij} \geq 0$, $\forall\, i, j$.

It follows from Corollary 1 of Theorem 2 that there exists an optimal solution for this problem such that for $1 \leq i, j \leq 5$, $P(s_i, s_j) = min(p_{1i}, p_{2i})$ for $i = j$ and zero otherwise. This partial solution reduces the size of the original problem from $12 \times 5 = 60$ to $7 \times 5 = 35$ variables as summarized in Table 9. The reduced problem can now be solved using any standard solver of transportation problems.

Table 7: Sampling Scheme for Example 3

Survey I | Survey II

| $s_m$ | $P_{1m}$ | $s_n$ | $P_{1n}$ |
|---|---|---|---|
| (1) | .15 | (1) | .40 |
| (2) | .018 | (2) | .15 |
| (3) | .012 | (3) | .05 |
| (4) | .24 | (4) | .30 |
| (5) | .04 | (5) | .10 |
| (1,4) | .30 | | |
| (1,5) | .05 | | |
| (2,4) | .036 | | |
| (2,5) | .006 | | |
| (3,4) | .024 | | |
| (3,5) | .004 | | |
| (0) | .12 | | |

Table 8: The Cost Matrix and Its Variation

| | 1 | 2 | 3 | 4 | 5 |
|---|---|---|---|---|---|
| 1 | 1+ | 0 | 0 | 0 | 0 |
| 2 | 0 | 1+ | 0 | 0 | 0 |
| 3 | 0 | 0 | 1+ | 0 | 0 |
| 4 | 0 | 0 | 0 | 1+ | 0 |
| 5 | 0 | 0 | 0 | 0 | 1+ |
| 6 | 1 | 0 | 0 | 0 | 1 |
| 7 | 1 | 0 | 0 | 1 | 0 |
| 8 | 0 | 1 | 0 | 0 | 1 |
| 9 | 0 | 1 | 0 | 0 | 1 |
| 10 | 0 | 0 | 1 | 1 | 0 |
| 11 | 0 | 0 | 1 | 0 | 1 |
| 12 | 0 | 0 | 0 | 0 | 0 |

Table 9: Sampling Scheme for the Reduced Problem

| Survey I | | Survey II | |
|---|---|---|---|
| $s_m$ | $P_{1m}$ | $s_n$ | $P_{1n}$ |
| (1, 4) | .30 | (1) | .25 |
| (1, 5) | .05 | (2) | .132 |
| (2, 4) | .036 | (3) | .038 |
| (2, 5) | .006 | (4) | .06 |
| (3, 4) | .024 | (5) | .06 |
| (3, 5) | .004 | | |
| (0) | .12 | | |

**Example 3**

This example establishes the optimality of Lahiri's selection scheme (1954). This scheme requires a serpentine ordering of the psu's as illustrated in Figure 1 below so that geographically contiguous units occur next to each other in the sampling frame.

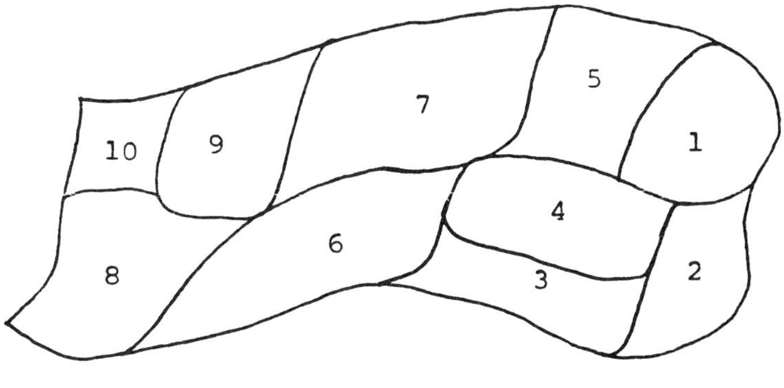


Fig. 1. Lahiri's Serpentine Ordering of PSU's

**Algorithm I (Northeast)**

```
begin
        p_ij := 0; 1 ≤ i ≤ m, 1 ≤ j ≤ n;
        i := 1; j := n;
        while (j > 0) or (i < m ) do
                if p[i] < q[j] then
                begin
                        p_ij := p[i];
                        q[j] := q[j] - p[i];
                        i := i + 1
                end
                else if p[i] ≥ q[j] then
                begin
                        p_ij := q[j];
                        p[i] := p[i] - q[j];
                        j := j - 1
                end
                else
                begin
                        p_ij := q[j];
                        i := i + 1;
                        j := j - 1
                end
end.
```

**Algorithm II (Positive Variation)**

```
begin
        for i := 1 to m do;
                for j := 1 to n do;
                begin
                        positive := true;
                        k := 1;
                        repeat
                                l := 1;
                                repeat
                                        if variation(i,j,k,l) < 0
                                                then positive :=
                                                false
                                        l := l + 1
```

```
                until (not positive) or
                      (l > n);
                k := k + 1
            until (not positive) or (k > m)
end.
```

## References

Aragon, J. and Pathak, P. K. (1990): A transportation algorithm for optimal integration of two surveys, to appear in *Sankhya.*

Arthanari, T. S. and Dodge, Y. (1981): *Mathematical Programming in Statistics*, Wiley, New York.

Causey, B. D., Cox, L. H., and Ernst, L. R. (1985): Applications of transportation theory to statistical problems, *J. Amer. Statist. Assoc.*, 80, 903-909.

Chvatal, V. (1983): *Linear Programming*, Freeman, New York.

Felligi, I. P. (1966): Changing the probabilities of selection when two units are selected with pps without replacement, *Proceedings of the Social Statistics Section, Amer. Statist. Assoc.*, 434-442.

Goodman, R. and Kish, L. (1950): Controlled selection—A technique in probability sampling, *J. Amer. Statist. Assoc.* 45, 350-372.

Hanson, R. J. and Hiebert, K. L. (1981): A sparse linear programming subprogram, *Sandia National Laboratory Report* SAND81-0297.

Hoffman, A. J. (1963): On simple linear programming problems, *Proc. Sympos. Pure Math.* 7, 317-327.

Keyfitz, N. (1951): Sampling with probability proportional to size: adjustments for changes in probabilities, *J. Amer. Statist. Assoc.* 46, 105-109.

Kish, L. and Scott, A. (1971): Retaining units after changing strata and probabilities, *J. Amer. Statist. Assoc.* 66, 461-470.

Krishnamoorthy, K. and Mitra, S. K. (1986): Cost robustness of an algorithm for optimal integration of surveys, *Sankhya B* 48, 233-245.

Lahiri, D. B. (1954): Technical paper on some aspects of the development of the sample design, *Sankhya* 14, 264-316.

Maczynski, M. J. and Pathak, P.K. (1980): Integration of surveys, *Scan. J. Statist.* 7, 130-138.

Mitra, S. K. and Pathak, P. K. (1984): Algorithms for optimum integration of two or three surveys, *Scan. J. Statist.* 11, 257-263.

Monge, G. (1781): Mémoire sur la théorie des déblais et des remblais, *Mémoires de L'Académie des Sciences* 19, 666-704.

Raj, D. (1957): On the method of overlapping maps in sample surveys, *Sankhya* 17, 89-98.

# THE MODEL BASED (PREDICTION) APPROACH TO FINITE POPULATION SAMPLING THEORY

Richard M. Royall, Department of Biostatistics,
The Johns Hopkins University

## Introduction

Estimating a finite population mean from a sample is equivalent to predicting the mean of the non-sample values. This view, that finite population inference problems are actually prediction problems, leads naturally to a theory in which prediction models, not sample selection probabilities, are central. This paper is an informal survey of that theory.

The first section describes the model-based approach and attempts to make clear how and why it differs from the prevailing (randomization-based) theory. This section is built around a simple example, which is used to illustrate various facets of the approach. The second section addresses the question "What has the model-based approach accomplished?" This is not an attempt to catalog significant contributions to model-based sampling theory, but to describe and interpret the general kinds of developments that have occurred. Finally, the third section consists of some brief observations on current research.

## What Is Model-Based Sampling Theory?

Model-based sampling theory begins by recognizing that problems of estimating finite population characteristics are naturally expressed as prediction problems (Kalbfleisch and Sprott, 1969; Geisser, 1986, p. 163). For example, Figure 1 shows the data for a sample of $n = 32$ hospitals. For each sample hospital we know the number of beds ($x$) and we have observed the number of patients discharged ($y$) during a given month. If we must estimate how many patients were discharged from another hospital, say one with x = 400 beds, we might fit the dotted line in Figure 1. The slope of that line, the ratio of total sample discharges to total sample beds, shows that in sample hospitals there were 3.1 patients discharged per bed. Thus we might estimate that there were about $3.1 \times 400 = 1240$ patients discharged from the other hospital. More generally, to estimate how many patients were discharged from a set r of non-sample hospitals having a total of $\Sigma_r x_i$ beds, we might use $3.1\ \Sigma_r x_i$. Then to estimate the patient total for the entire population composed of the thirty-two hospitals in the sample s as well as those in r, we would simply add the observed total for the thirty-two sample hospitals, $\Sigma_s y_i$ to our estimate for those not observed, $3.1\ \Sigma_r x_i$.

Clearly this estimate of the population total is reasonable only if it is reasonable to assume that the hospitals in r are "like" the ones in s: if the sample hospitals are in the eastern United States while the r-hospitals are in France, then this estimate is certainly questionable. How can we formalize this reasoning,

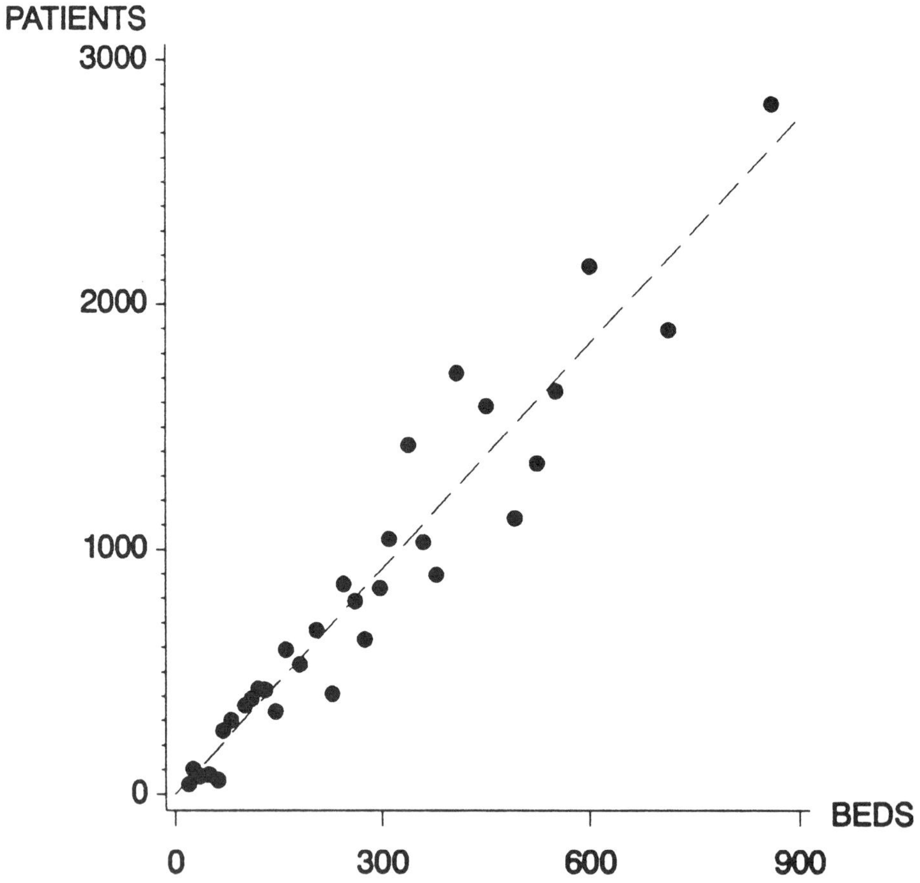


Figure 1. Number of patients discharged and number of beds in 32 short-stay U.S. Hospitals, June 1968.

exposing and clarifying the underlying assumptions and explaining when the estimate is a good one, and when it is not?

A natural way to express the assumptions is through a probability model for the numbers of patients discharged from each of the hospitals, both those in the sample $s$ and those in $r$. The model represents these numbers, $y_1, y_2, \ldots, y_N$, as realized values of independent random variables $Y_1, Y_2, \ldots, Y_N$, where $N$ is the total number of hospitals.

**Model M.** $$E(Y_i) = \beta x_i \, , \; var(Y_i) = \sigma^2 x_i \, ,$$

$$cov(Y_i, Y_j) = 0 \, , \; i \neq j.$$

Under model M the $y$'s will tend to be roughly proportional to the $x$'s, with more variability about the expected value, $\beta x$, in large hospitals than in small ones. This model is consistent with the thirty-two observations shown in Figure 1, but is not unique in this respect. We must be alert to the possibility that other models might be more appropriate. Nevertheless, analysis under model M can explain much of what our informal look at the problem has already suggested.

First we note that the model represents a link between the two sets of numbers, $\{y_i;\ i \in s\}$ and $\{y_i;\ i \in r\}$, that enables us to learn about the second set by studying the first. Now the problem of estimating $T = \Sigma_s y_i + \Sigma_r y_i$ is evidently equivalent to the problem of predicting the value, $\Sigma_r y_i$, of the random variable, $\Sigma_r Y_i$. The estimate that we derived intuitively, $\hat{T}_R = \Sigma_s y + b\Sigma_r x$, where $b = \Sigma_s y / \Sigma_s x = 3.1$, is the best linear unbiased (BLU) estimator of $T$ under model M, because $b\Sigma_r x$ is the BLU predictor of $\Sigma_r y$. Note that this is actually the popular *ratio estimator*, $(\Sigma_s y / \Sigma_s x)\Sigma_1^N x$. The reason that we would not use this estimate if the non-sample hospitals were in France is that we would be unwilling to apply the same model (with the same value of the expected number of patients per bed, $\beta$) to both the sample and non-sample facilities. Note that this conclusion would apply even if we had decided at random which ones to exclude from the sample and had chosen the overseas hospitals by bad luck. Our reluctance to use the sample ratio, 3.1 discharges per bed, to estimate for those not in the sample arises from unwillingness to make the assumptions expressed in the model, not from the process used to choose which hospitals to put in the sample and which ones to leave out.

The model also provides guidance in sampling. For a given split of the population into sample $s$ and non-sample $r$ hospitals, the estimation (prediction) error in the ratio estimate of $T$ is $\hat{T}_R - T = b\Sigma_r x - \Sigma_r y$. Its expected value under $M$ is zero and its variance is $var(\hat{T}_R - T) = (N/f)(1 - f)(\bar{x}\bar{x}_r/\bar{x}_s)\sigma^2$, where $\bar{x}$ is the population mean, $\bar{x}_s$ and $\bar{x}_r$ are sample and non-sample means, $n$ is the sample size, and $f$ is the sampling fraction, $n/N$. This variance decreases as $\bar{x}_s$ increases, so it is minimized when $s$ consists of the $n$ largest hospitals. Equally important, it is *maximized* when $s$ consists of the $n$ smallest. Although we will find that robustness considerations imply that it is often unwise to choose the largest units for $s$, the smallest units represent the worst possible sample under a wide variety of conditions.

Another role of the model is to validate large sample confidence intervals: if the population is enlarged, so that both sets of hospitals, $s$ and $r$, grow in a stable way, then $(\hat{T}_R - T)/[var(\hat{T}_R - T)]^{1/2}$ converges in distribution to the standard normal. Because $v = \Sigma_s(y_i - bx_i)^2/nx_i$ is a consistent estimator of $\sigma^2$, an approximate confidence interval for $T$ is given by $\hat{T}_R \pm z[(N/f)(1 - f)(\overline{x}\overline{x}_r/\overline{x}_s)v]^{1/2}$ when $n$ and $N - n$ are both large (Royall and Cumberland, 1978).

Although the ratio estimator is BLU under model M, other estimators might also be considered, because of robustness, simplicity, or other criteria. Analysis under M remains critical: the estimator we choose must at least have reasonable properties under this model if it is to be appropriate for estimating the total number of patients discharged. For example, the simple expansion estimator $\hat{T}_E = \Sigma_s y + (N - n)\overline{y}_s = N\overline{y}_s$, which estimates the non-sample mean $\overline{y}_r$ by the sample mean $\overline{y}_s$, would be inappropriate here in any sample $s$ of hospitals whose mean size $\overline{x}_s$ is not very close to the population mean $\overline{x}$. This is because the estimator is biased under M:

$$E(\hat{T}_E - T) = N\beta(\overline{x}_s - \overline{x}).$$

This expression shows that the expansion estimator will tend to underestimate $T$ if the average size of sample hospitals, $\overline{x}_s$, is smaller than the population average, $\overline{x}$, and to overestimate when $\overline{x}_s$ is larger. By contrast, the linear *regression estimator* $\hat{T}_{RG} = N[\overline{y}_s + b_1(\overline{x} - \overline{x}_s)]$ where $b_1 = \Sigma_s(x_i - \overline{x}_s)y_i/\Sigma_s(x_i - \overline{x}_s)^2$, is, like the ratio estimator, unbiased under M in any sample $s$: $E(\hat{T}_{RG} - T) = 0$.

Thus we can evaluate estimators in terms of bias and variance under M, study how these properties are affected by characteristics of the sample, like $\overline{x}_s$, and find approximating distributions for setting confidence intervals. If M were known to be true, then this body of theoretical results might be satisfactory for guiding us in selecting a sample and making inferences from observations.

But M is not true. A sufficiently large sample of hospitals would surely reveal that M, like any mathematical model, is at best an approximation. Although we have adopted M as a working model for this population, we remain skeptical, aware that theoretical results derived under M have practical value only if they are robust in the face of plausible departures from this model.

Robustness to departures from M can be studied by changing the model. For example if we generalize by relaxing the restriction that $var(Y_i)$ be proportional to $x_i$, we see that the ratio estimator remains unbiased and consistent. But the large-sample confidence interval is no longer valid, because the estimator of $var(\hat{T}_R - T)$ is no longer consistent. Fortunately there are variance estimators that are consistent under the generalized model, providing robust large sample confidence intervals (Royall and Cumberland, 1981a).

To study the effects of errors in the working model's regression function, $E(Y_i) = \beta x_i$, we might consider a sequence of generalizations, first adding a constant term, then a quadratic, etc. Each term added to the regression model introduces bias in the ratio estimator. For example, if $E(Y_i) = \alpha + \beta x_i$, then the bias is $E(\hat{T}_R - T) = N\alpha(\bar{x} - \bar{x}_s)/\bar{x}_s$. Protection against this bias can be achieved by choosing a sample that is balanced on $x$: $\bar{x}_s = \bar{x}$. Protection against a quadratic term's bias can be achieved by balancing on $x^2$ as well: $\Sigma_s x^2/n = \Sigma_1^N x^2/N$. And balancing on other powers of $x$ protects against bias caused by the presence of corresponding terms in the true regression function (Royall and Herson, 1973).

Thus in order to protect against the bias that can be caused by departure from the working model's regression function, we might choose a balanced sample in preference to the optimal (minimum variance) sample composed of the $n$ largest hospitals. The same type of trade-off, efficiency for robustness, might apply to other aspects of the problem as well, such as the choice of an estimator for $T$ and an estimator of the error variance, $var(\hat{T} - T)$. The model-based theory does not assume that a particular model is correct and proceed blindly under that assumption: alternative models are used to examine the key practical issue of robustness.

The main features of model-based sampling theory have appeared in our look at the hospital discharge population:

(i) representing the unknown numbers of interest as realized values of observable random variables,

(ii) recognizing that estimating a population value from an observed sample is a prediction problem, and

(iii) using probability models as the formal basis for prediction and for determining the primary statistical properties of samples and predictors.

The use of probability models as the basis for inference from sample to population, (iii), is the critical feature distinguishing the model-based theory from the prevailing one. Although a random sampling plan may be used for choosing which hospitals will be observed (and for which hospitals the number of discharges must be estimated), the basic inference framework is the probability model, not the random sampling plan. By contrast, the prevailing theory of finite population sampling reverses the priority, avoiding probability models in favor of distributions created by random sampling plans as the formal basis for inference.

Conventional theory defines bias, for example, with respect to the probability distribution generated by the random sampling plan. Thus the expansion estimator, $N\bar{y}_s$, is an unbiased estimator of $T$ if every set of $n$ hospitals is given the same probability of being selected as the sample. But the same estimator is biased if the sample is chosen by another selection scheme. The bias in $N\bar{y}_s$ is determined, not by relationships between the hospitals in $s$ and those not in *s*, but by the probabilities with which other samples might have been selected. Recall that the model-based theory under model M said that this estimator has a positive bias if the sample consists of hospitals that are larger, on average, than those not in the sample, a negative bias if the sample hospitals are smaller, and no bias only if the sample is balanced on size. Although both definitions of bias are mathematically valid, for the purpose of inference from a given sample of hospitals the model-based one is clearly relevant and informative while the conventional one is misleading.

Conventional theory defines variance also as an average value over all possible samples. Again this is in contrast to model-based theory, which, because it defines the variance for a specific sample with respect to a prediction model for the unobserved variates, conditions on the characteristics of the sample actually observed as well as on those of the non-sample units whose values must be predicted.

Model-based theory, by insisting that inferences should be based on prediction models, not on probability distributions created by randomly choosing which units to observe, does not preclude the use of random sampling plans. It is not the presence or absence, but the *role*, of random sampling that distinguishes model-based from conventional finite population sampling theory. The terminology invites misunderstanding on this point: because the word *sampling* in the name suggests only the design phase–choosing samples – *model-based sampling theory* is easily misinterpreted as signifying a theory for choosing samples using models, whereas the critical feature is the use of models in inference.

There are other model-based approaches. The one sketched above is developed in terms of bias, variance, and approximate normality under linear models. Alternatives include approaches based on fiducial (Kalbfleisch and Sprott, 1969), likelihood (Royall, 1976b), and Bayesian prediction models. Ericson (1988) has recently surveyed the Bayesian theory. We will focus on the linear prediction approach, because it has seen the most vigorous development, empirical testing, and critical discussion.

### What Has the Model-Based Approach Accomplished?

The model-based approach has bridged the gap between finite population problems and the rest of statistics. Before the model-based approach, finite population sampling was an eccentric realm where many of the basic concepts and tools of statistics were curiously inapplicable. Statisticians skilled in designing experiments and in applying linear models to make inferences from experimental and observational data found that finite population problems were apparently beyond the scope of their techniques. Although there were some familiar-looking formulas, such as the linear regression estimator shown in Section 1, these statistics lacked the familiar rationale and properties. Not only was the linear regression estimator biased (and therefore certainly not a BLU estimator) it was not even linear, because the random choice of observation points turned the denominator of the estimated slope into a random variable. To make matters appear utterly hopeless to one interested in statistical theory, Godambe (1955) proved that the BLU estimator for a finite population average does not exist and furthermore (1966) showed that the likelihood function generated by a random sample from a finite population is, for all practical purposes, totally uninformative. Attempts to fill the theoretical vacuum were uniformly unsuccessful (e.g., Godambe, 1966; Hanurav, 1968; Hartley and Rao, 1969; Royall, 1969).

The prediction approach revealed that the problem was rooted, not in esoteric aspects of finite population problems that invalidated the methods applicable to the rest of statistics, but in the attachment of those who worked in finite population sampling theory to a restrictive statistical doctrine based on a dubious principle. This is the Randomization Principle, proclaimed and then renounced by Fisher (1935 §21, 1960 §21.1), which asserts that the only probability distributions appropriate for statistical inference are those created by deliberate randomization.

A particularly clear statement of the Randomization Principle in the finite population setting was given by Stuart (1962):

> If you feel at times that the statistician, in his insistence on random sampling methods, is merely talking himself into a job, you should chasten yourself with the reflection that in the absence of random sampling, the whole apparatus of inference from sample to population falls to the ground, leaving the sampler without a scientific basis for the inferences which he wishes to make.

This Principle has had its champions in experimental statistics (Kempthorne, 1955), where it underlies the curious claim that no valid statistical inferences are possible in observational studies. (This last point is discussed in Royall (1976a), with references.) But in that area the Principle faced strong opposition, from "Student" (1937) and Neyman and Pearson (1937, p. 384) for example, and it never held sway. The Principle's unchallenged domination of finite population theory is thus curious; it is doubly curious because this domination is credited to Neyman (1934) (ref. Smith, 1976; O'Muircheartaigh and Wong, 1981).

The theoretical vacuum in finite population sampling was an inevitable consequence of the Randomization Principle. If the Principle is applied in other areas of statistics, entirely analogous results follow: if all inferences must be based on the probability distribution created by artificial randomization, so that all variables that have not been *made* random by the experimenter's actions must be treated as fixed (possibly unknown) constants, then the likelihood function for randomized comparative experiments is just like the finite population likelihood function—uninformative (Cornfield, 1966). Likewise, if inferences about regression coefficients must be based on the distribution created by using deliberate randomization to select material for observation, then the

Gauss-Markov theorem can justify least-squares estimators only in those cases where at each value of the regressor $x$ the average response $\bar{y}$ over all units actually available for observation falls precisely on the regression line: thus the Principle would imply the non-existence of BLU estimators in essentially all real-world applications, certainly including all problems where each potential sample unit is characterized by a unique vector of regressor values.

Deliberate randomization is a valuable statistical tool (for protecting against unconscious bias, for example). Few statisticians would deny this. But the Randomization Principle claims much more: the only biases, standard errors, significance levels, and confidence coefficients acceptable for inference are those defined and justified in terms of deliberate randomization. The model-based prediction approach to finite population sampling consists of nothing more radical than taking the concepts, techniques, and tools that form the familiar core of applied statistics and using them where previously they had been precluded by acceptance of the Randomization Principle. This has had several important effects:

(i) providing techniques for systematic study of some finite population sampling problems that the randomization approach is ill-equipped to address,

(ii) bringing an alternative theoretical perspective to finite population methods that have been analyzed previously in terms of randomization theory,

(iii) revitalizing conventional randomization-based finite population theory,

(iv) providing a new context for studying the model-based methods that are standard outside of finite populations, and

(v) testing general statistical concepts and principles in a new setting.

Examples in the first category – problems that are difficult to address in terms of deliberate randomization alone – include non-response (Särndal, 1981; Little, 1982; Chiu and Sedransk, 1986), small area estimation (Laake, 1979; Holt, Smith, and Tomberlin, 1979; Royall, 1979), and inference from non-random samples (Smith, 1983; Kott, 1984). This is not to say that there was no methodology for these problems before the model-based approach came along. There were various techniques that had been derived intuitively and developed by trial and error. What models did was to provide a theoretical framework for studying the methods (such as *synthetic estimates* for small area estimation) and for describing the implicit assumptions behind them, as well as for suggesting alternatives.

Of greater theoretical interest are activities of the second type – applications of the model-based approach to problems where the old randomization approach had already generated a body of results. In some cases the prediction approach simply provided a new explanation and interpretation of conclusions that had been reached by conventional sampling theory. An example is the finding that the Yates-Grundy estimator is better than the Horvitz-Thompson estimator for the variance of the mean-of-ratios statistics, $N\bar{x}\Sigma_s(y_i/x_i)/n$, in samples chosen by *a probability-proportional-to-x* sampling plan (Cumberland and Royall, 1981).

In other cases the prediction approach revealed a clear preference for one of two procedures where the randomization approach had been noncommittal. One example is in post-stratification, where some followers of randomization theory had chosen to condition the variance on the actual stratum sample sizes, while others had chosen to use the unconditional variance. The deadlock was described by Holt and Smith (1979), whose prediction theory analysis made clear the need to condition.

Variance estimation for the ratio estimator provides another example of the activities in category (ii). Randomization theory had been unable to choose between two proposed variance estimators, yet model-based analyses revealed that the more popular of the two has a severe conditional bias. This bias is positive in some samples, leading to overly conservative confidence intervals, and negative in others, producing undercoverage. The second statistic is free of these biases. It is worth noting that empirical comparisons of these two variance estimators had also been inconclusive, because the investigators, guided by randomization theory, had averaged the results over all of the values of the conditioning variable, and had thereby averaged out the biases (Rao and Rao, 1971). Empirical studies guided by prediction theory exposed the biases clearly enough to inspire efforts to accommodate the conditional results within randomization theory (Fuller, 1981; Robinson, 1987).

The model-based approach has stimulated conventional sampling theory in other ways as well. For example, model-based results on variance estimation (Royall and Cumberland, 1981a) have inspired significant developments in conventional theory (Wu and Deng, 1983; Deng and Wu, 1987). At a more general level, the model-based approach has forced those who object to it to examine and articulate the reasons for their opposition (e.g., Hansen, Madow, and Tepping, 1983) and to extend and adapt the conventional theory to accommodate those model-based results that they find compelling (e.g., the above-cited attempts to develop a conditional randomization theory for the ratio estimator). Another general effect on conventional sampling theory has been to create a greater awareness of models and willingness to use them in analyses. Very important work has been done in studying the effects of using standard computer packages (i.e., analyses based on simple models) to analyze sample survey data when the models do not adequately describe the process generating the observations. Some of this work has been model-based and some has been based on random sampling distributions, but stimulated by the model-based activity, and using models in the analysis (e.g., Holt, Smith, and Winter, 1980; Skinner, Holmes, and Smith, 1986).

Developments in category (iv) are of very general importance. The model-based approach brings new statistical methods to finite populations, methods that are widely used in other areas of statistics. These new applications represent important tests cases for the methods, which are now used in real samples from real populations that can be examined *in toto* to determine exactly how large the estimator error is, whether the true mean actually lies within the confidence interval, etc. Studying statistical methods in finite populations entails a degree of realism and relevance to real-world phenomena that is hard to achieve in other contexts, where the object of estimation is an unobservable (usually purely conceptual) model parameter, or where the test data are generated artificially.

This is illustrated by the finite-population tests of the standard variance estimates in linear regression models (Royall and Cumberland, 1981a, b). These empirical studies showed that the estimates are much more sensitive to errors in the models' variance structure than had been generally acknowledged (see e.g. Efron, 1979 §7). This suggests that more attention should be paid to bias-robust alternatives. But further finite population studies have produced frightening examples showing that confidence intervals based on bias-robust estimates, although better than those based on the standard variance estimates, can also perform very poorly under conditions that, though not uncommon, are difficult to recognize when they occur (Royall and Cumberland, 1985).

Finally, the model-based approach to finite population sampling has also helped to clarify the basic concepts and principles of statistics. Stimulated by the good advice "Look at the data," along with exciting computer capabilities for display and analysis of samples, statisticians now rely heavily on the data to suggest and criticize models. Finite population studies have helped to emphasize the limitations of this sort of empiricism: model failure that is not apparent in the sample can produce seriously misleading inferences (e.g., Royall and Cumberland, 1981a; Rubin, 1983). Thus, robustness is vitally important even when the model fits the observed data well. Other important general issues that have been emphasized and illustrated in the model-based approach to finite population sampling include the critical distinction between probabilistic and inferential validity and the need for conditioning on ancillary statistics to achieve the latter (see Royall, 1976a, for discussion; ref. also Hinkley, 1983), the inferential inadequacy of probability distributions generated by artificial randomization, and the fundamental importance of likelihoods (Royall, 1976b, discusses the last two points).

### Some Current Developments

The role of randomization in a model-based approach to finite population sampling is a subject of continuing research. Randomization is certainly valuable at the sampling stage. For example, it can ensure that the chances are good that the sample selected will be well balanced, so that in that sample a given estimator is robust with respect to variables that are not adequately accounted for by the prediction model (Royall and Herson, 1973). But just when and how random sampling probabilities should influence inferences from a given sample has proved to be a difficult issue. On one hand, the set of labels identifying the sampled units is an ancillary statistic, so that the Conditionality Principle evidently precludes any role for the random sampling distribution in inference (Basu, 1971). On the other hand, the expected balance associated with simple random sampling is a characteristic whose statistical relevance does not seem to vanish entirely when the perspective shifts from (i) choosing which units to observe to (ii) making estimates from an observed sample (ref. Royall, 1976a, p. 471). Thus there are continuing efforts to formalize and explain the precise role of random sampling in finite population inference (e.g., Sugden and Smith, 1984; Pfefferman and Holmes, 1985; Cumberland and Royall, 1988; Kott, 1988; and Tam, 1988) and to reconcile the prediction and randomization approaches (Brewer, Hanif, and Tam, 1988).

But recent progress in model-based theory has not been limited to the interface with randomization theory. Tam (1986) has given an elegant extension and unification of earlier work on robust estimation. Chambers (1988) has contributed both theoretical and empirical results on model-based estimation for domains within a larger population. And Valliant has used the prediction approach to analyze the statistical properties of a widely-used method of variance estimation (1987a), to discover critical conditional properties of estimators in stratified samples (1987b), and to study an important problem in economic statistics (1988).

**References**

Basu, D. (1971): An Essay on the Logical Foundations of Survey Sampling, Part I, in *Foundations of Statistical Inference*, eds. Godambe and Sprott, Holt, Rinehart and Winston of Canada, Toronto, 203-233.

Brewer, K. R. W., Hanif, M., and Tam, S. M. (1988): How nearly can model-based prediction and design-based estimation be reconciled?, *Journal of the American Statistical Association* 83, 128-131.

Chambers, R. L. (1988): Limited information estimation of domain means, manuscript submitted for publication.

Chiu, H. Y., and Sedransk, J. (1986): A Bayesian procedure for imputing missing values in sample surveys, *Journal of the American Statistical Association* 81, 667-676.

Cornfield, J. (1966): A note on the likelihood function generated by randomization over a finite set, *Bulletin of the International Statistical Institute* 41, Book 1, 79-80.

Cumberland, W. G., and Royall, R. M. (1981): Prediction models and unequal probability sampling, *Journal of the Royal Statistical Society*, Ser. B, 43, 353-367.

Cumberland, W. G., and Royall, R. M. (1988): Does simple random sampling provide adequate balance?, *Journal of the Royal Statistical Society*, Ser. B, 50, 118-124.

Deng, L.-Y., and Wu, C. F. J. (1987): Estimation of variance of the regression estimator, *Journal of the American Statistical Association* 82, 568-576.

Efron, B. (1979): Bootstrap methods: another look at the jackknife, *Annals of Statistics* 7, 1-26.

Ericson, W. A. (1988): Bayesian inference in finite populations, Chapter 9 in *Handbook of Statistics, Volume 6, Sampling*, eds. Krishnaiah and Rao, North-Holland, New York, 213-246.

Fisher, R. A. (1935): *The Design of Experiments*, first edition, Oliver and Boyd, Edinburgh.

Fisher, R. A. (1960): *The Design of Experiments*, seventh edition, Hafner Publishing Company, New York.

Fuller, W. A. (1981): Comment, *Journal of the American Statistical Association* 76, 76-80.

Geisser, S. (1986): Predictive analysis, in *Encyclopedia of Statistical Sciences*, Volume 7, eds. Kotz and Johnnson, John Wiley & Sons, Inc., New York.

Godambe, V. P. (1955): A unified theory of sampling from finite populations, *Journal of the Royal Statistical Society*, Ser. B, 17, 267-278.

Godambe, V. P. (1966): A new approach to sampling from finite populations I: Sufficiency and linear estimation, *Journal of the Royal Statistical Society*, Ser. B, 28, 310-219.

Hansen, M. H., Madow, W. G., and Tepping, B. J. (1983): An evaluation of model-dependent and probability-sampling inferences in sample surveys, *Journal of the American Statistical Association* 78, 776-793.

Hanurav, T. V. (1968): Hyper-admissibility and optimum estimators for sampling finite populations, *Annals of Mathematical Statistics* 39, 621-642.

Hartley, H. O., and Rao, J. N. K. (1968): A new estimation theory for sample surveys, *Biometrika* 55, 547-557.

Hinkley, D. V. (1983): Can frequentist inferences be very wrong? A conditional yes, in *Scientific Inference, Data Analysis, and Robustness*, eds. Box, Leonard, and Wu, Academic Press, New York, 85-103.

Holt, D., and Smith, T. M. F. (1979): Post stratification, *Journal of the Royal Statistical Society*, Ser. A, 142, 33-46.

Holt, D., Smith, T. M. F., and Tomberlin, T. J. (1979): A model-based approach to estimation for small subgroups of a population, *Journal of the American Statistical Association* 74, 405-410.

Holt, D., Smith, T. M. F., and Winter, P. D. (1980): Regression analysis of data from complex surveys, *Journal of the Royal Statistical Society*, Ser. A, 143, 474-487.

Kalbfleisch, J. D., and Sprott, D. (1969): Applications of likelihood and fiducial probability to sampling finite populations, in *New Developments in Survey Sampling*, eds. Johnson and Smith, John Wiley & Sons, New York, 358-389.

Kempthorne, O. (1955): The randomization theory of experimental inference, *Journal of the American Statistical Association* 50, 946-967.

Kott, P. S. (1986): Some asymptotic results for the systematic and stratified sampling of a finite population, *Biometrika* 73, 485-491.

Kott, P. S. (1988): Model-based finite population correction for the Horvitz-Thompson Estimator, *Biometrika* 75, to appear.

Laake, P. (1979): A predictive approach to subdomain estimators in finite populations, *Journal of the American Statistical Association* 74, 355-358.

Little, R. J. A. (1982): Models for nonresponse in survey samples, *Journal of the American Statistical Association* 77, 237-250.

Neyman, J. (1934): On the two different aspects of the representative method: the method of stratified sampling and the method of purposive selection, *Journal of the Royal Statistical Society* 97, 558-606.

Neyman, J., and Pearson, E. S. (1937): Note on some points in 'Students' paper on 'Comparison between balanced and random arrangements of field plots', *Biometrika* 29, 380-388.

O'Muircheartaigh, C., and Wong, S. T. (1981): The impact of sampling theory on survey sampling practice: A review, *Bulletin of the International Statistical Institute* 49, 465-493.

Pfefferman, D., and Holmes, D. J. (1985): Robustness considerations in the choice of a method of inference for regression analysis of survey data, *Journal of the Royal Statistical Society*, Ser. A, 148, 268-278.

Rao, P. S. R. S., and Rao, J. N. K. (1971): Small sample results for ratio estimators, *Biometrika* 58, 625-630.

Robinson, J. (1987): Conditioning ratio estimates under simple random sampling, *Journal of the American Statistical Association* 82, 826-831.

Royall, R. M. (1968): An old approach to finite population sampling theory, *Journal of the American Statistical Association* 63, 1269-1279.

Royall, R. M. (1976a): Current advances in sampling theory: Implications for human observational studies, *American Journal of Epidemiology* 104, 463-474.

Royall, R. M. (1976b): Likelihood functions in finite population sampling theory, *Biometrika* 63, 605-614.

Royall, R. M. (1979): Prediction models in small area estimation, in *Synthetic Estimates for Small Areas*, ed. J. Steinberg, NIDA Monograph 24, National Institute on Drug Abuse, Washington, D.C., 63-87.

Royall, R. M., and Cumberland, W. G. (1978): Variance estimation in finite population sampling, *Journal of the American Statistical Association* 73, 351-358.

Royall, R. M., and Cumberland, W. G. (1981a): An empirical study of the ratio estimator and estimators of its variance, *Journal of the American Statistical Association* 76, 66-77.

Royall, R. M., and Cumberland, W. G. (1981b): The finite-population linear regression estimator and estimators of its variance—An empirical study, *Journal of the American Statistical Association* 76, 924-930.

Royall, R. M., and Cumberland, W. G. (1985): Conditional coverage properties of finite population confidence intervals, *Journal of the American Statistical Association* 80, 355-359.

Royall, R. M., and Herson, J. H. (1973): Robust estimation in finite populations I, *Journal of the American Statistical Association* 68, 880-889.

Rubin, D. B. (1983): A case study of the robustness of Bayesian methods of inference: Estimating the total of a finite population using transformations to normality, in *Scientific Inference, Data Analysis, and Robustness*, eds. Box, Leonard, and Wu, Academic Press, New York, 213-244.

Särndal, C. E. (1981): Frameworks for inference in survey sampling with applications to small area estimation and adjustment for nonresponse, *Bulletin of the International Statistical Institute* 49, Book 1, 494-513.

Skinner, C. J., Holmes, D. J., and Smith, T. M. F. (1986): The effect of sample design on principal component analysis, *Journal of the American Statistical Association* 87, 789-798.

Smith, T. M. F. (1976): The foundations of survey sampling: A review, *Journal of the Royal Statistical Society*, Ser. A, 139, 183-195.

Smith, T. M. F. (1983): On the validity of inferences from non-random samples, *Journal of the Royal Statistical Society*, Ser. A, 146, 394-403.

Sugden, R. A., and Smith, T. M. F. (1984): Ignorable and informative designs in survey sampling inference, *Biometrika* 71, 495-506.

Stuart, A. (1962): *Basic Ideas of Scientific Sampling*, Hafner Publishing Company, New York.

"Student" (1937): Comparison between balanced and random arrangements of field plots, *Biometrika* 29, 363-379.

Tam, S M. (1986): Characterization of best model-based predictors in survey sampling, *Biometrika* 73, 232-235.

Tam, S. M. (1988): Asymptotically design-unbiased predictors in survey sampling, *Biometrika* 75, 175-177.

Valliant, R. (1987a): Some prediction properties of half-sample variance estimators in single-stage sampling, *Journal of the Royal Statistical Society*, Ser. B, 49, 68-81.

Valliant, R. (1987b): Conditional properties of some estimators in stratified sampling, *Journal of the American Statistical Association* 82, 509-519.

Valliant, R. (1988): Estimation of the Laspeyres price indexes using the prediction approach for finite population sampling, *Journal of Business and Econometric Statistics* 6, 189-196.

Wu, C. F. J., and Deng, L. Y. (1983): Estimation of variance of the ratio estimation: An empirical study, in *Scientific Inference, Data Analysis, and Robustness*, eds. Box, Leonard, and Wu, Academic Press, New York, 245-277.

## SAMPLING THEORY USING EXPERIMENTAL DESIGN CONCEPTS

Jaya Srivastava, Colorado State University

and

Zhao Ouyang, Colorado State University

### Abstract

In this paper, we consider the application of concepts of Statistical Experimental Design to Sampling Theory. As is well-known, because of its inherent nature, Experimental Design Theory involves a relatively heavy amount of Combinatorial Mathematics. It turns out that, over the years, relatively speaking, it is this combinatorial aspect of Design, that has found much application in Sampling. We present a brief review of the same, including some of the latest work in the field.

### Introduction

The subject of *sampling using experimental design concepts* has attracted more and more attention in recent years. A very explicit connection was made by M.C. Chakrabarti (1963) who indicated that balanced incomplete block designs (BIBD's) could be used as sampling schemes. At first, it was shown that a BIBD procedure has properties similar to SRSWOR (simple random sampling without replacement). But later on it was found that a BIBD corresponds, in a sense, to *controlled sampling*, which was proposed by Goodman and Kish in 1950, and to which further contributions were made by Avadhani and Sukhatme (1965, 1968, 1973).

Consider an agricultural survey. Suppose we use SRSWOR to draw a sample of $n$ counties from a population of $N$ counties. It may happen that the $n$ counties in our sample are spread out in an undesirable or inconvenient manner. As pointed out by Avadhani and Sukhatme (1973), "this may not only increase considerably the expenditure on travel, but the quality of data collected is also likely to be seriously affected by non-sampling errors, particularly non-response and investigator bias, since in such cases organizing close supervision over the field work would generally be fraught with administrative difficulties". Such a sample is considered as *non-preferred*. Hence the total set of $\binom{N}{n}$ samples can be classified into two classes: *preferred* samples and *non-preferred* samples (Goodman and Kish, 1950). Hence, our objective is to design a sampling procedure which reduces the probability of drawing a non-preferred sample as much as possible, and at the same time *resembles* SRSWOR (assuming no stratification, clustering, etc. is present, and there are no auxiliary variables).

The problem of controlled sampling was first proposed by Goodman and Kish (1950). This method involves stratified sampling and emphasizes the minimization of the probability of the selection of the non-preferred samples. But, as discussed by Avadhani and Sukhatme (1973), this method may lose precision in estimation. In their three papers (1965, 1968, 1973), Avadhani and Sukhatme discuss the problem of minimizing the chance of selection of non-preferred samples without losing efficiency relative to SRSWOR.

We recall some useful notation from Srivastava (1985). Let $U$ denote a population with $N$ units denoted by the integers 1, 2,...,$N$. Let $y$ be the variable of interest, and let $y_i$ ($i = 1,\ldots,N$) be the value of $y$ for the unit $i$ in $U$. Let $Y \equiv \sum_1^N y_i$ be the *population total.* The class of all subsets of U is denoted by $2^U$, and any $\omega \in 2^U$ is called a sample of $U$. (This includes the empty sample.) For any set $K$, let $|K|$ denote the number of elements in $K$. For any $\omega \in 2^U$, let $(\omega{:}n)$ be the class of all $n$-element subsets of $\omega$; if $|\omega| < n$, then this class is empty. A sampling measure, denoted by $p(\cdot)$, is a probability density $\{p(\omega)\}$ defined on $2^U$. For a given $p(\cdot)$, let

$$\pi_i = \sum_{\omega:\, i \in \omega} p(\omega), \quad i = 1,\ldots,N. \tag{1}$$

Then, $\pi_i$ ($i = 1,\ldots,N$) is the probability that the unit $i$ is included in the sample. For any non-empty sample $\omega$, let $\bar{y}_\omega$ denote the sample mean. Consider a sampling measure $p$ for which all inclusion probabilities $\pi_i$ ($i = 1,\ldots,N$) equal $(n/N)$. Then, Avadhani and Sukhatme define $p$ to be admissible if (i) $N\bar{y}_\omega$ is an unbiased estimator of $Y$, and (ii) $Var_p(N\bar{y}_\omega) \leq Var_{SRS}(N\bar{y}_\omega)$, where $Var_p$ and $Var_{SRS}$ denote the variance respectively under the measure $p$, and the measure $q$ induced by SRSWOR with sample size $n$. (Note that, for all $\omega \in 2^U$, $q(\omega) = \{1/\binom{N}{n}\}$, if $|\omega| = n$, and $q(\omega) = 0$, otherwise.)

Let $3 \leq n \leq N-3$. The following results are given by Avadhani and Sukhatme (1973).

**Theorem 1**

Let $S \subset (U{:}\ n)$, and let $|S| = b$. Then the sampling measure which selects each $\omega \in S$ with probability $(1/b)$ is admissible if and only if $|\{\omega{:}\ \omega \in S,\ i,j \in \omega,\ i \neq j\}|$, are the same for all $i \neq j$, $i,j = 1,\ldots,N$. For such a measure, $|\{\omega{:}\ \omega \in S,\ i \in \omega\}|$ are the same for all $i = 1,\ldots,N$.

Under the condition of Theorem 1, let

$$\lambda = |\{\omega{:}\ \omega \in S,\ i,j \in \omega,\ i \neq j\}| \tag{2}$$

$$r = bn/N. \tag{3}$$

It is easy to see that the existence of $S$ in Theorem 1 is equivalent to the existence of a BIBD with parameters $(N, b, r, n, \lambda)$, such that $N$ is the number of treatments, $b$ the number of blocks, $r$ the number of replications for each treatment, $\lambda$ the number of blocks which contain any given pair of treatments, and $n$ the block size. In fact, such a $S$ is a BIBD with the above parameters. But, when $N$ and $n$ are large, such a BIBD may be hard to identify. So, the next two theorems are useful.

**Theorem 2**

The measure induced by the following (two-part) sampling procedure is admissible:

(i) Split the population randomly into $k$ subpopulations with fixed sizes $N_i$ $(i = 1,\ldots,k)$ such that $\sum_{i=1}^{k} N_i = N$,

(ii) For $i = 1,\ldots,k$, select $n_i$ units from the $i^{th}$ subpopulation by using an admissible sampling measure (with inclusion probability $(n_i/N_i)$). The selection of the units from the different subpopulations should be done independently.

**Corollary 1**

The measure induced by the following procedure is admissible:

(i) Draw a sample of size $n' > n$ from the population by SRSWOR.

(ii) From the sample selected in (i), draw a sample of size $n$ by using an admissible measure with inclusion probability $n/n'$ for each unit.

In view of the above, Avadhani and Sukhatme suggest that the following steps may be followed for controlled sampling:

(i) Let $N_1 + N_2 + \ldots + N_g = N$. Divide the original population randomly into $g$ subpopulations, which have sizes $N_1, N_2,\ldots,N_g$ respectively.

(ii) Let $n_1 + n_2 + \ldots + n_g = n$. For $i = 1, 2,\ldots,g$, select an integer $n_i'$ such that $n_i \leq n_i' < N_i$ and also select a BIBD with parameters $(n_i', b_i, r_i, n_i, \lambda_i)$. (It is preferred that $n_i$ be much smaller than $n_i'$.) Use SRSWOR to select (independently for each $i$) a sample of size $n_i'$ from the $i^{th}$ subpopulation of size $N$.

(iii) For each sample of size $n_i'$ $(i = 1,\ldots,g)$ drawn in step (ii), collect the information on all the preferred subsamples of size $n_i$ and then find a

BIBD with parameters $(n_i', b_i, r_i, n_i, \lambda_i)$ such that the number of the blocks which correspond to the preferred subsamples of size $n_i$ is as large as possible. Then draw one block with probability $1/b_i$ from the $i^{th}$ BIBD independently for $i = 1, \ldots, g$. In this way, we get a sample of total size $n_1 + \ldots + n_g = n$.

An example of controlled sampling using BIBD will be given in the last section in this paper.

**Other Works on Sampling Using Concepts of Experiment Design**

In the first section, we discussed the use of BIBD in controlled sampling. It is very clear that for a BIBD with parameters $(N, b, r, n, \lambda)$, $N$ corresponds to the number of units in the population, $b$ corresponds to the (maximum possible) number of distinct samples, and $n$ corresponds to the size of the sample. With this interpretation, it is easy to see that the parameters $r$ and $\lambda$ in the BIBD correspond respectively to the first order and the second order inclusion probabilities. So, for some time, the use of BIBD in sampling has been discussed widely.

As early as 1963, Chakrabarti pointed out the equivalence between SRSWOR and BIBD in the sense of having the same first order and second order inclusion probabilities. It is clear that the smaller the support of (i.e., the number of distinct blocks in) the BIBD, the better is the possibility of adapting it for a given situation of controlled sampling. Thus, BIBD's with a small support size have importance in sampling theory. Because of this, the work of Hedayat and others in the field of BIBD's with small supports is useful.

In 1977, Wynn showed that for each sampling measure $p_1$ there is a measure $p_2$, which gives rise to the same first and second order inclusion probabilities as $p_1$, and whose support size is not greater than $N(N-1)/2$. For the case of SRSWOR, he showed that no BIBD with support size less than $N$ can be equivalent to SRSWOR in the above sense. Hence, with the help of BIBD's we can reduce the support size from SRSWOR's $\binom{N}{n}$ to something between $\binom{N}{2}$ and $N$.

Besides BIBD, Fienberg and Tanur (1985) listed some parallel concepts in Design of Experiments and Sampling. These include randomization in design and random sampling, blocking in design and stratification in sampling, Latin square in design and lattice sampling, split-plot design and cluster sampling, and covariance adjustment in design and post-stratification in sampling. By using some similar parallel concepts in design and sampling, Meeden and Ghosh (1983) found some admissible strategies in sampling and Cheng and Li (1983) showed that Rao-Hartley-Cochran and Hansen-Hurwitz strategies are approximately minimax under some models. Brewer et al. (1977) discussed use of experimental design in the planning of sample surveys, and Sedransk (1967) discussed the use of experimental design in the analysis of sample surveys. But, even though experimental design and sampling have so many parallel concepts and similar structure, sampling has been developed separately from experimental design.

Smith and Snyder (1985) pointed out the main distinction between experimental design and sampling from their nature of inference. They concluded that "the differences between survey and experiments are as important as the similarities, and that each will continue to develop in its own way". An excellent discussion of experimental design and sample surveys, both with respect to their similarities and differences, was given by Fienberg and Tanur (1985).

Hedayat (1979) gave a method for finding a sampling design which has the same first and second order inclusion probabilities, but has a reduced support size than SRSWOR. (In other words, he gave a general method for obtaining BIBD's with relatively small support sizes.) Let $M$ denote the incidence matrix of all the pairs $(i, j)$ versus all the samples of $U$ with size $n$, where $i, j \in U$. Thus, $M$ is a $\left(\binom{N}{2} \times \binom{N}{n}\right)$ zero-one matrix. Suppose all the samples of $U$ with size $n$ are arranged in a list in an arbitrary but fixed order. Consider a BIBD (with block size n) in which $f_k$ denotes the frequency of the $k^{th}$ sample in the above list. Let $\underline{f} = (f_1, f_2, \ldots, f_{\binom{N}{n}})$. Consider a sampling measure $p$ which assigns probability $(f_k / \sum_i f_i)$ to the $k^{th}$ sample. Then, $p$ has the same first and second order inclusion probabilities as SRSWOR of size $n$ iff $M\underline{f} = \lambda\underline{1}$, where $\lambda$ is a positive integer and $\underline{1}$ is a column vector with all entries equal to 1. So each feasible solution of the system

$$M\underline{f} = \lambda\underline{1},\ \underline{f} \geq 0 \tag{4}$$

gives a sampling measure equivalent to SRSWOR of size $n$. Notice that there is always a solution for the system. So we can introduce another quantity, for example, the number of non-zero entries in $\underline{f}$, and find a feasible solution of the system to minimize the quantity. The algorithm of mathematical programming can be used to get such a solution. In other papers in combinatorics, Hedayat and others give further results.

In Hedayat and Pesotan (1983), $(R \times L)$ *triply balanced matrices* was discussed. The $(R \times L)$ triply balanced matrices arise in estimating the mean square error of nonlinear estimators in sampling. Briefly, a $(R \times L)$ triply balanced matrix is $\Delta = (\delta_{ij})$ with entries +1 or −1 such that $\sum_{r=1}^{R} \delta_{rh} = 0$, $\sum_{r=1}^{R} \delta_{rh}\delta_{rs} = 0$, $\sum_{r=1}^{R} \delta_{rh}\delta_{rs}\delta_{rt} = 0$, where the $h$, $s$, $t$ are distinct and $h, s, t = 1, \ldots, L$. It was proved that a $(R \times L)$ triply balanced matrix $\Delta$ is an orthogonal array of strength 3 and 2 symbols.

In Hedayat, Rao, and Stufken (1988), balanced sampling plans excluding contiguous units are discussed. In some situations, the $N$ units of the population are arranged in a natural order. In this case it may happen that contiguous units provides us *similar* information so that it seems more reasonable to select a sampling plan such that the contiguous unit cannot appear in the sample. Here the term *balanced* means that the first and second order inclusion probabilities are fixed. The condition of the existence of such a sampling measure is given in this paper, and a method of constructing such a sampling measure is also proposed.

**Use of $t$-Design**

Suggested by the usefulness of BIBD with sampling, the use of $t$-design in sampling was proposed by Srivastava and Saleh (1985). A BIBD, which has the same inclusion probabilities (of individual units, and pairs of units) as SRSWOR, has the same moments as SRSWOR up to order two. Generalizing this, Srivastava and Saleh showed that a $t$-design has the same moments as SRSWOR up to order $t$, because it has the same inclusion probabilities as SRSWOR up to order $t$ (i.e. every set of $i$ units ($i = 1,\ldots,t$) has the same inclusion probability, say $q_i$). Also, as for the BIBD, the sample space under a $t$-design can be much smaller than the sample space under SRSWOR. Thus, using $t$-designs we can try to avoid non-preferred samples, and still maintain resemblance to SRSWOR up to moments of order $t$.

For later use, define $a_{i\omega}$ ($i \in$ U, $\omega \in 2^U$) by

$$a_{i\omega} = 1, \text{ if } i \in \omega$$
$$= 0, \text{ otherwise.} \tag{5}$$

Let $1 \leq k \leq N$. For any sampling measure $\{p(\omega)\colon \omega \in 2^U\}$ define

$$\pi(i_i,\ldots, i_k) = \sum_{\omega} p(\omega) a_{i_1\omega} a_{i_2\omega} \cdots a_{i_k\omega}, \tag{6}$$

where $i_1, i_2,\ldots,i_k \in U$.

In this section, we suppose the sample size is always equal to $n$, a fixed integer. We are interested in estimating the population total $Y$.

The following results from Srivastava and Saleh (1985) are useful in the studies on using $t$-design theory in sampling.

**Lemma 1**

Let $2 \leq k \leq n$. Suppose $i_1,\ldots,i_k$ are distinct elements of $U$. Then we have

$$\sum_{i_k=1}^{N} \pi(i_1,\ldots,i_k) = (n - k + 1)\pi(i_1,\ldots,i_{k-1}),$$
$$i_k \neq i_1,\ldots,i_{k-1} \tag{7}$$

$$\sum_{i=1}^{N} \pi(i) = n. \tag{8}$$

This lemma says that for $2 \leq k \leq n$, the inclusion probabilities of order $j(1 \leq j \leq k-1)$ are determined by the inclusion probabilities of order $k$.

**Theorem 3**

Suppose there are two different sampling measures on $2^U$. Let $t$ be a positive integer. Then these two sampling measures give the same inclusion probabilities of order $t$ if and only if these two sampling measures give the same values of $E(\overline{y}_\omega^k)$, $k = 1,\ldots, t$, for all possible values of $(y_1,\ldots,y_n)$.

Let

$$\psi_q(\omega) = \sum_{i \in \omega} (y_i - \overline{y}_\omega)^9 \tag{9}$$

$$s_\omega^2 = \frac{1}{n-1} \sum_{i \in \omega} (y_i - \overline{y}_\omega)^2 = \frac{1}{n-1} \psi_2(\omega). \tag{10}$$

Then, we have

**Theorem 4**

Consider two sampling measures on $2^U$. Consider the following four conditions:

(i) For all possible values of $\underline{y} = (y_1,\ldots,y_n)'$, $E(\overline{y}_\omega)$ is the same under these two sampling measures,

(ii) For all possible values of $\underline{y}$, $E(\overline{y}_\omega^2)$, or $E(s_\omega^2)$, or $V(\overline{y}_\omega)$ is the same under these two sampling measures,

(iii) For all possible values of $\underline{y}$, $cov(\overline{y}_\omega, s_\omega^2)$ is the same under these two sampling measures,

(iv) For all possible values of $\underline{y}$, $V(s_\omega^2)$ is the same under these two sampling measures.

Let $t$ be an integer such that $1 \leq t \leq 4$. Then the above conditions (i), (ii), up to ($t$) are true if and only if these two sampling measures have the same inclusion probabilities of order $t$.

One can generalize Theorem 4 to higher order. But the most important case is order 4. In this case, we can characterize the mean and the variance of a linear estimator, and characterize the variance of a quadratic estimator of the variance of the linear estimator.

Now consider a $t$-design $D(N, n, t, b)$ where $N$ is the number of varieties, $n$ the block size, $b$ the number of blocks (which may or may not be distinct), and where every combination of $t$ varieties $(t \leq u)$ occurs in $b\binom{u}{t}\Big/\binom{N}{n}$ blocks.

Consider a sampling measure (called a *t-design sampling measure*) which selects each block of $D(N, n, t, b)$ with probability $1/b$. When $b = \binom{N}{n}$ and each block in $D\left(N, n, t, \binom{N}{n}\right)$ is distinct, this sampling measure becomes SRSWOR. In this case SRSWOR is a $t$-design $D\left(N, n, t, \binom{N}{n}\right)$, where $t$ can take any value from 1 to $n$.

For the $t$-design sampling measure mentioned above, for distinct $i_1,\ldots,i_t \in U$, we have

$$\pi(i_1,\ldots,i_t) = \frac{1}{b}\, b\binom{n}{t}\Big/\binom{N}{t} = \binom{n}{t}\Big/\binom{N}{t}. \tag{11}$$

Hence we have the following theorem.

**Theorem 5**

SRSWOR (with sample size $n$) and the $t$-design sampling measure have the same inclusion probabilities of order $t$ and hence have the same moments up to order $t$.

For any $t$-design $D(N, n, t, b)$, the number of distinct blocks is not greater than $\binom{N}{n}$, and usually is much less than $\binom{N}{n}$. This makes a $t$-design useful in controlled sampling. In fact, a BIBD is a 2-design. Because we need to estimate $V(\bar{y}_\omega)$, we need to consider up to the fourth moments; the first two moments are not enough. In view of this, Srivastava and Saleh assert that it would be much better to use 4-designs rather than BIBD's, since the former gives rise to the same moments as SRSWOR up to order 4.

**Connection with Arrays**

The theory of factorial designs constitutes a major part of the whole subject of experimental design. Furthermore, the modern theory of factorial designs is largely built around the concept of arrays. Indeed, arrays constitute a very important tool in all of design theory, since for example, BIBD's, PBIBD's and $t$-designs, etc. may (through their *incidence matrices*) be studied in terms of arrays. Because of this, in this section, we discuss the application of arrays in sampling theory. An array is a matrix whose elements come from a finite set. Suppose the finite set has m elements in it. Without loss of generality, we use the integers $0, 1,\ldots,m-1$ to denote the elements of the finite set. In this case, an array is a matrix whose elements belong to the set $\{0, 1,\ldots,m-1\}$. When $m = 2$, such an array becomes (0, 1) matrix which is of special importance.

A special case of a (0, 1) matrix is the incidence matrix of a class of subsets of a given finite set. The rows of an incidence matrix correspond to the elements of the given finite set and the columns correspond to the subsets of the given finite set. In sampling, an incidence matrix is $\Omega_U$ which is a $(N \times 2^N)$ (0, 1)-matrix such that its columns correspond to the elements of $2^U$, and rows to the elements of $U$. In order to simplify the discussion, and without loss of generality, we assume that the $i^{th}$ row of $\Omega_U$ corresponds to the element $i$ of $U$, and the $j^{th}$ column of $\Omega_U$ corresponds to the $j^{th}$ element of $2^U$ such that the elements of $2^U$ are arranged in the following standard order:

(i) If $\omega_1, \omega_2 \in 2^U$ and $|\omega_1| < |\omega_2|$, then $\omega_1$ precedes $\omega_2$;

(ii) If $|\omega_1| = |\omega_2|$ but there exists a $k \in U$ such that $|\{1,\ldots,k\} \cap \omega_1| > |\{1,\ldots,k\} \cap \omega_2|$ and $|\{1,\ldots,\ell\} \cap \omega_1| = |\{1,\ldots,\ell\} \cap \omega_2|$ for $0 \leq \ell < k$, then $\omega_1$ precedes $\omega_2$.

In this way, the elements of $2^U$ are arranged as

$$\{\omega(0), \omega(1),\ldots,\omega(2^N - 1)\} \tag{12}$$

and $i \in \omega(j)$ if and only if the $i^{th}$ coordinate of the $j^{th}$ column of $\Omega_U$ is equal to 1. For $N = 3$, $\Omega_U$ is equal to

$$\begin{bmatrix} 0 & 1 & 0 & 0 & 1 & 1 & 0 & 1 \\ 0 & 0 & 1 & 0 & 1 & 0 & 1 & 1 \\ 0 & 0 & 0 & 1 & 0 & 1 & 1 & 1 \end{bmatrix}. \tag{13}$$

Now, given any sampling measure $\{p(\omega): \omega \in 2^U\}$, we can rewrite it as a vector which is called the vector form of sampling measure

$$\underline{p}' = \big(p(\omega(0)), p(\omega(1)),\ldots,p(\omega(2^N - 1))\big). \tag{14}$$

Combining $\Omega_U$ and $\underline{p}'$, we have a matrix $\pi_U(p)$ where

$$\pi_U(p) = \begin{bmatrix} \Omega_U \\ \hline \underline{p}' \end{bmatrix}. \tag{15}$$

Thus, $\pi_U(p)$ presents a sampling measure in a matrix form. Now suppose all the $p(\omega(j))$ are rational numbers and $p(\omega(j)) = v_j/v$ such that $v_j$ is a non-negative integer, $v = \Sigma v_j$ (where the sum runs over all $j$), also suppose there is no common factor other than 1 among the $v_j$ ($j = 0, 1,\ldots,2^n-1$). Suppose

$$\Omega_U = [\underline{c}_0, \underline{c}_1,\ldots,\underline{c}_{2^n-1}]. \tag{16}$$

Now we introduce another matrix $\Delta_U(p)$ such that

$$\Delta_U(p) = \left[\underline{c}_0 1'_{v_0} \middle| \underline{c}_1 1'_{v_1} \middle| \cdots \middle| \underline{c}_{2^n-1} 1'_{v_{2^n-1}}\right] \tag{17}$$

where $1'_k$ is the $(1 \times k)$ vector containing 1 *everywhere*, and where if for any $j$, we have $v_j = 0$, then the columns $(\underline{c}_j)$ do not appear in $\Delta_U(p)$. Now, drawing a column from $\Delta_U(p)$ with probability $(1/v)$ is equivalent to drawing a column from $\Omega_U$ with probability measure $\{p(\omega(j)) = v_j/v, j = 0, 1,\ldots,2^N-1\}$. So, the matrix $\Delta_U(p)$ represents the sampling measure in the form of an array and it is called *sampling array* in Srivastava (1988), wherein the following result is proved.

**Theorem 6**

For any vector form of sampling measure $\underline{p}$ and $\epsilon > 0$, there exists a vector form of sampling measure $\underline{p}^*$ whose elements are rational such that $(\underline{p} - \underline{p}^*)'(\underline{p} - \underline{p}^*) < \epsilon$. (Note that every sampling measure can be expressed in the vector form.)

Although this theorem seems simple it has an important interpretation in that we can replace a sampling measure by a rational sampling measure as closely as we want. On the other hand, by using a rational sampling measure we get a sampling array. So the above theorem connects sampling theory to the theory of arrays in a fundamental manner, and hence to factorial and other experimental designs.

Now consider the problem of estimating the population total $Y$ by a general linear estimator $\hat{Y}_G$ ($G$ means general), where

$$\hat{Y}_G = \sum_{i \in \omega} c_{i\omega} y_i = \sum_{i=1}^{N} c_{i\omega} a_{i\omega} y_i, \tag{18}$$

and where $c_{i\omega}$ are known real numbers which depend on $i$ and $\omega$ for all $i \in U$, $\omega \in 2^U$. Define

$$\phi_{ic} = \sum_{\omega} c_{i\omega} a_{i\omega} p(\omega) \tag{19}$$

$$\phi^0_{iic} = \sum_{\omega} c^2_{i\omega} a_{i\omega} p(\omega),\ \phi^0_{ijc} = \sum_{\omega} c_{i\omega} c_{j\omega} a_{i\omega} a_{j\omega} p(\omega) \tag{20}$$

$$\underline{\phi}_c = (\phi_{1c},\ldots,\phi_{Nc})',\ \Phi^0_c = (\phi^0_{ijc})_{N \times N} \tag{21}$$

$$\Phi_c = \Phi^0_c - (J_{N1}\phi'_c + \underline{\phi}_c J_{1N}) + J_{NN} \tag{22}$$

where $J_{mn}$ is a $m \times n$ matrix which elements are equal to 1. It is easy to check that

$$\Phi_c = \sum_{\omega} p(\omega) \underline{U}_{c_\omega} \underline{U}'_{c_\omega} \tag{23}$$

where

$$\underline{U}_{c_\omega} = (c_{1\omega} a_{1\omega} - 1,\ c_{2\omega} a_{2\omega} - 1,\ldots, c_{N\omega} a_{N\omega} - 1). \tag{24}$$

We have the following theorem:

**Theorem 7**

The mean square error of $\hat{Y}_G$ as an estimator of $Y$ denoted by $MSE(\hat{Y}_G)$ is

$$MSE(\hat{Y}_G) = \underline{Y}'\Phi_c\underline{Y}. \tag{25}$$

Notice that the matrix $\Phi_c$ is known when the sampling measure and the estimator $\hat{Y}_G$ are selected. The matrix $\Phi_c$ in sampling theory is similar to the information matrix in the theory of experimental design.

## A General Estimator

In this section, we consider an estimator proposed in Srivastava (1985). There is an interesting history relevant here. First, in 1985, Srivastava observed the connection between combinatorial arrays and sampling theory, discussed in the last section. This appeared to open up a quite new theoretical field, in which variable sample size appeared to be inherent. Thus, there seemed to be a need for a general estimator in which sample size was not necessarily fixed. Now, most estimators in sampling theory relate to fixed size. In many ways, the most general estimator (which, among other things, allows variable sample size) existing in 1985 was the Horvitz-Thompson estimator. But this is entirely dependent on the sampling measure, which is of course decided upon *before* the sample is drawn. In an attempt to be able to utilize the new knowledge (*independent* of the sample, but obtained during the course of actual sampling) the concepts of the *sample weight function* (discussed below), and the estimator of this section, were discovered. This estimator is extremely general, in that most of the known estimators turn out to be its special cases.

The most important concept in this estimator is the introduction of the *sample weight function* $r$, defined on $2^U$, such that for all $\omega \in 2^U$, $r(\omega)$ is a finite real number. For every $K \subset U$, and $k \in (1, 2,\ldots,N)$, let

$$(K: k) = \{\omega: w \subset K, |\omega| = k\}.$$

Clearly, if $\underline{i} \in (U: k)$, then $\underline{i}$ is a $k$-tuple, with $k$ distinct elements from $U$. From here on, $\sum_{\underline{i}}$ will denote the sum over all $\underline{i} \in (U: k)$, $\sum_{\omega\underline{i}}$ will denote the sum over all $\omega \in 2^U$ such that $\underline{i} \subset \omega$, and $\sum_{\underline{i}\omega}$ will denote the sum over all $\underline{i} \in (w: k)$. Note that the last sum could be empty. In this section, we always look upon $\underline{i} = (i_1,\ldots,i_k)$ as an unordered set $\{i_1,\ldots,i_k\}$. Let

$$\pi_r(\underline{i}) = \sum_{\omega\underline{i}} p(\omega)r(\omega). \tag{26}$$

For $k \in (1, 2,\ldots,N-1)$, $t \in (1, 2,\ldots,N)$, and $\underline{i} \in (U: k)$, let $T_r(\underline{i}, t)$ be the class of all unordered sets $\underline{j} = (j_0,\ldots,j_{t-1})$ such that $\underline{j} \in (U: t)$ and $\underline{i} \subset \underline{j}$

where $\underline{i} = (i_1, \ldots, i_k)$, and $\pi_r(\underline{j}) \neq 0$. Let

$$\nu_r(\underline{i}, t) = | T_r(\underline{i}, t)|. \tag{27}$$

Also, let $\alpha(\underline{i}, t)$ be real numbers which satisfy the following two conditions:

$$\alpha(\underline{i}, t) = 0, \text{ if } \nu_r(\underline{i}, t) = 0, \text{ and} \tag{28}$$

$$\sum_{t=1}^{N} \alpha(\underline{i}, t) = 1. \tag{29}$$

For all $\omega \in 2^U$, $\underline{i} \in (U\colon k)$, define

$$\beta_r(\underline{i}, \omega) = r(\omega) \sum_{t=1}^{N} \alpha(\underline{i}, t) [v_r(\underline{i}, t)]^{-} \left\{ \sum{}^{*} [\pi_r(\underline{j})]^{-} \right\} \tag{30}$$

where $a^{-} = a^{-1}$ if $a \neq 0$ and $a^{-} = 0$ if $a = 0$, and $\Sigma^*$ runs over all $\underline{j} \in (U\colon t)$ such that $\underline{i} = (i_1, \ldots, i_k) \subset \underline{j}$ and $\underline{j} \subset \omega$. Now, consider the estimation of the following *symmetric linear population function* $Q(\psi)$ where

$$Q(\psi) = \sum_{\underline{i}} \psi(\underline{i}), \tag{31}$$

where $\psi$, defined over $(U\colon k)$, is such that for all $\underline{i} \in (U\colon k)$, $\psi(\underline{i})$ is a real number. Notice when $\underline{i} \subset \omega$ and $\omega$ is selected, $\psi(\underline{i})$ can be calculated. Thus, once a sample $\omega$ is drawn, we can compute $\hat{Q}^{sr}(\psi)$ where

$$\hat{Q}^{sr}(\psi) = \sum_{\underline{i}\omega} \psi(\underline{i}) \beta_r(\underline{i}, \omega). \tag{32}$$

Here in $\hat{Q}^{sr}$, $s$ means that we are estimating a symmetric function, and $r$ means that the sample weight function $r$ is being used.

### Theorem 8

The statistic $\hat{Q}^{sr}(\psi)$ is an unbiased estimator of $Q(\psi)$, if and only if for every $\underline{i} \in (U\colon k)$ with $\psi(\underline{i}) \neq 0$, there exists a $t$ such that $1 \leq t \leq N$, $\nu_r(\underline{i}, t) \neq 0$.

For the case of $\pi_r(\underline{i}) \neq 0$ for all $\underline{i} \in (U\colon k)$, let $\alpha(\underline{i}, k) = 1$ and $\alpha(\underline{i}, t) = 0$ for all $t \neq k$. Then we have

$$\beta_r(\underline{i}, \omega) = r(\omega)[\pi_r(\underline{i})]^{-1} \text{ and} \tag{33}$$

$$\hat{Q}^{sr}(\psi) = r(\omega) \sum_{\underline{i}\omega} \psi(\underline{i})/\pi_r(\underline{i}). \tag{34}$$

By using Theorem 8, it can be checked that (34) is unbiased for $Q(\psi)$.

The variance of $\hat{Q}^{sr}(\psi)$ and an unbiased estimator of the variance of $\hat{Q}^{sr}(\psi)$ were also obtained in Srivastava (1985). Now we turn to an estimator of the population total.

Let $k = 1$, $\underline{i} = 1$, $\psi(\underline{i}) = y_i$. Then $Q(\psi) = \sum_{i=1}^{N} y_i = Y$. Then, (34) gives

$$\hat{Q}^{sr}(\psi) = r(\omega) \sum_{i \in \omega} y_i/\pi_r(i) \equiv \hat{Y}_{sr1} \text{ say.} \tag{35}$$

By Theorem 8, if $\pi_r(i) \neq 0$ for all $i \in U$, then $\hat{Y}_{sr1}$ is an unbiased estimator of $Y$. When

$$r(\omega) = 1, \text{ for all } \omega \in 2^U, \tag{36}$$

$\pi_r(i)$'s become $\pi_i$'s where $\pi_i$ is the probability such that the unit $i$ is included in the sample. At this time, $\hat{Y}_{sr1}$ becomes the well known Horvitz-Thompson estimator $\hat{Y}_{HT}$ where

$$\hat{Y}_{HT} = \sum_{i\omega} y_i/\pi_i. \tag{37}$$

The variance of $\hat{Y}_{sr1}$ is given in the following theorem.

**Theorem 9**

Suppose $\pi_r(i) \neq 0$, $i = 1,\ldots,N$. Then

$$Var(\hat{Y}_{sr1}) = \sum_{i=1}^{N} y_i^2 \left[ \frac{\pi_{r2}(i) - \left(\pi_r(i)\right)^2}{\left(\pi_r(i)\right)^2} \right] + \sum_{i \neq j,=1}^{N} y_i y_j \left[ \frac{\pi_{r2}(i,j) - \pi_r(i)\pi_r(j)}{\pi_r(i)\pi_r(j)} \right] \tag{38}$$

where

$$\pi_{r^2}(i) = \sum_{\omega i} p(\omega)[r(\omega)]^2; \; i = 1,\ldots,N \tag{39}$$

$$\pi_{r^2}(i,j) = \sum_{\omega ij} p(\omega)[r(\omega)]^2, \; i \neq, \; i,j = 1,\ldots,N. \tag{40}$$

It is easy to see that

$$\pi_{r^2}(i)/(\pi_r(i))^2 \geq 1/\pi_i,\ i = 1,\ldots,N. \tag{41}$$

So the term in $y_i^2$ in $Var(\hat{Y}_{sr1})$ is always larger than the correspondent term for $\hat{Y}_{HT}$. But we can choose $r(\omega)$ such that the cross product terms of $\hat{Y}_{sr1}$ are small so that $Var(\hat{Y}_{sr1})$ is small. Examples are given in Srivastava (1985).

**Balanced Array Sampling**

We have defined arrays in the fourth section. Let $K(a \times b)$ and $\underline{k}(a \times 1)$ be a matrix and a vector with elements from $\sigma_s$, where $\sigma_s$ is a finite set whose elements are $(0, 1,\ldots,s-1)$. The symbol $\lambda(\cdot,\cdot)$ is defined as a counting operator, such that $\lambda(\underline{k}, K)$ is equal to the number of times $\underline{k}$ occurs as a column of $K$. Let $\psi_s$ be the permutation group over $\sigma_s$. For $\psi \in \psi_s$, and $j \in \sigma_s$, let $\psi(j)$ be the image of $j$ when the permutation $\psi$ is applied. Similarly, we define $\psi(\underline{k}) = (\psi(k_1),\ldots,\psi(k_a))$ if $\underline{k} = (k_1,\ldots,k_a)$ is a $(a \times 1)$ array over $\sigma_s$.

**Definition 1**

Let $K$ be a $(a \times b)$ array over $\sigma_s$. Then $K$ is a *balanced array* (B-array, or BA) of strength $t$ if and only if

$$\lambda(\underline{k}_0, K_0) = \lambda\ (\psi(\underline{k}_0), K_0) \tag{42}$$

where $\underline{k}_0$ is any $(t \times 1)$ array over $\sigma_s$, $K_0$ is any $(t \times b)$ subarray of $K$ and $\psi$ is any permutation in $\psi_s$.

Balanced arrays play an important role in factorial experimental design and coding theory. For $\underline{i} = (i_1,\ldots,i_k) \in (U: k)$, define

$$\pi(i_1,\ldots, i_k) = \pi(\underline{i}) = \sum_{\omega \underline{i}} p(\omega). \tag{43}$$

When $k = 1$ or 2, the following customary notations will be used instead of $\pi(i)$, $\pi(i,j)$

$$\pi_i = \pi(i),\ \pi_{ij} = \pi(i,j). \tag{44}$$

**Definition 2**

Let $p(\cdot) = \{p(\omega): \omega \in 2^U\}$ be a sampling measure. Then $p(\cdot)$ corresponds to *balanced array sampling with strength* $k$ iff $\pi(i_1,\ldots,i_g)$ is fixed, for all possible $(i_1,\ldots,i_g) \in (U: g)$. Here, $g = 0, 1,\ldots,k$.

Thus, if $p(\cdot)$ corresponds to balanced array sampling with strength $k$, then there exists $\theta_1,\ldots,\theta_k$ such that

$$\pi_r(i_1,\ldots,i_g) = \theta_g, \tag{45}$$

for $(i_1,\ldots,i_g) \in (U\colon g)$, and $g = 0, 1,\ldots,k$.

**Theorem 10**

Suppose, the measure $p(\cdot)$ corresponds to BA sampling with strength $k$. Then there exists a sampling measure $p^*(\cdot)$ whose sampling array is $\Delta_U(p^*)$ such that $\Delta_U(p^*)$ is a B-array of strength $k$, and $p^*(\cdot)$ is arbitrarily close to $p(\cdot)$. (In the sense of Theorem 6.)

**Theorem 11**

Suppose $\Delta_U^*(p)$ is $(N \times v)$ B-array of strength $k$. Let $p(\cdot)$ be a sampling measure such that it gives a probability $(1/v)$ to each column of $\Delta_U^*(p)$ for being selected. Then $p(\cdot)$ corresponds to balanced array sampling with strength $k$.

Let $\delta_1$ and $\delta_2$ be the mean and the variance of the sample size under the measure $p(\cdot)$, i.e.,

$$\delta_2 = \sum_\omega p(\omega)|\omega| \tag{46}$$

$$\delta_2 = \sum_\omega p(\omega)(|\omega| - \delta_1)^2. \tag{47}$$

Then we have the following theorem.

**Theorem 12**

Consider BA sampling whose inclusion probability is given by (45). Then

$$\hat{Y}_{HT} = \theta_1^{-1}|\omega|\bar{y}_\omega \tag{48}$$

$$V(\hat{Y}_{HT}) = S^2\left[\frac{N}{\delta_1}\left\{(N - \delta_1) - \frac{\delta_2}{\delta_1}\right\}\right] + \frac{N^2\delta_2}{\delta_1^2}\,\bar{Y}^2 \tag{49}$$

$$= S^2N^2\left(\frac{1}{\delta_1} - \frac{1}{N}\right) + \frac{\delta_2}{\delta_1}\,[N^2\bar{Y}^2 - S^2]$$

where $\bar{Y} = \frac{1}{N}\sum_{i=1}^{N} y_i$, $S^2 = \frac{1}{N-1}\sum_{i=1}^{N}(y_i - \bar{Y})^2$, are respectively of the population mean and variance. (The significance of this result lies in the fact that if we have some idea of the value of $\bar{Y}$, we can reduce the variance below that of SRSWOR. This may happen, for example, in recursive sampling.)

**Definition 3**

Let $p(\cdot)$ be a sampling measure. Then, $p(\cdot)$ corresponds to *proportional array sampling with strength k* (or, briefly, *proportional sampling*) iff for all integer $g$ such that $1 \leq g \leq k$, and all $(i_1,\ldots,i_g) \in (U\colon g)$ we have

$$\pi(i_1,\ldots,i_g) = \pi(i_1)\ldots\pi(i_g). \tag{50}$$

Notice that when $\pi_i$ is fixed, say $\theta$, for all $i \in U$, then the *proportional array sampling with strength k* is also *balanced array sampling with strength k*. In this case we call it *balanced proportional sampling with strength k*.

In order to construct a $p(\cdot)$ which corresponds to proportional array sampling with strength $k$, we need the definition of orthogonal array (OA).

**Definition 4**

Let $K$ be a $(a \times b)$ array over $\sigma_s$. Then $K$ is an orthogonal array of strength $t$ if and only if

$$\lambda(\underline{k}_0, K_0) = b \times s^{-t} \tag{51}$$

where $\underline{k}_0$ is any $(t \times 1)$ array over $\sigma_s$, $K_0$ is any $(t \times b)$ subarray of $K$. It is easy to see that an OA with strength $t$ is a BA with strength $t$.

Let $L(N \times b) = (\underline{\ell}_1,\ldots,\underline{\ell}_N)'$ be an OA of strength $k$ over $\sigma_s$ where $s$ is a prime number. Let $s_i$ be an integer satisfying $1 \leq s_i \leq s$, $i = 1,\ldots,N$. In $\underline{\ell}_i'$, replace the $(s_i - 1)$ symbols $\{2, 3,\ldots,s_i\}$ by 1, leave the original 1 unchanged, and replace the other symbols (if any) by 0. Notice when $s_i = s$, then the symbol $s_i$ is the same as symbol 0. Let $L(N \times b)$ be the array obtained by the above replacement.

**Theorem 13**

Consider a sampling measure $p(\cdot)$ such that it has $L(N \times b)$ as a sampling array. Then $p(\cdot)$ corresponds to proportional sampling of strength $k$, such that the inclusion probability of unit $i$ is equal to $s_i/s$, for $i = 1,\ldots,N$.

**Theorem 14**

We have

$$var(\hat{Y}_{HT}) = \sum_{i=1}^{N} y_i^2\left(\frac{1}{\pi_i} - 1\right) \tag{52}$$

for proportional sampling and

$$var(\hat{Y}_{HT}) = \left(\frac{1}{\delta_1} - 1\right)\left[(N-1)s^2 + N\bar{Y}^2\right]. \tag{53}$$

for balanced proportional sampling.

We can use BA sampling with strength 4 to imitate SRSWOR up to the $4^{th}$ moments. Notice that the binomial sampling referred to in the literature, is a balanced proportional sampling with strength $N$. It is clear that it should be adequate enough to use balanced proportional sampling with strength 4 instead of using binomial sampling.

**Weight Balanced Sampling**

Now we introduce an estimator of $Y$ called $\hat{Y}_{s2}$ which is a special case of $\hat{Y}_{s1}$ when

$$r(\omega) = |\omega|^{-1}, \text{ for all } \omega \in 2^U, \omega \neq \phi. \tag{54}$$

Let

$$\pi'_i = \sum_{\omega} \frac{p(\omega) a_{i\omega}}{|\omega|}, \; i = 1,\ldots,N \tag{55}$$

$$\pi''_i = \sum_{\omega} \frac{p(\omega) a_{i\omega}}{|\omega|^2}, \; i = 1,\ldots,N \tag{56}$$

$$\pi''_{ij} = \sum_{\omega} \frac{p(\omega) a_{i\omega} a_{j\omega}}{|\omega|^2}, \; i \neq j, \; i,j = 1,\ldots,N. \tag{57}$$

where we assume that empty samples are not allowed.

**Theorem 15**

Suppose $\pi_i > 0$ for $i = 1,\ldots,N$. Then

$$\hat{Y}_{s2} = |\mathrm{w}|^{-1} \sum_{i \in \omega} y_i/\pi'_i, \tag{58}$$

$$E(\hat{Y}_{s2}) = Y, \tag{59}$$

$$var(\hat{Y}_{s2}) = \sum_{i=1}^{N} y_i^2 \left( \frac{\pi''_i}{(\pi'_i)^2} - 1 \right) + \sum_{i \neq j} y_i y_j \left( \frac{\pi''_{ij}}{\pi'_i \pi'_j} - 1 \right). \tag{60}$$

Notice that when the sample size is fixed, $\hat{Y}_{s2} = \hat{Y}_{HT}$.

**Definition 5**

A sampling measure $p(\cdot)$ corresponds to *weight-balanced* (WB) sampling, if and only if $(\pi''_i/(\pi'_i)^2)$ and $(\pi''_{ij}/\pi'_i\pi'_j)$ are constants for $i \in U$ and $i \neq j$, $i,j \in U$ respectively.

Let

$$\pi''_i/(\pi'_i)^2 = \beta_1, \text{ for all } i \tag{61}$$

$$\pi''_{ij}/\pi'_i\pi'_j = \beta_2, \text{ for all } i \neq j,\ i,j \in U. \tag{62}$$

We have the following corollary of Theorem 15.

**Corollary 1**

Under WB sampling, we have

$$V(\hat{Y}_{sr1}) = (N-1)S^2(\beta_1 - \beta_2) + N\bar{Y}^2[(\beta_1 - \beta_2) + N(\beta_2 - 1)]. \tag{63}$$

**Definition 6**

A sampling measure $p(\cdot)$ corresponds to *strongly weight-balanced* (SWB) sampling if and only if $\pi'_i$, $\pi''_i$, $\pi''_{ij}$ are constants for $i \in$ U and $i \neq j$, $i,j \in U$ respectively.

Let

$$\pi'_i = \beta_3 \quad i \in U, \text{ and} \tag{64}$$

$$\beta_0 = \sum_{\omega} p(\omega)/|\omega| = \sum_{j=1}^{N} \pi''_j \tag{65}$$

**Theorem 16**

For SWB sampling, we have

$$var(\hat{Y}_{s2}) = N^2 S^2\left(\beta_0 - \frac{1}{N}\right). \tag{66}$$

Suppose $q(n) > 0$, $n = 1,\ldots,N$ and $\sum_{n=1}^{N} q(n) = 1$. Suppose we draw a sample in this way: firstly select the sample size $n$ with probability $q(n)$, then use SRSWOR to draw a sample of size $n$. Then use $N\bar{y}_\omega$ to estimate the population total $Y$. In this way, we select a particular sample of size $n$ with probability $q(n)/\binom{N}{n}$. We have

$$var(N\bar{y}_\omega) = \sum_{n=1}^{N} q(n)\left[(N\bar{y}_\omega - Y)^2 \,\middle|\, |\omega| = n\right] \tag{67}$$

$$= \sum_{n=1}^{N} q(n) N^2 S^2 \left(\frac{1}{n} - \frac{1}{N}\right)$$

$$= N^2 S^2 \left(\beta_0 - \frac{1}{N}\right).$$

Hence, the technique of using $\hat{Y}_{s2}$ to estimate the population total in SWB sampling is a technique which imitates SRSWOR.

An estimator $\hat{Y}_G$ is said to be location invariant if and only if

$$\hat{Y}_G \text{ (given that } \underline{y} = \underline{y}^*) = -y_0 N + \hat{Y}_G \text{ (given that } \underline{y} = \underline{y}^* + y_0 J_{1N}) \quad (68)$$

for all real $y_0$, when $\underline{y} = (y_1, \ldots, y_N)'$. It is easy to see that $\hat{Y}_G = \left(\sum_{i=1}^{N} C_{i\omega} a_{i\omega} y_i\right)$ is location invariant iff $\sum_{i=1}^{N} c_{i\omega} a_{i\omega} = N$ for all $\omega \in 2^U$.

**Theorem 17**

Under SWB sampling, $\hat{Y}_{s2}$ is location invariant.

The material in this section comes from Srivastava (1987), where examples of WB are given. From an unpublished paper of Srivastava and Ouyang (1988), we know that $\hat{Y}_{s2}$ is an admissible linear estimator of $Y$, and has a variance formula which is similar to the Yates-Grundy variance formula for $Var(\hat{Y}_{HT})$ when the sample size is fixed.

**An Example of Controlled Sampling and BA Sampling**

Now we discuss an example given by Avadhani and Sukhatme (1973) in controlled sampling. Let $N = 7$, and suppose these seven units are located as in the diagram below:

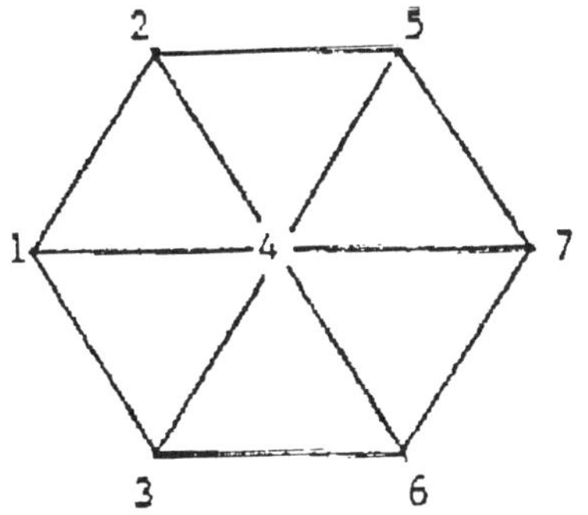


Here, any two units which are connected by a line are considered as neighbors. We are going to get a sample of size 3 from these 7 units. In order to reduce the travel cost, we hope the sample we get consists of neighboring units. So a sample $\omega = \{i_1, i_2, i_3\}$ is considered preferred if and only if after a suitable permutation, there is a line between $i_1$ and $i_2$ and also there is a line between $i_2$ and $i_3$. So the total number of preferred samples is 21, and the total number of possible samples is $\binom{7}{3} = 35$.

Consider a BIBD with parameter $N = 7$, $k = 3$, $b = 7$, $r = 3$, $v = 1$:

$$T = \begin{bmatrix} 2 & 2 & 3 & 1 & 3 & 1 & 1 \\ 5 & 4 & 6 & 4 & 4 & 2 & 5 \\ 7 & 6 & 7 & 7 & 5 & 3 & 6 \end{bmatrix} \tag{69}$$

Now only the block correspond to column 7 is not preferred. Hence if we use probability 1/7 to draw a column from $T$, we reduce the probability of drawing a non-preferred sample greatly, and at the same time we have the same first two moments as SRSWOR. But this technique does not avoid the nonpreferred samples totally. To avoid the nonpreferred samples totally, consider a balanced array approach as follows. We have a list of 16 samples: {147}, {246}, {543}; {125}, {257}, {576}, {763}, {631}, {321}, {15}, {27}, {56}, {73}, {61}, {32}; {5}. With probability (1/11) we draw any one of the first three samples, and with probability (1/22) we draw any one of the remaining samples. Hence we avoid the nonpreferred samples. But we use some subsamples of the preferred samples.

The problem of controlled sampling may be approached through the concepts of array sampling as follows.

(i) Decide the preferred and nonpreferred samples.

(ii) Decide whether fixed sample size should be used or not.

(iii) Consider using BA sampling or WB sampling.

(iv) Suppose BA sampling is used. Then we need to find a BA whose columns consist of the preferred samples. If we fail to get such a BA, then consider subsamples of these samples. Sometimes we have to change the decision in step (ii) to consider using some non-preferred samples in this step (with minimal probability).

**References**

Avadhani, M. S. and Sukhatme, B. V. (1963): Controlled simple random sampling, *J. India Soc. Agricultural Stat.* 17, 34-42.

Avadhani, M. S. and Sukhatme, B. V. (1967): Controlled sampling with varying probability with and without replacement, *Austral. J. Stat.* 9(1), 8-15.

Avadhani, M. S. and Sukhatme, B. V. (1973): Controlled sampling with equal probabilities and without replacement, *International Stat. Review* 41(2), 175-182.

Brewer, K. R. W., Foreman, E. K., Mellor, R. W. and Trewin, D. J. (1977): Use of experimental design and population modelling in survey sampling, *Bulletin of the International Stat. Inst.* 3, 173-190.

Brewer, K. R. W. and Hanif, M. (1983): *Sampling with Unequal Probabilities*, Springer-Verlag, New York, Heidelberg, Berlin.

Brunk, M. E. and Federer, W. T. (1953): Experimental designs and probability sampling in marking research, *JASA* 48, 440-452.

Cassel, C. M., Sarndal, C. E. and Wretman, J. H. (1977): *Foundations of Inference in Survey Sampling*, Wiley, New York.

Chakrabarti, M. C. (1963): On the use of incidence matrices of designs in sampling from finite populations, *J. Indian Stat. Soc.* 1, 78-85.

Cheng, C. S. and Li, K. C. (1983): A minimax approach to sample surveys, *Annals of Statistics* 11, 552-563.

Cochran, W. G. (1977): *Sampling Techniques*, 3rd edition, Wiley, New York.

Cox, D. R. (1956): A note on weighted randomization, *Annals of Math. Stat.* 27, 1144-1151.

Cox, D. R. (1958): *Planning of Experiments*, Wiley, New York.

Cumberland, W. G. and Royal, R. M. (1981): Prediction models and unequal probability sampling, *J. Roy. Stat. Soc.*, Ser. B, 43(3), 353-367.

Deming, W. E. (1953): On the distinction between enumerative and analytic survey, *JASA* 48, 244-255.

Dharmadhikari, S. W. (1982): Connectedness and zero variance in sampling designs, *Stat. and Prob.*, 221-225.

Dwyer, P. S. (1972): Moment functions of sample moment functions, in *Symmetric Functions in Statistics*, ed. Derrick S. Tracy, University of Windsor, Windsor, Ontario, 11-51.

Fellegi, I. P. (1963): Sampling with varying probabilities without replacement: Rotating and rotating samples, *JASA* 58, 183-201.

Kalton, G. (1983): Models in the practice of survey sampling, *International Stat. Review* 51, 175-188.

Kiefer, J. (1961): Optimum designs in regression problems. II, *Annals of Math. Stat.* 321, 298-325.

Kempthorne, O. (1952): *The Design and Analysis of Experiments*, Wiley, New York.

Kempthorne, O. (1955): The randomization theory of experimental inference, *JASA* 50, 946-967.

Kish, L. (1965): *Survey Sampling*, Wiley, New York.

Kruskal, W. H. and Mosteller, F. (1980): Representative sampling IV: The history of the concept in statistics. 1815-1939, *International Stat. Review* 48, 169-195.

Lacayo, H., Pereina, C. A. de, Proschan, F. and Sarndal, C. F. (1982): Optimal sample depends on optimality criterion, *Scand. J. Stat.* 9, 47-48.

Little, R. J. A. (1983): Estimating a finite population mean from unequal probability sample, *JASA* 78(3), 596-604.

Lin, T. P. and Thompson, M. E. (1983): Journal of quadratic finite population: The batch approach, *Annals of Stat.* II(1), 275-285.

Mahalanobis, P. C. (1944): On large-scale sample surveys, *Philosophical Transactions of the Roy. Soc.*, London, 231(B), 329-451.

Meeden, G. and Ghosh, M. (1983): Choosing between experiments: Applications to finite population sampling, *Annals of Stat.* 11, 296-305.

Mikhail, N. N. and Ali, M. M. (1981): Unbiased estimates of the generalized variance for finite population, *J. Indian Stat. Assoc.* 19, 85-92.

Mukhopadhyay, P. (1975): An optimum sampling design to the HT method of estimating a population total, *Metrika* 22, 119-127.

Mukhopadhyay, P. (1982): Optimal strategies for estimating the variance of a finite population under a superpopulation model, *Metrika* 29, 143-158.

Murthy, M. N.: Sampling theory and methods, *Stat. Pub. Soc.*, Calcutta.

Padmawar, V. R. (1981): A note on the comparison of certain sampling strategies, *J. Roy. Stat. Soc.*, Ser. B, 43(3), 321-326.

Pathak, P. K. (1964): On inverse sampling with unequal probabilities, *Biometrika* 51, 185-193.

Pereira, C. A. B. and Rodriques, J. (1983): Robust linear prediction in finite populations, *International Stat. Review* 51, 293-300.

Prasad, N. G. N. and Srivenkataramana, T (1980): A modification to the Horvitz-Thompson estimator under the Midzuno sampling scheme, *Biometrika* 67, 709-711.

Raj, D. (1958): On the relation accuracy of some sampling techniques, *JASA* 53, 98-101.

Raj, D. (1972): *Sampling Theory*, McGraw-Hill Book Company, New York.

Ramakrisknan, M. K. (1975a): Choice of an optimum sampling strategy–I, *Annals of Stat.* 3(3), 669-679.

Ramakrishnan, M. K. (1975b): A generalization of the Yates-Grundy variance estimator, *Sankhya*, Ser. C, 37, 204-206.

Rao, C. R. (1975): Some problems of sample surveys, *Adv. Appd. Prob.* 7 (Supplement), 50-61.

Robinson, P M. (1982): On the convergence of the Horvitz-Thompson estimator, *Austral. J. Stat.* 24, 234-238.

Rosen, B. (1972): Asymptotic theory for successive sampling with varying probabilities without replacement I and II, *Annals of Math. Stat.* 43, 7, 373-392, 748-776.

Roy, J. (1957): A note on estimation of variance components in multistage sampling with varying probabilities, *Sankhya* 17, 367-372.

Rubin, D. B. (1978): Bayesian inference for causal effects: The role of randomization, *Annals of Stat.* 6, 34-58.

Sedransk, J. (1967): Designing some multi-factor analytical studies, *JASA* 62, 1121-1139.

Srivastava, J. (1985): On a general theory of sampling, using experimental design. Concept I: Estimation, *Bulletin of International Stat. Inst.*, Vol. 51, Book 2, 10.3-1 - 10.3-16.

Srivastava, J. (1988): On a general theory of sampling, using experimental design. Concepts II: Relation with arrays, in *Probability and Statistics*, ed. J. N. Srivastava, North-Holland.

Srivastava, J. and Ouyang, Z. (1988a): Studies on the general estimator in sampling theory, based on the sample weight function, unpublished paper.

Srivastava, J. and Ouyang, Z. (1988b): Optimal properties of balanced array sampling and weight balanced sampling, unpublished paper.

Srivastava, J. and Saleh, F. (1985): Need of t-designs in sampling theory, *Utilitas Mathematica* 28, 5-17.

Stenger, H. and Gabler, S. (1981): On the completeness of the class of fixed size sampling strategies, *Annals of Stat.* 9, 229-232.

Strauss, I. (1982): On the admissibility of estimators for the finite population variance, *Metrika* 29, 195-202.

Sukhatme, P. V. and Sukhatme, B. V. (1976): Sampling theory of surveys with applications, New Dehli, *Indian Soc. of Agricultural Stat.*

Tepping, B. J., Hurvitz, W. N. and Deming, W. E. (1943): On the efficiency of deep stratification in block sampling, *JASA* 38, 93-100.

Vardeman, S. and Meeden, G. (1983): Admissible estimators in finite population sampling employing types of prior information, *JSPI* 7, 329-341.

Wilk, M. B. and Kempthorne, O. (1956): Some aspects of the analysis of factorial experiments in a completely randomized design, *Annals of Math. Stat.* 27, 950-985.

Wynn, H. P. (1976): Optimum designs for finite populations sampling, in *Statistical Decision Theory and Related Topics*, eds. S. S. Gupta and D. S. Moore, Academic Press, New York.

Wynn, H. P. (1977a): Minimax purposive survey sampling design, *JASA* 72(359), 655-657.

Wynn, H. P. (1977b): Convex set of finite population plans, *Annals of Stat.* 5, 414-418.